S0-CFN-505

BUSINESS MATHEMATICS

FOURTH EDITION

BUSINESS MATHEMATICS

FOURTH EDITION

BURTON S. KALISKI
New Hampshire College

ROBERT L. DANSBY
Columbus Area Vocational-Technical School

HBJ

Harcourt Brace Jovanovich, Publishers
and its subsidiary, Academic Press

San Diego New York Chicago Austin Washington, D.C.
London Sydney Tokyo Toronto

ISBN: 0-15-505644-1
Library of Congress Catalog Card Number: 86-81630
Printed in the United States of America

Preface

In our fast-moving, rapidly changing world, skills once considered desirable have now become indispensable. Although a mere acquaintance with arithmetic may have satisfied the professional and personal needs of past generations, today a thorough mastery of the principles and techniques of business mathematics is essential for success in your career and everyday activities as consumer and citizen.

Business Mathematics, Fourth Edition, will provide you with the principles you must know and use to solve business problems. It will also make you more aware of the importance of business mathematics in your personal life. The topics you will study are those recommended by more than 150 experienced college instructors. Although one volume cannot cover every area associated with the term "business mathematics," all the standard topics are included, as well as those of special current interest, including individual retirement plans, charge account plans, home mortgages, and NOW accounts.

The arithmetic of business bears a close relationship to the tax laws of our country. As you most likely know, a major revision of income tax laws has just occurred in late 1986 which will affect some of the problems shown in the text. Annual updates will be provided to the users of this text to indicate which problems need to be modified in order to conform to projected and yet to be determined changes in the laws.

Because of its importance to business, we have chosen the income statement as the focal point around which the book has been organized. The text is divided into six units and one appendix.

Unit 1 thoroughly reviews the fundamental processes.

Unit 2 deals with buying, selling, and inventory, the first factors considered in calculating a firm's profit or loss.

Unit 3 discusses operating expenses that reduce the profit or increase the loss from buying and selling activities, and the actual calculation of net profit or loss.

Unit 4 considers interest as a form of additional income or expense.

Unit 5 deals with investment in stocks, bonds, mutual funds, and other areas as still another source of income or expense.

Unit 6 discusses procedures for summarizing and analyzing data.

The appendix explains basic techniques for operating the pocket calculator, enabling you to use this tool as desired.

In *Business Mathematics,* Fourth Edition, the step-by-step explanation of principles—profusely illustrated by numerous sample problems with solutions—will make it easy for you to grasp each new concept and technique. "*Test Your Ability*" practice problems appear at intervals throughout each chapter. Since they are organized in the same sequence as the sample problems, you can always find a practice problem to do immediately after studying a sample problem. Answers to "*Test Your Ability*" items appear at the back of the book, immediately following the glossary.

In all units except Unit 1, end-of-chapter exercises are divided into Group A and Group B. Each Group A problem matches a corresponding "*Test Your Ability*" problem; however, the items appear in different sequences to enable you to check your grasp of the principles and procedures learned in the chapter. Group B problems require a deeper understanding of the material. Each chapter also includes questions titled "*Check Your Reading,*" which enable you to verify your understanding of the terminology and concepts.

The majority of problems require only simple calculations, and the answers work out evenly. However, to reflect real business practice, some problems include odd amounts, requiring more extensive computation for their solution. Since most business problems are solved by using simple arithmetic, we have deliberately minimized calculation by formula. In addition, because tables are used extensively in business practice, they are also included here wherever possible. Some of the tables include data in actual use as of mid-1986, such as those for computing withholding tax; others are sample tables, such as those for stocks, bonds, and insurance.

The review section at the end of each unit lists the important terms and reinforces the major principles learned in the unit. The summary problems in the "*Unit Review*" are intended as self-tests and should serve as effective review for examinations.

The Fourth Edition of *Business Mathematics* represents a thorough restructuring and updating of the Third Edition, in line with the recommendations of faculty consultants. The text material throughout has been reworked, revised, and reordered to provide rigorous, comprehensive coverage of the standard topics included in a business mathematics course. Interest rates, wage rates, and tax deductions have been updated. New topics have been added—inventory, ACRS depreciation, and shortcut methods—and the metric system and additional first-year depreciation have been deleted. The original Chapter 1 on fundamentals has been divided into three chapters for deeper understanding.

Business Mathematics, Fourth Edition, is dedicated to our families for their love, encouragement, and patience throughout the preparation of this book. Our thanks, therefore, go to the Kaliskis—Janice, Burton, Jr., Kristen, John, Karen, and Michael—and to the Dansbys—Barbara and Champ. Special thanks to Michael Kaliski for checking answers and pasting up the manuscript.

We also wish to thank the staff members at Harcourt Brace Jovanovich and Academic Press who are responsible for this book—specifically, Susan Mease—for their professionalism and unflagging devotion to the project.

BURTON S. KALISKI
ROBERT L. DANSBY

Contents

2 *Fractions* / 22

3 *Decimals* / 35

4 *Base, Rate, and Percentage* / 48

UNIT 3 The Mathematics of Operating / 113

8 Payroll / 115

9 Insurance / 136

10 Depreciation, Other Expenses, and Business Profits / 162

UNIT 1

Fundamentals of Arithmetic

Business depends upon numbers. To enable us to solve number problems in business, we apply the fundamentals of arithmetic to both routine and unusual business situations. Unit 1 covers these fundamentals in order to prepare you for the work of the following units.

Chapter 1 opens with a brief discussion of how numbers are written. It then reviews the basic processes of addition, subtraction, multiplication, and division in relation to whole numbers. An important application of the basic skills is the process of bank reconciliation, the major business application covered in this unit. Chapters 2 and 3 review the four basic processes in relation to fractions and decimals, respectively. All the problems in this textbook are solved by using one or more of these skills.

Chapter 4 presents one of the most common applications of business mathematics: finding base, rate, and percentage.

CHAPTER 1

The Basic Skills

Because you have been working with arithmetic since you started going to school, you may now be very adept at it. If so, skim through Chapter 1, pausing at the "Test Your Ability" sections to try the problems and check your knowledge.

If you are "rusty" or find new ideas to be learned in Chapter 1, study them very carefully. Mastery of the skills in this chapter is essential to your successful completion of this course.

Your Objectives

After completing Chapter 1, you should be able to

1. Perform the fundamental processes of arithmetic.
2. Add, subtract, multiply, and divide whole numbers.
3. Add, subtract, multiply, and divide whole numbers using shortcuts.
4. Prepare a bank reconciliation statement.

The Decimal System

Our number system is the *decimal system* because it is based on 10 (*decem* is Latin for ten). Any number can be written using the proper choice and combination of ten symbols: 0, 1, 2, 3, 4, 5, 6, 7, 8, and 9. Since these numbers form all other numbers in our number system, they are referred to as *whole numbers, natural numbers* or *integers.*

The decimal system is a *place value* system. This means that the value of each digit depends in part on its position. As the diagram on page 4 shows, each position is labeled, including the *decimal point* that separates whole numbers from their parts, or *fractions.* As you move to the left in a decimal number, each place is worth ten times the previous place; as you move to the right, each place is worth one-tenth of the previous place. We can analyze the decimal number 6,471.307 as follows:

6	4	7	1	.	3	0	7
thousands	*hundreds*	*tens*	*units*	*decimal point*	*tenths*	*hundredths*	*thousandths*

The number 6,471.307 is fully expressed as *six thousands, four hundreds, seven tens, one unit, three tenths, no hundredths, and seven thousandths*—or read in the more conventional way as *six thousand, four hundred seventy-one, and three hundred seven thousandths.* Notice where the "and" was placed in reading the above number. The "and" is pronounced only after all whole numbers. In other words, the "and" is used to separate whole decimal numbers from decimal fractions. To take another example, 155.5 is read "one hundred fifty-five and five tenths," not "one hundred and fifty-five and five tenths."

Even large numbers can be conveniently expressed using the place value system.

SAMPLE PROBLEM 1

Read aloud the following number: 1,256,761.513

Solution

One million, two hundred fifty-six thousand, seven hundred sixty-one *and* five hundred thirteen thousandths.

TEST YOUR ABILITY

Read aloud, or write, the following numbers:

1. 75
2. 275.5
3. 6.7
4. 99.99
5. 356.205
6. 1,986.225
7. 10,975.01
8. 78,222.065
9. 122,918.2
10. 4,589.002

Adding Whole Numbers

Addition is an arithmetic operation in which two or more numbers are combined into a single equivalent quantity. The numbers that are combined together are referred to as the *addends,* and the equivalent quantity is referred to as the *sum* or the *total.*

SAMPLE PROBLEM 2

Add the following numbers: 45 + 50 + 18

Solution

```
11 ←———————— carry
 45 ┐
 50 ├   addends
 18 ┘
———
113     sum
```

Notice, in the preceding example, that the total of the units column is 13, which means 3 is written below this column and 1 is carried to the next (tens) column. The total of the tens column is 10, and adding the 1 that was carried from the units column makes 11. This procedure is always followed. Thus, units are added to units, tens are added to tens, hundreds are added to hundreds, and so on, with numbers carried as needed.

SAMPLE PROBLEM 3

Add the following numbers: 322 + 84 + 276 + 88 + 730

Solution

$$\begin{array}{r} 322 \\ 84 \\ 276 \\ 88 \\ \underline{730} \\ 1{,}500 \end{array}$$

Checking for Accuracy

Accuracy should be checked in any arithmetic operation. One way to check the accuracy of addition is to add in reverse order. Thus, if a column of figures were added downward (which is usually the case) the accuracy of the operation can be checked by adding the same figures upward. Adding upward creates different number combinations which could result in detecting an error that was made when the numbers were added downward.

SAMPLE PROBLEM 4

Adding upward, check the accuracy of the addition in Sample Problem 3.

Solution

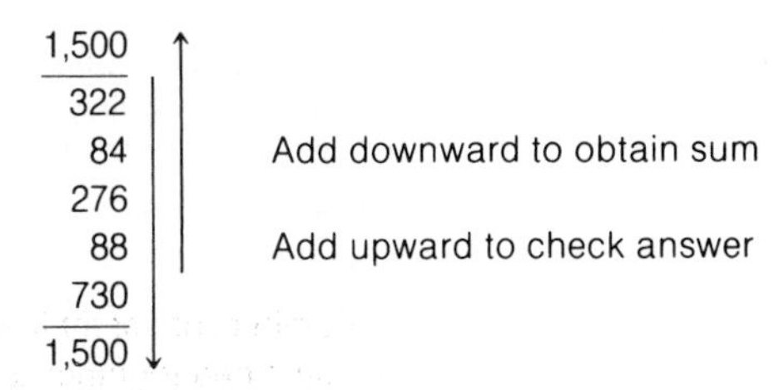

TEST YOUR ABILITY

Add the following numbers. Check the accuracy of each sum by adding in reverse order.

1.	2.	3.	4.	5.
663	450	221	222	772
356	45	220	388	923
902	230	876	645	619
818	925	564	836	417

6.	7.	8.	9.
6,289	7,435	10,567	256,000
7,821	9,040	11,988	217,800
9,289	5,678	16,222	353,344

Shortcut Addition

Recognizing certain combinations of numbers can save time and reduce errors when adding. To illustrate, add the following numbers:

$$\begin{array}{r} 387 \\ 723 \\ 256 \\ 858 \\ 444 \\ \hline \end{array}$$

What technique did you use in your addition? You may have said to yourself: "7 + 3 = 10; 10 + 6 = 16; 16 + 8 = 24; 24 + 4 = 28, write 8 and carry 2 to the next column." If you proceeded in this manner, you arrived at the correct total, but you did not make the best use of time.

The goal of the successful math student is to work with accuracy and speed. Speed in adding is attained by looking for combinations of digits, instead of adding a digit at a time. Train yourself to recognize combinations that total 10:

$$\begin{array}{c} 1 + 9 = 10 \\ 2 + 8 = 10 \\ 3 + 7 = 10 \\ 4 + 6 = 10 \\ 5 + 5 = 10 \end{array}$$

Or, in reverse order:

$$\begin{array}{c} 9 + 1 = 10 \\ 8 + 2 = 10 \\ 7 + 3 = 10 \\ 6 + 4 = 10 \\ 5 + 5 = 10 \end{array}$$

Look for these and similar combinations. Then add the column of numbers above in the following manner. (It may look somewhat complicated in print but it's fast in practice!)

Start in the far right column. You see 7 + 3 and 6 + 4 (or 2 tens) + an 8, which equals 28. Write 8 and carry the 2. In the second column you see 8 + 2 and 5 + 5 (or 2 tens), then the carried 2, +4 equals 26. Write 6 and carry 2. In the third column you see 3 + 7 and 2 + 8 (or 2 tens), and the carried 2, +4 equals 26. The complete procedure is diagrammed below.

2 ←	2 ←		Carry
3, 7) 10	8, 2) 10	7, 3) 10	
2, 8) 10	5, 5) 10	6, 8, 4) 10	
4	4		
10 + 10 + 2 + 4	10 + 10 + 2 + 4	10 + 10 + 8	
26	26	28	
26	6	8	= 2,668

TEST YOUR ABILITY

Add the following numbers. Watch for combinations that total 10.

1.	2.	3.	4.	5.
273	388	243	398	298
827	288	296	286	372
565	432	864	109	498
545	672	553	721	612

6.	7.	8.	9.	10.
346	598	937	444	398
764	309	173	566	498
882	45	897	589	612
282	561	213	111	712

Horizontal Addition

Thus far the numbers we have added have all involved *vertical addition* because we added down or up. Other problems call for adding across, or *horizontal addition.* Vertical and horizontal additon are often used together in order to double-check each other.

SAMPLE PROBLEM 5

Add the following numbers horizontally: 8 + 2 + 9 + 6 + 4

Solution

$$8 + 2 + 9 + 6 + 4 = 29$$

(running totals: 10, 19, 25)

When more than one digit is to be added horizontally, proceed in the same manner as in vertical addition. That is, add units to units, tens to tens, hundreds to hundreds, and so on until a sum is reached.

Let's look at Sample Problem 6, in which horizontal addition is used to find each sales representative's sales while vertical addition is used to find each day's sales. The results can be double-checked by (1) adding the vertical totals horizontally and (2) adding the horizontal totals vertically. The two grand totals should be equal. A difference in the answers shows that an error in addition has been made somewhere in the problem.

SAMPLE PROBLEM 6

Complete the following sales summary for Dionne's Dairy Products for the current week:

Solution

Dionne Dairy Products, Inc.
Sales Summary
July 10–14, 1988

Sales Representative	Monday	Tuesday	Wednesday	Thursday	Friday	Total
Marcia Lang	$ 725	$ 504	$ 631	$ 898	$ 646	*$ 3,404*
Jeffrey Engel	805	940	528	807	735	*3,815*
Lenny Wilson	377	624	339	981	427	*2,748*
Selma Ford	437	382	681	427	481	*2,408*
Marie Landa	847	704	401	207	565	*2,724*
TOTAL	*$3,191*	*$3,154*	*$2,580*	*$3,320*	*$2,854*	***$15,099***

TEST YOUR ABILITY

Add the following numbers vertically and horizontally:

1. 54 + 67 + 56 + 43 = ____
   45 + 24 + 66 + 24 = ____
   78 + 48 + 39 + 28 = ____
   66 + 34 + 56 + 98 = ____
   67 + 59 + 78 + 56 = ____
   __ __ __ __ = ____

2. 48 + 49 + 28 + 37 = ____
   78 + 27 + 49 + 33 = ____
   89 + 41 + 21 + 88 = ____
   90 + 43 + 77 + 22 = ____
   58 + 17 + 69 + 84 = ____
   __ __ __ __ = ____

3. Complete the following inventory summary for Barrett Department Store by using vertical and horizontal addition:

Barrett Department Store
Inventory Summary
October, 1988

Dept.	Units in Counters	Units in Stockroom	Units on Order	Total
Clothing	35	206	375	____
Jewelry	57	125	390	____
Pet	26	194	240	____
Lamp	18	206	126	____
Records	41	25	84	____
TOTAL				

Subtracting Whole Numbers

Subtraction is the second of the basic skills. It is an arithmetic operation that involves deducting one number from another. A number from which another number is deducted is called the *minuend.* The number that is deducted from the minuend is called the *subtrahend.* And the number that results from deducting the subtrahend from the minuend is called the *difference.*

SAMPLE PROBLEM 7

Subtract 372 from 884

Solution

```
 884   minuend
-372   subtrahend
 512   difference
```

Subtraction problems such as the one shown above give us little trouble because each number in the subtrahend is smaller than the corresponding number in the minuend. But to subtract, for example, 387 from 836, we must use the technique of *borrowing* from columns to the left. The "1" borrowed from 3 in step (a) below, because of the place value of the 3, is really ten. When we add this ten to the 6, what we are doing is breaking 36 into 20 and 16.

```
  7 ¹2  1
  8  3  6
 -3  8  7
  4  4  9
```

(a) Borrow 1 from the 3, making it 2 and changing the 6 to 16. $16 - 7 = 9$

(b) Borrow 1 from the 8, making it 7 and changing the 2 to 12. $12 - 8 = 4$

(c) $7 - 3 = 4$

Checking the Accuracy of Subtraction

The accuracy of a subtraction problem can be checked by simply adding the amount of the difference to the subtrahend. The sum of these two will, assuming your work is accurate, equal the minuend.

SAMPLE PROBLEM 8

Subtract 678 from 921 and check your answer.

Solution

```
 921
-678
----
 243
```

Accuracy check

```
 678
+243
----
 921
```

TEST YOUR ABILITY

Subtract each of the following and check for accuracy:

```
1.   545     2.   498     3.   385     4.  1,246
    -469         -419         -296        -   58

5. 10,566    6.   609     7.   958     8.   498
  - 8,760        -281         -789         -294

9.  3,458   10.  18,988
   -1,467       -     29
```

Horizontal Subtraction

In business, it is often necessary to do *horizontal subtraction.* The process is the same as vertical subtraction in that we subtract ones from ones, tens from tens, hundreds from hundreds, and so on.

SAMPLE PROBLEM 9

Horizontally subtract the following numbers. Check your answer by vertically adding the columns.

98 −	64 =	*34*
45 −	33 =	*12*
67 −	47 =	*20*
93 −	72 =	*21*
303	*216*	*87*

In the above problem, the difference between the column totals—303 and 216—is 87; and the sum of the differences between the individual amounts is 87. Thus, our accuracy has been verified.

Sample Problem 10 shows the use of both vertical addition, to find the three totals; and horizontal subtraction, to find the net figure on each line and provide a check on the grand total.

SAMPLE PROBLEM 10

Complete the following sales summary:

Solution

Foye Furnishings, Inc.
Branch Sales Summary
May, 1988

Branch	Sales	–	Sales Return	=	Net Sales
Los Angeles	$13,207		$ 737		*$12,470*
Boston	4,561		412		*4,149*
Houston	15,294		1,112		*14,182*
Hempstead	8,307		375		*7,932*
Seattle	11,934		799		*11,135*
Phoenix	9,136		804		*8,332*
TOTAL	*$62,439*		*$4,239*		***$58,200***

TEST YOUR ABILITY

Subtract the following horizontally and add the columns vertically:

1. 45 – 33 = _____
   87 – 76 = _____
   44 – 31 = _____
   98 – 79 = _____
   65 – 45 = _____
   __ – __ = _____

2. 6,788 – 4,566 = _____
   5,677 – 3,455 = _____
   6,899 – 4,688 = _____
   5,222 – 4,321 = _____
   9,765 – 6,860 = _____
   ____ – ____ = _____

3. Complete the following inventory summary by addition and subtraction:

Leslie Glass Corporation
Inventory Summary
July, 1988

Item	Balance 7/1	+	Receipts	=	Total	–	Issues	=	Balance 7/31
test tubes	76		52		_____		39		_____
juice glasses	68		51		_____		71		_____
jam jars	47		18		_____		24		_____
storage jars	29		26		_____		31		_____
baby bottles	103		14		_____		105		_____
milk bottles	74		19		_____		72		_____
TOTAL									

Multiplying Whole Numbers

Multiplication is the third, and most widely used, of the four basic skills in business mathematics. Multiplication involves finding the *product* of two numbers, the *multiplicand* and the *multiplier.* Let's look at a multiplication problem and review its terminology:

SAMPLE PROBLEM 11

Multiply 475 times 12

Solution

```
   475     multiplicand
×   12     multiplier
 -----
   950
  475
 -----
 5,700     product
```

To find the product of 475 and 12, you would proceed as follows:

Step 1: Write the smaller number under the larger with units under units, tens under tens, hundreds under hundreds, and so on.

Step 2: Multiply the multiplicand by the units figure in the multiplier, and write the partial product directly below the answer line ($475 \times 2 = 950$).

Step 3: Multiply the multiplicand by the tens figure in the multiplier, and write the product starting one place to the left of the previous partial product ($475 \times 1 = 475$).

Step 4: Add the partial products to obtain the product of the two numbers.

Checking the Accuracy of Multiplication

A popular method for checking the accuracy of a multiplication problem is to reverse the numbers and multiply the multiplier by the multiplicand.

SAMPLE PROBLEM 12

Multiply 55×34 and check.

Solution

```
    55
  × 34
 -----
   220
  165
 -----
 1,870
```

Check

```
    34
  × 55
 -----
   170
  170
 -----
 1,870
```

Another method for checking the accuracy of a multiplication problem is to divide the product by the multiplier to obtain the multiplicand, or divide the product by the multiplicand to obtain the multiplier.

TEST YOUR ABILITY

Multiply in each problem. Check your answers by reversing multiplier and multiplicand.

1. 26×45
2. 37×19
3. 24×38
4. 56×94
5. 32×81
6. 206×47
7. 311×84
8. 29×519
9. 412×637
10. 519×804

Shortcut Multiplication A knowledge of shortcut multiplication techniques can improve your speed and accuracy. Let's look at some popular shortcut techniques that are simple to perform and easy to remember.

Multiplying by a Power of 10. To multiply by 10, or a "power" of 10 (100, 1,000, and so on), simply annex (add on) the number of zeroes in the multiplier to the multiplicand.

SAMPLE PROBLEM 13

Find the product of each of the following:

456 × 10
520 × 100
300 × 1,000

Solution

456 × 10 = 4,560 (annex one zero)
520 × 100 = 52,000 (annex two zeroes)
300 × 1,000 = 300,000 (annex three zeroes)

In a multiplication problem such as 509 × 30, first deal with the zeroes. That is, simply annex a zero and then multiply by 3. For example:

509 × 30 = 509 × 10 × 3 = 5,090 × 3 = 15,270

This procedure helps to increase the number of problems we can do mentally. Some problems, however, require a written procedure. To multiply 864 by 600 most efficiently:

```
    864
  × 600 ←— (a) Set zeroes to right
  -----
     ↓↓
518,400 ←— (b) Copy zeroes on answer line
 ↑——————— (c) Multiply 6 × 864 = 5,184
```

Multiplying by 25 and 50. When a number is multiplied by 25, the product can be found quickly by annexing two zeroes (multiplying by 100) to the multiplicand, and then dividing by four, since 25 is one-fourth of 100. Similarly, when a number is multiplied by 50, the product can be found quickly by annexing two zeroes and then dividing by 2, since 50 is one-half of 100.

SAMPLE PROBLEM 14

Using the shortcut method, multiply the following:

400 × 25
650 × 25
800 × 50

Solution

400 × 25 = 40,000 ÷ 4 = 10,000
650 × 25 = 65,000 ÷ 4 = 16,250
800 × 50 = 80,000 ÷ 2 = 40,000

Another way to improve speed is to be sure that you know your multiplication tables. If you are slow at multiplying numbers such as 5 × 6 or 8 × 9, you can improve your speed by preparing and practicing tables like the one here. Table 1.1 can be expanded to include other common operations, such as 12 × 11, if you wish. Notice that only half of the table need be shown since, for example, 3 × 6 is the same as 6 × 3.

Table 1.1

Multiplication Facts

×	1	2	3	4	5	6	7	8	9
1	1	2	3	4	5	6	7	8	9
2		4	6	8	10	12	14	16	18
3			9	12	15	18	21	24	27
4				16	20	24	28	32	36
5					25	30	35	40	45
6						36	42	48	54
7							49	56	63
8								64	72
9									81

TEST YOUR ABILITY

Multiply each of the following using shortcuts:

1. 215 × 10
2. 313 × 40
3. 411 × 100
4. 130 × 1,000
5. 345 × 50
6. 6,750 × 100
7. 8,000 × 25
8. 5,400 × 10
9. 7,200 × 200
10. 9,400 × 300
11. 4,815 × 100
12. 9,000 × 50
13. 6,000 × 25
14. 8,800 × 3,000
15. 2,440 × 1,000

Dividing Whole Numbers

Division is an arithmetic process that determines how many times one number is contained in another number. The number into which another number is divided is called the *dividend*, and the number that is divided into the dividend is called the *divisor*. The number that results from dividing one number into another number is called the *quotient*.

SAMPLE PROBLEM 15

Divide 288 by 4

Solution

$$\begin{array}{lrl} & 72 & \text{quotient} \\ \text{divisor} & 4\overline{)288} & \text{dividend} \end{array}$$

This is an example of short division in that we can divide the dividend by the divisor without further calculations. Long division is used when a problem is too involved for short division. Long division problems are relatively rare in business, but when they occur, you should know how to solve them.

First, let's consider a problem in which there is no remainder: Divide 5040 by 24.

```
       210      (a) 24 goes into 50 two times
  24)5040
     48         (b) 2 × 24 = 48; 50 − 48 = 2; bring
     --             down the next digit, 4
      24
      24        (c) 24 goes into 24 once
      --
      00        (d) 1 × 24 = 24; 24 − 24 = 0; bring
                    down the next 0
                (e) 24 goes into 00 zero times
```

5,040 (*the dividend*) ÷ 24 (*the divisor*) = 210 (*the quotient*)

Now, look at the division of 603 by 32. Notice that there is a remainder in the answer.

```
(a) 32 goes into 60 once ──→  18     (c) 32 goes into 283 eight times
                          32)603     (d) 8 × 32 = 256; 283 − 256 = 27
(b) 1 × 32 = 32; 60 − 32 =    32
    28; bring down the 3 ──→  283
                              256
                              ---
                               27
```

Should anything further be done with this answer? That depends on the directions given for the problem. If the answer is to be expressed with a fractional remainder, the answer is $18\frac{27}{32}$.

If the answer is to be expressed in tenths, we must carry the division two places further (to hundredths). Write 603 as 603.00 in order to divide to two places. The full problem will then appear as follows:

```
     18.84
 32)603.00
    32
    --
    283
    256
    ---
     270
     256
     ---
      140
      128
      ---
       12
```

Rounded to tenths, 18.84 becomes 18.8.

Checking the Accuracy of Division Problems

The most popular way to check the accuracy of a division problem is to multiply the quotient by the divisor. If your math is correct, the product will equal the dividend. Should a remainder exist, it is added to the product of the quotient and divisor.

SAMPLE PROBLEM 16

Divide 27,125 by 60 and check for accuracy

Solution

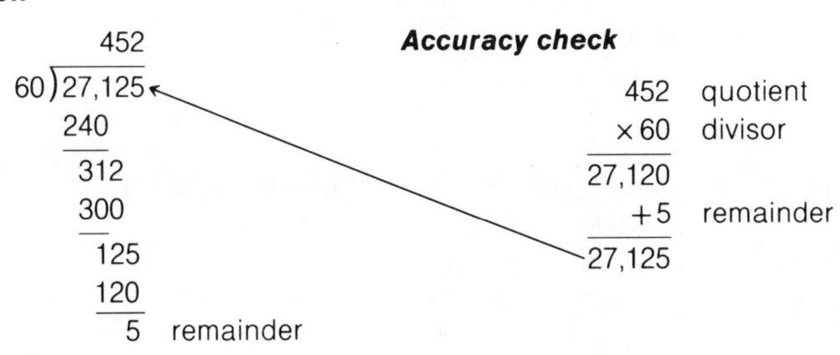

Divide each of the following and check your answer by multiplication:

TEST YOUR ABILITY

1. 24)34,560
2. 18)36,720
3. 50)78,400
4. 17)34,340
5. 19)41,800
6. 30)36,030
7. 445)378,250
8. 340)680,000
9. 250)500,000

Shortcut Division

Just as there are certain shortcuts that can save you time when multiplying, there are shortcuts that can save you time when dividing. Consider these two popular shortcut methods for dividing:

Dividing by a Power of 10. To divide by 10, or a power of 10 (100, 1,000, and so on), simply move the decimal point one place left in the dividend for each zero in the divisor.

SAMPLE PROBLEM 17

Divide each of the following:

2,400 ÷ 10
788.8 ÷ 100
244.6 ÷ 1,000

Solution

2,400 ÷ 10 = 240 (move the decimal point one place to the left)
788.8 ÷ 100 = 7.888 (move the decimal point two places to the left)
244.6 ÷ 1,000 = .2446 (move the decimal point three places to the left)

Division by 25. When a number is divided by 25, the quotient can be found quickly by multiplying the dividend by 4; then dividing by 100 (moving the decimal point two places to the left).

SAMPLE PROBLEM 18

Find the quotient in the following problem: 10,400 ÷ 25

Solution

10,400 × 4 = 41,600
41,600 ÷ 100 = 416

Division by 50. When a number is divided by 50, the quotient can be found quickly by multiplying the dividend by 2; then dividing by 100.

SAMPLE PROBLEM 19

Find the quotient in the following problem: 62,000 ÷ 50

Solution

62,000 × 2 = 124,000
124,000 ÷ 100 = 1,240

TEST YOUR ABILITY

Using shortcuts, find the quotient in each of the following problems:

1. 450 ÷ 10	5. 5,690 ÷ 1,000
2. 600 ÷ 100	6. 8,875 ÷ 25
3. 800 ÷ 25	7. 9,450 ÷ 50
4. 750 ÷ 100	8. 12,950 ÷ 50

The Bank Reconciliation: An Application of the Basics

One common form of analysis is the comparison of two different numbers to see if they agree. Another word for the procedure is *reconciliation*—the resolution of differences. It is all too common for the bank's records and the firm's records of the same account to disagree. It then becomes the accountant's job or—in this case—your job to reach an agreement—or, if not, to find out why not.

Let's explore very briefly the operations of a checking account and then return to the process of reconciliation, a process which applies the basic skills.

The Checking Account

The majority of business transactions are paid for by check. A *check* is an order by the *drawer,* who writes the check on the *drawee* (the drawer's bank), to pay the *payee.* Since it is *negotiable,* it can be transferred by *endorsement.*

A model check appears below:

NO. 47

70-2915
711

June 7 1988

PAY TO THE ORDER OF Henry Stevens $ 250.00

Two hundred-fifty and 00/100 DOLLARS

FIRST NATIONAL BANK
Centerville, State 09077

Jane Adams

⑆0711⑈2915⑆ 255⑈200803⑈

The *drawer* is Jane Adams; the *drawee,* the First National Bank; and the *payee,* Henry Stevens. Let's follow this check around.

Stevens receives the check and cashes it in his own bank, United Trust. United Trust sends the check to a check-sorting center, called the *clearinghouse,* which returns the check to the drawee. When First National receives the check, it subtracts the $250 from Adams's account. The check's total travel time is about a week.

Once a month, First National returns to Adams all the *canceled checks,* which it has deducted from her account. It also includes a *bank statement* showing the bank's record of Adams's deposits, checks deducted, and balance.

Adams properly handles her own *checkbook records* by adding all deposits as she makes them and subtracting all checks as she draws them. She also has a record of her balance.

In theory, the bank's records and Adams's records should agree. However, it is rare that they do. Why? Look at the following possibilities:

1. Checks that Adams has written and deducted from her records may not yet have reached her bank. If Adams draws a check on July 27 that reaches her bank on August 3, a bank statement as of July 31 will not show this *outstanding check.*
2. *A late deposit* made by Adams on the last day of the month will be on her own records, but not recorded by the bank on the month-end statement.
3. Although this is a changing trend, many checking accounts are subject to a *service charge* per check (such as 15¢) or a flat monthly fee ($2.50, for example). The bank always includes the charge while Adams may have been waiting until she receives her bank statement before doing so.
4. The bank may have collected a note for Adams, a fact that she does not know until the bank statement arrives. Or a check from a customer, perhaps Robert Harris, may *bounce*—that is, when it reaches the drawer's bank, the account has *insufficient funds* to pay the check. In such a case, the check is not accepted and is returned to her for action against Harris. Finally, it is possible that either Adams or the bank made an *error.*

In order to know exactly how much she has in her account, Adams prepares a *bank reconciliation statement,* a form used to show whether a depositor's records and the bank's records are *actually in agreement.*

Preparing the Statement

To reconcile an account, begin by listing the checkbook balance in one column and the bank statement balance in a parallel column. List any item that has been recorded by the depositor *but not by the bank* on the bank's side. Treat it as the depositor treated it on the depositor's own records. If it was *added* on the depositor's records, *add* it to the bank's balance. If it was *deducted* on the depositor's records, *deduct* it from the bank's balance.

List any item that has been recorded by the bank *but not by the depositor* on the depositor's side and treat it as the bank treated it.

Correct any *error* on the side of the party who made the error.

SAMPLE PROBLEM 20

From the data presented below, prepare a bank reconciliation for the Groat Company on April 30, 1988.

Bank statement balance	$2,328.55
Checkbook balance	1,979.75
Outstanding checks	318.25
Late deposit	165.20
Service charge	4.25
Collection of note by bank	200.00

Solution

Groat Company
Bank Reconciliation Statement
April 30, 1988

Checkbook balance	$1,979.75	Bank statement balance	$2,328.55
Less: Service charge	4.25	Less: Outstanding checks	318.25
	$1,975.50		$2,010.30
Add: Collection of note	200.00	Add: Late deposit	165.20
Adjusted balance	$2,175.50	Available balance	$2,175.50

Notes

(1) The statement contains two adjustments on each side. This is not always the case; there might have been many additional transactions to record.

(2) If the final balances do not agree, recheck. Look for an error in the checkbook. If there is none, then the bank should be notified.

(3) If two items appear under any heading, the proper form is:

Less:	Service charge	$ 5.00	
	Insufficient funds	100.00	105.00

(4) Why is the service charge on the checkbook side? Because the bank recorded it and Groat has not. Why is it subtracted? Because the bank subtracted it. Follow this logic for every item.

(5) A different bank may use a different form for reconciliation.

(6) If there was a series of outstanding checks, the proper way to list them is by number and amount, such as:

Less:	Outstanding checks		
	#105	$ 40.00	
	107	12.00	
	111	110.00	162.00

TEST YOUR ABILITY

1. Prepare a bank reconciliation statement for Paul Shaw from the following information on May 31, 1988:

Bank statement balance	$1,166.56
Checkbook balance	1,046.20
Outstanding checks	209.91
Service charge	3.05
Late deposit	86.50

2. Prepare a bank reconciliation statement for the Windsor Chemical Corporation from the following data, as of July 31, 1988:

Bank statement balance	$10,192.56
Checkbook balance	10,239.32
Outstanding checks	857.19
Service charge	8.95

Late deposit	955.00
Collection of note by bank	125.00
Check of customer marked "Insufficient Funds"	65.00

3. Alice Warner's bank statement shows a balance of $2,041.00. She finds that a deposit made on July 31 for $210.50 has not been recorded on the bank statement, and she determines that checks totaling $147.19 are outstanding. The bank has deducted a $3.15 service charge and a $75.00 credit card advance from her account. A check for $68.00 returned by the bank was recorded as $86.00 in her own checkbook, which shows a balance of $2,164.46 on July 31, 1988. Prepare a bank reconciliation statement for Warner.

CHECK YOUR READING

1. Write out the following numbers:
   (a) 555.05
   (b) 22.025
   (c) 3,456.78
2. Explain the difference between vertical and horizontal addition.
3. In a problem using vertical and horizontal addition, how is each used to check the other?
4. Multiply each of the following numbers by 10:
   (a) 45
   (b) 72.1
   (c) 6.37
   (d) .854
5. Multiply each of the following numbers by 100:
   (a) 6
   (b) 26
   (c) 37.55
   (d) 9.276
6. Using shortcuts, multiply the following numbers as indicated:
   (a) 450 × 25
   (b) 300 × 50
   (c) 950 × 25
   (d) 200 × 50
7. Divide the following numbers by 10:
   (a) 50
   (b) 60
   (c) 800
   (d) 925
8. Divide the following numbers by 100:
   (a) 678.25
   (b) 222.665
   (c) 2,450.65
   (d) 10,895
9. Using shortcuts, divide the following:
   (a) 600 by 25
   (b) 900 by 50
   (c) 26,550 by 50
   (d) 12,400 by 25

10. Wilson gives a check to Phillips. Wilson's bank is Bank A; Phillips's, Bank B. Identify the drawer, the drawee, and the payee.
11. Distinguish between a *canceled check* and an *outstanding check*.
12. State whether each of the following will be (a) added to the bank statement balance; (b) subtracted from the bank statement balance; (c) added to the checkbook balance; (d) subtracted from the checkbook balance.
    (1) Outstanding checks
    (2) Collection made by bank for depositor
    (3) Late deposits not recorded by bank
    (4) Service charge made by the bank
    (5) Insufficient funds check charge made by the bank
13. The Sally Seasoning Company has written checks 127 through 145 this month. Checks 127, 128, 129, 130, 131, 132, 133, 135, 136, 137, 138, 139, 141, and 144 are returned. Which checks are outstanding?
14. Why do the bank statement balance and the checkbook balance rarely agree?

EXERCISES

1. Add the following. Check for accuracy by adding in reverse order.

(a)	(b)	(c)	(d)
24,290	38,245	456,384	22,190
54,789	36,498	281,394	757
47,374	22,382	37,593	18,353
58,385	8,725	3,727	77
9,364	29,483	24,516	2,213

2. Add the following numbers horizontally.
   (a) $8 + 5 + 6 + 9 + 7 =$
   (b) $4 + 6 + 9 + 3 + 8 =$
   (c) $18 + 22 + 67 + 93 + 98 =$
   (d) $222 + 888 + 473 + 633 + 907 =$
3. Complete the following summary of sales by branch, using vertical and horizontal addition:

Andersen Clothing Co., Inc.
Branch Sales Summary
October–December, 1988

Branch	October	November	December	Total
Fall River	$2,060	$3,000	$3,900	____
Providence	1,850	2,600	3,200	____
Salem	1,910	2,700	4,000	____
Bedford	2,100	3,500	2,900	____
Laconia	1,400	2,200	3,100	____
Manchester	960	1,500	1,600	____
TOTAL	____	____	____	____

4. Subtract the following. Check for accuracy by adding.

(a)	(b)	(c)	(d)	(e)
34,560	345,669	444,230	34,600	3,276
22,198	45,091	2,564	228	176

5. Subtract the following numbers horizontally and then add the columns vertically:

4,780 − 2,395 = ____
3,565 − 3,458 = ____
4,550 − 2,560 = ____
5,245 − 4,300 = ____
3,568 − 2,458 = ____
26,450 − 8,455 = ____
9,467 − 7,548 = ____
____ − ____ = ____

6. Multiply the following. Check your answer by reverse multiplication.
   (a) 408 × 18 (b) 3,457 × 22 (c) 3,200 × 125
   (d) 9,885 × 228 (e) 16,552 × 525 (f) 200 × 10
7. Multiply the following using shortcuts.
   (a) 846 × 20 (d) 5,600 × 600 (g) 75,600 × 400
   (b) 344 × 40 (e) 777 × 100 (h) 3,569 × 200
   (c) 3,450 × 4,000 (f) 4,895 × 900 (i) 25,455 × 10
8. Multiply the following using shortcuts.
   (a) 400 × 25
   (b) 900 × 100
   (c) 750 × 50
   (d) 90 × 10
   (e) 3,200 × 25
9. Divide the following. State remainders as fractions.
   (a) 75)155,250 (c) 60)244,800 (e) 155)345,805
   (b) 600)987,600 (d) 250)600,000 (f) 44)572
10. Divide the following using shortcuts.
   (a) 600 by 10 (e) 200 by 25
   (b) 900 by 100 (f) 300 by 50
   (c) 875 by 1,000 (g) 3,500 by 25
   (d) 2.25 by 10 (h) 35,900 by 1,000
11. Jim Nichol's bank statement balance is $14,265.37. He discovers that a deposit of $509.56 made on August 31, 1988, does not appear on the bank statement. Checks of $1,146.83 are outstanding. The bank has collected a note for $656.50 on behalf of Nichols and has deducted $17.50 for a service charge. Check 475, recorded in Nichols's checkbook for $467.12, was discovered actually to be for $467.21. Prepare a bank reconciliation statement to determine if Nichols's checkbook balance of $12,989.19 is correct.
12. Using the following data, prepare a bank reconciliation statement as of October 31, 1988 for the Kobler Chemical Company.

Bank statement balance	$15,391.84
Checkbook balance	15,105.10
Outstanding checks	1,374.95
Service charge	19.45
Late deposit	318.51
Note paid by bank to payee	555.50
Customer's check marked "Insufficient Funds"	194.75

CHAPTER **2**

Fractions

The word "fraction" comes from the Latin word *fractos,* which means "to break." Thus, a fraction is a division of a whole unit into parts. Every business day, we are constantly confronted with fractions. In newspaper, radio, and TV ads, for example, we often see fractions used; "Gigantic $\frac{1}{2}$ price sale"; "beautiful house for sale on $\frac{3}{4}$ acre lot, featuring $2\frac{1}{2}$ baths with $\frac{1}{10}$ down and an assumable mortgage rate of only $10\frac{1}{4}$%"; "Coca Cola stock was actively traded today at $64\frac{1}{4}$"; "come see Florida's beautiful beaches this weekend, only $3\frac{1}{2}$ hours from Atlanta." These and other uses of fractions are so commonplace, that you need a firm foundation in the use and application of common fractions.

Your Objectives

After studying Chapter 2, you should be able to

1. Reduce a fraction to lowest terms.
2. Raise a fraction to higher terms.
3. Convert improper fractions to mixed numbers.
4. Convert mixed numbers to improper fractions.
5. Add, subtract, multiply, and divide fractions.

Working with Fractions

A fraction is expressed as two whole numbers separated by a bar, for example, $\frac{4}{7}$. The fraction bar is either horizontal (–) or slanted (/). The number above the horizontal bar—or to the left of the slanted bar—is called the *numerator.* It shows how many parts of a unit are expressed. The number below the horizontal bar—or to the right of the slanted bar—is called the *denominator.* The denominator shows how many parts the unit is divided into. The numerator and the denominator are the *terms* of the fraction.

A *proper fraction* is one in which the numerator is smaller than the denominator. Thus, $\frac{1}{2}$, $\frac{3}{4}$, $\frac{4}{5}$, and $\frac{7}{9}$ are all examples of proper fractions. An *improper fraction* is one in which the numerator is equal to or greater than the denominator. Thus, $\frac{3}{2}$, $\frac{2}{2}$, and $\frac{5}{4}$ are examples of improper fractions. A *mixed number* is composed of both whole numbers and fractions. Examples of mixed numbers are $2\frac{1}{4}$, $3\frac{1}{8}$, $15\frac{1}{2}$, and $20\frac{1}{8}$.

Reducing to Lowest Terms

Multiplying or dividing both terms of a fraction by the same number (factor)—other than zero—does not change the value of the fraction. Thus, a fraction stated in higher terms can often be reduced to lower terms to make calculations easier. We say that a fraction has been *reduced to lowest terms* when there is no factor in common by which the numerator and the denominator can be divided.

To reduce a fraction to lowest terms, divide both the numerator and the denominator by common factors until further division is impossible. For example, to reduce the fraction $\frac{450}{750}$ to lowest terms:

1. Divide by the obvious factor that is common to both terms, 10.

$$\frac{450}{750} = \frac{450 \div 10}{750 \div 10} = \frac{45}{75}$$

2. Divide by the next obvious factor, 5.

$$\frac{45}{75} = \frac{45 \div 5}{75 \div 5} = \frac{9}{15}$$

3. Divide by 3.

$$\frac{9}{15} = \frac{9 \div 3}{15 \div 3} = \frac{3}{5}$$

Because the numbers 3 and 5 have no factor in common, the fraction $\frac{450}{750}$ has been reduced to lowest terms.

TEST YOUR ABILITY

Reduce each of the following to lowest terms:

1. $\frac{16}{18}$	6. $\frac{140}{180}$	11. $\frac{675}{1000}$
2. $\frac{34}{72}$	7. $\frac{250}{600}$	12. $\frac{700}{1200}$
3. $\frac{19}{35}$	8. $\frac{800}{900}$	13. $\frac{400}{1200}$
4. $\frac{40}{50}$	9. $\frac{175}{250}$	14. $\frac{150}{3000}$
5. $\frac{50}{100}$	10. $\frac{255}{415}$	15. $\frac{600}{600}$

Determining the Greatest Common Divisor

In reducing the above problems to lowest terms, we looked for a factor (divisor) that would reduce both terms of the fraction. We continued this procedure until further division was impossible. In some cases, the largest factor—referred to as the *greatest common divisor*—is obvious. For example, in reducing the fraction $\frac{40}{50}$ we could mentally observe that dividing by 10 would reduce 40 to 4, and reduce 50 to 5, thus reducing the fraction to its lowest terms, $\frac{4}{5}$. In other fractions, the greatest common divisor is not as obvious. For example, reduce $\frac{104}{152}$ to lowest terms. First, find the greatest common divisor, as follows:

1. Divide the denominator by the numerator. If the denominator is equally divisible by the numerator, the numerator is the greatest common divisor. If it is not equally divisible, continue to Step 2.
2. Divide the remainder into the previous divisor.
3. Continue dividing the next remainder into the previous divisor until it goes evenly, or there is no remainder. The last divisor used, unless it is 1, will be the greatest common divisor. Should you get a remainder of 1, the fraction is in its lowest terms.

SAMPLE PROBLEM 1

Find the greatest common divisor and reduce $\frac{104}{152}$ to lowest terms.

Solution

To find the greatest common divisor:

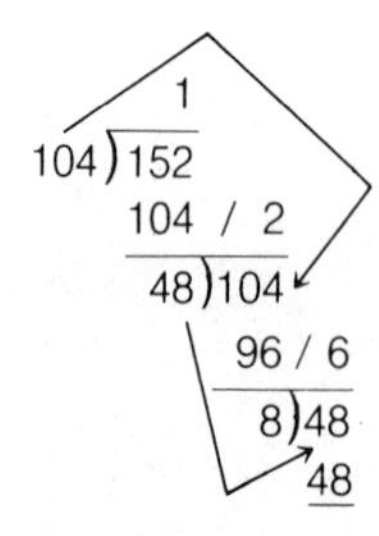

The greatest common divisor is 8.

To reduce to lowest terms:

$$\frac{104 \div 8}{152 \div 8} = \frac{13}{19}$$

TEST YOUR ABILITY

Find the greatest common divisor in each of the following fractions:

1. $\frac{109}{254}$
2. $\frac{245}{390}$
3. $\frac{90}{630}$
4. $\frac{54}{540}$
5. $\frac{496}{690}$
6. $\frac{242}{488}$
7. $\frac{320}{970}$
8. $\frac{1,442}{3,688}$
9. $\frac{2,880}{4,620}$
10. $\frac{1,425}{1,550}$

Raising a Fraction to Higher Terms

A fraction can be raised to higher terms by multiplying both the numerator and the denominator by the same number. This process is accomplished by dividing the original denominator into the desired denominator, then multiplying the numerator by this figure.

SAMPLE PROBLEM 2

Change the following fraction to the term indicated:

$$\frac{3}{4} = \frac{?}{12}$$

Solution

Divide original denominator into the desired denominator:

$$4\overline{)12}^{\,3} \qquad \frac{3 \times 3}{4 \times 3} = \frac{9}{12}$$

TEST YOUR ABILITY

Change the following fractions to the terms indicated.

1. $\frac{2}{3} = \frac{?}{9}$
2. $\frac{4}{5} = \frac{?}{15}$
3. $\frac{6}{8} = \frac{?}{32}$
4. $\frac{5}{7} = \frac{?}{42}$
5. $\frac{2}{15} = \frac{?}{45}$
6. $\frac{1}{10} = \frac{?}{100}$
7. $\frac{45}{90} = \frac{?}{180}$
8. $\frac{2}{9} = \frac{?}{54}$
9. $\frac{5}{8} = \frac{?}{128}$

Changing Improper Fractions to Whole or Mixed Numbers

Final answers involving improper fractions are usually stated as whole or mixed numbers. An improper fraction is changed to a whole or mixed number simply by dividing the numerator by the denominator and stating the remainder as a numerator over the original denominator.

SAMPLE PROBLEM 3

Change $\frac{5}{4}$ to a mixed number.

Solution

$$\begin{array}{r} 1 \\ 4\overline{)5} \\ 4 \\ \hline 1 \end{array} \text{ remainder} \qquad = 1\tfrac{1}{4}$$

TEST YOUR ABILITY

Change the following to whole or mixed numbers, reduced to lowest terms:

1. $\frac{40}{15}$
2. $\frac{60}{20}$
3. $\frac{90}{35}$
4. $\frac{75}{5}$
5. $\frac{44}{20}$
6. $\frac{388}{300}$
7. $\frac{350}{50}$
8. $\frac{125}{25}$
9. $\frac{400}{125}$
10. $\frac{275}{5}$
11. $\frac{2,000}{50}$
12. $\frac{1,650}{40}$
13. $\frac{2,488}{60}$
14. $\frac{4,550}{12}$
15. $\frac{1,200}{20}$

Changing Mixed Numbers to Improper Fractions

Some operations with fractions call for the use of improper fractions. A mixed number is changed to an improper fraction by following these steps:

1. Multiply the whole number by the denominator.
2. Add the numerator to the product obtained in Step 1.
3. State the sum obtained in Step 2 as the numerator over the original denominator.

SAMPLE PROBLEM 4

Change $3\frac{2}{5}$ to an improper fraction:

Solution

$5 \times 3 = 15 + 2 = \frac{17}{5}$

TEST YOUR ABILITY

Change each of the following mixed numbers to improper fractions:

1. $1\frac{1}{3}$	6. $17\frac{1}{4}$	11. $240\frac{1}{4}$
2. $9\frac{2}{3}$	7. $18\frac{2}{9}$	12. $125\frac{1}{2}$
3. $6\frac{1}{7}$	8. $35\frac{1}{2}$	13. $260\frac{4}{5}$
4. $8\frac{3}{4}$	9. $14\frac{2}{3}$	14. $320\frac{7}{8}$
5. $7\frac{1}{2}$	10. $40\frac{7}{9}$	15. $900\frac{5}{6}$

Adding Fractions

All four basic arithmetic processes used with whole numbers—addition, subtraction, multiplication, and division—can also be used with fractions.

To add fractions with the same denominator, simply add the numerators while keeping the common denominator.

$$\frac{1}{8} + \frac{3}{8} = \frac{4}{8}$$

Since 4 and 8 are both divisible by 4, reduce the answer to its lowest terms, $\frac{1}{2}$.

Fractions can only be added when all fractions in the problem have the same denominator. For example, we cannot add $\frac{1}{4}$ and $\frac{1}{3}$ because fourths and thirds are not equivalent terms. We must have a denominator that is common to both 4 and 3. To minimize later reductions, the denominator we use in solving such a problem should be the *lowest common denominator*. In many cases, the lowest common denominator can be obtained simply by observing the problem. For example, what is the lowest number that is common to 4 and 3? The answer is 12, since it is a multiple of both 4 and 3. Thus, we can add $\frac{1}{4}$ and $\frac{1}{3}$ as follows:

$$\frac{1 \times 3}{4 \times 3} = \frac{3}{12}$$

$$\frac{1 \times 4}{3 \times 4} = \frac{4}{12}$$

$$\frac{7}{12}$$

Finding the Lowest Common Denominator

For many problems, it is difficult to determine the lowest common denominator by observation. For such problems, the *prime number method* can be used to determine the lowest common denominator. A *prime number* is a number that is evenly divisible only by itself and 1. Examples of prime numbers are 1, 2, 5, 7, 11, 13, 17, 19, 23, and so on. The prime number method works by setting up the denominators of a problem in a scheme, and then factoring out the prime numbers. To illustrate this process, let's look at Sample Problem 5:

SAMPLE PROBLEM 5

Find the lowest common denominator of $\frac{2}{3}$, $\frac{8}{9}$, $\frac{11}{12}$, and $\frac{5}{18}$:

Solution

1. Place all denominators in a scheme, as follows:

   $\underline{)3 - 9 - 12 - 18}$

2. Select a prime number that will evenly divide into *at least two of the denominators*. For this problem, let's start with 3.

   $\underline{3)3 - 9 - 12 - 18}$

3. Divide the prime number you selected—3 in this case—into each of the denominators. If it does not divide evenly, simply bring down the denominator.

$$\begin{array}{r|l} 3 & \cancel{3} - \cancel{9} - \cancel{12} - \cancel{18} \\ \hline & 1 - 3 - 4 - 6 \end{array}$$

4. Continue dividing by prime numbers until you cannot divide a prime number into at least two of the denominators.

$$\begin{array}{r|l} 3 & \cancel{3} - \cancel{9} - \cancel{12} - \cancel{18} \\ \hline 3 & 1 - \cancel{3} - 4 - \cancel{6} \\ \hline 2 & 1 - 1 - \cancel{4} - \cancel{2} \\ \hline & 1 - 1 - 2 - 1 \end{array}$$

5. Find the least common denominator by multiplying the numbers along the edge of the scheme:

$$3 \times 3 \times 2 \times 1 \times 1 \times 2 \times 1 = 36$$

Having found our lowest common denominator of 36, we can now add this problem, as follows:

$$\begin{aligned} \tfrac{2}{3} &= \tfrac{24}{36} \\ \tfrac{8}{9} &= \tfrac{32}{36} \\ \tfrac{11}{12} &= \tfrac{33}{36} \\ \tfrac{5}{18} &= \tfrac{10}{36} \\ \hline \tfrac{99}{36} &= 2\tfrac{27}{36} = 2\tfrac{3}{4} \end{aligned}$$

Adding Mixed Numbers

To add two or more mixed numbers, proceed as follows: (1) add the fractions; (2) add the whole numbers; and (3) add the sum of the fractions to the sum of the whole numbers. Sample Problem 6 shows the addition of $12\frac{1}{2}$ and $10\frac{1}{4}$.

SAMPLE PROBLEM 6

Add $12\frac{1}{2} + 10\frac{1}{4}$

Solution

The lowest common denominator is 4:

$$\begin{aligned} 12\tfrac{1}{2} &= 12\tfrac{2}{4} \\ 10\tfrac{1}{4} &= 10\tfrac{1}{4} \\ \hline & \quad 22\tfrac{3}{4} \end{aligned}$$

TEST YOUR ABILITY

Find the lowest common denominator, and add each of the following sets of fractions:

1. $\frac{3}{4} + \frac{5}{8} + \frac{7}{9} + \frac{1}{12}$
2. $\frac{3}{5} + \frac{1}{8} + \frac{2}{3} + \frac{4}{9}$
3. $\frac{1}{2} + \frac{6}{7} + \frac{1}{3} + \frac{7}{8}$
4. $\frac{9}{11} + \frac{6}{9} + \frac{2}{3} + \frac{1}{4}$
5. $\frac{2}{7} + \frac{1}{7} + \frac{3}{14} + \frac{1}{28}$
6. $\frac{14}{21} + \frac{12}{42} + \frac{15}{63}$
7. $15\frac{4}{5} + 12\frac{2}{3}$
8. $\frac{12}{15} + \frac{6}{45} + \frac{41}{90}$
9. $\frac{15}{40} + \frac{11}{30} + \frac{17}{60}$
10. $25\frac{7}{8} + 18\frac{2}{9}$
11. $\frac{150}{300} + \frac{90}{150}$
12. $145\frac{11}{12} + 18 + 220\frac{1}{6}$

Subtracting Fractions

As in addition, fractions must have the same denominator to be subtracted. To subtract fractions, first make the denominators identical; then find the difference between the numerators and place this difference over the common denominator.

To subtract $\frac{5}{16}$ from $\frac{9}{16}$:

$$\frac{9}{16} - \frac{5}{16} = \frac{4}{16} = \frac{1}{4}$$

To subtract $\frac{2}{3}$ from $\frac{4}{5}$:

$$\frac{4}{5} - \frac{2}{3} = \frac{4 \times 3}{5 \times 3} - \frac{2 \times 5}{3 \times 5} = \frac{12}{15} - \frac{10}{15} = \frac{2}{15}$$

Subtracting Mixed Numbers

A mixed number is subtracted from another mixed number in two steps, as follows: (1) find a common denominator and subtract the fractional portion of the subtrahend from the fractional portion of the minuend and (2) subtract the whole number portion of the subtrahend from the whole number portion of the minuend.

SAMPLE PROBLEM 7

Find the difference between $45\frac{1}{2}$ and $12\frac{1}{3}$.

Solution

The lowest common denominator is 6.

$$\begin{array}{r} 45\frac{1}{2} = 45\frac{3}{6} \\ -12\frac{1}{3} = 12\frac{2}{6} \\ \hline 33\frac{1}{6} \end{array}$$

In the above mixed number, the fractional portion of the minuend was larger than the fractional portion of the subtrahend. Thus, to complete the subtraction process, it was not necessary to borrow from the whole number portion of the minuend. In the subtraction of some mixed numbers, however, it is necessary to borrow a whole unit from the whole number portion of the minuend. The whole unit borrowed is always expressed as a fraction with a denominator equal to the minuend. Notice in Sample Problem 8, below, that the whole unit borrowed is equal to $\frac{9}{9}$.

SAMPLE PROBLEM 8

Find the difference between $15\frac{1}{3}$ and $5\frac{7}{9}$

Solution

$$\begin{array}{r} 15\frac{1}{3} = 15\frac{3}{9} = 14\frac{12}{9} \\ 5\frac{7}{9} = 5\frac{7}{9} = 5\frac{7}{9} \\ \hline 9\frac{5}{9} \end{array}$$

Since the fractional portion of the minuend, $\frac{3}{9}$, was smaller than the fractional portion of the subtrahend, $\frac{7}{9}$, we must borrow a whole unit, or $\frac{9}{9}$, from the whole number portion of the minuend, 15. This changes $15\frac{3}{9}$ to $14\frac{12}{9}$ which permits the subtraction of the fractional portion of the subtrahend.

TEST YOUR ABILITY

Subtract the following:

1. $\frac{1}{2} - \frac{1}{4}$
2. $\frac{1}{8} - \frac{1}{9}$
3. $\frac{7}{8} - \frac{3}{7}$
4. $\frac{2}{3} - \frac{1}{3}$
5. $\frac{8}{9} - \frac{1}{2}$
6. $14\frac{1}{2} - 7\frac{1}{3}$
7. $10\frac{1}{4} - 2\frac{1}{8}$
8. $75\frac{3}{7} - 9\frac{1}{6}$
9. $81\frac{1}{2} - 7\frac{1}{3}$
10. $40\frac{7}{8} - 4\frac{5}{9}$
11. $38\frac{1}{2} - 6\frac{7}{8}$
12. $18\frac{2}{3} - 16\frac{3}{4}$
13. $101\frac{3}{4} - 58\frac{7}{9}$
14. $245\frac{4}{7} - 12\frac{6}{7}$
15. $340\frac{3}{5} - 19\frac{3}{8}$

Multiplying Fractions

To multiply fractions, multiply like terms: numerator by numerator; denominator by denominator. (Reduce all answers to lowest terms.) To multiply $\frac{4}{9}$ by $\frac{7}{12}$:

$$\frac{4}{9} \times \frac{7}{12} = \frac{4 \times 7}{9 \times 12} = \frac{28}{108}$$

You can now reduce $\frac{28}{108}$ by the factor 4, as follows:

$$\frac{28}{108} = \frac{28 \div 4}{108 \div 4} = \frac{7}{27}$$

To multiply a whole number by a fraction, change the whole number to a fraction by placing the number over 1. To multiply 4 by $\frac{1}{5}$:

$$4 \times \frac{1}{5} = \frac{4}{1} \times \frac{1}{5} = \frac{4 \times 1}{1 \times 5} = \frac{4}{5}$$

Dividing Fractions

To divide one fraction by another, *invert* (reverse terms of) the divisor (the second fraction) and multiply. To divide $\frac{3}{5}$ by $\frac{2}{3}$:

$$\frac{3}{5} \div \frac{2}{3} = \frac{3}{5} \times \frac{3}{2} = \frac{9}{10}$$

To divide a whole number by a fraction, write the whole number in fraction form, invert the second fraction, and multiply.

$$12 \div \frac{2}{3} = \frac{12}{1} \div \frac{2}{3} = \frac{12}{1} \times \frac{3}{2} = \frac{36}{2} = \frac{18}{1} = 18$$

TEST YOUR ABILITY

1. Multiply. Reduce answers to lowest terms.
   (a) $\frac{3}{5} \times \frac{2}{3}$
   (b) $\frac{5}{6} \times \frac{1}{3}$
   (c) $\frac{4}{9} \times \frac{3}{5}$
   (d) $4 \times \frac{2}{13}$
   (e) $10 \times \frac{1}{2}$
2. Divide. Reduce answers to lowest terms.
   (a) $\frac{2}{3} \div \frac{3}{4}$
   (b) $\frac{1}{3} \div \frac{5}{6}$
   (c) $\frac{4}{9} \div \frac{3}{5}$
   (d) $7 \div \frac{7}{12}$
   (e) $9 \div \frac{2}{3}$

Cancellation in Multiplication and Division

Multiply $645 \times 4\frac{2}{5}$.

$$645 \times 4\frac{2}{5} = \frac{645}{1} \times \frac{22}{5} = \frac{14190}{5} = 2{,}838$$

This procedure is correct, but having to work with a number as large as 14,190 increases the likelihood of making errors. You can use a simpler procedure, called *cancellation,* in which the fractions to be multiplied are reduced by common factors *before* multiplying. To multiply $\frac{3}{4}$ by $\frac{8}{3}$, for example, set up the problem as follows:

$$\frac{3}{4} \times \frac{8}{3} = \frac{3 \times 8}{4 \times 3}$$

Instead of multiplying 3 × 8 and 4 × 3, stop and look for factors that are common to the numerator and denominator. The number 3, for instance, is a factor of both; therefore, dividing both 3's by 3 reduces them to 1.

$$\frac{\overset{1}{\cancel{3}} \times 8}{4 \times \underset{1}{\cancel{3}}}$$

The number 4 is a factor of both 8 and 4. First, divide 8 and 4 by 4; then multiply the reduced numbers.

$$\frac{\overset{1}{\cancel{3}} \times \overset{2}{\cancel{8}}}{\underset{1}{\cancel{4}} \times \underset{1}{\cancel{3}}} = \frac{2}{1} = 2$$

Let's return to the earlier problem of $645 \times 4\frac{2}{5}$. Reduce the factor 5 prior to multiplying.

$$\frac{645}{1} \times \frac{22}{5} = \frac{\overset{129}{\cancel{645}} \times 22}{1 \times \underset{1}{\cancel{5}}} = 2{,}838$$

The figures are still large, but 645 ÷ 5 is easier than dividing 14,190 by 5.

Ratios

A *ratio* is a comparison of one number to another. A 3 to 1 ratio—written as either 3/1 or 3:1—means that the first number is 3 times as large as the second number. If the second number is 100, the first number is 300. The ratio of the second number to the first number is 1/3 or 1:3.

The most common use of ratios in business mathematics is the comparing of a part to a whole or one part to another. For example, if F. T. Hall, an insurance corporation, has 500 employees, 200 of whom are typists and 300 of whom are clerks:

The Ratio of	Is
clerks to employees	300/500 or 3/5 or 3:5 to .6:1
typists to employees	200/500 or 2/5 or 2:5 or .4:1
clerks to typists	300/200 or 3/2 or 3:2 or 1.5:1
typists to clerks	200/300 or 2/3 or 2:3 or .7:1

Notice the many ways in which a ratio can be shown. The final ratio on each line, *something:1,* is a common expression. To change 3:5 into .6:1, both numbers—3 and 5—are divided by the second number, 5.

Let's consider a variation of this problem. If a firm has 600 employees, with a 3:2 ratio of blue-collar to white-collar workers, how many of each group of workers are employed?

A 3:2 ratio means 3 "parts" blue-collar to 2 "parts" white-collar. There are altogether 5 parts, 3 of which ($\frac{3}{5}$) are blue-collar, and 2 of which ($\frac{2}{5}$) are white-collar. Thus, $\frac{3}{5} \times 600 = 360$ blue-collar; $\frac{2}{5} \times 600 = 240$ white-collar. Since the sum of the parts is equal to the whole, you can check your results by adding all the parts and seeing that they equal the original figure. Since $360 + 240 = 600$, the answer is correct.

SAMPLE PROBLEM 9

Venus Cigarettes employs 400 sales workers and 1,200 production workers.

(a) What fraction of the staff is in sales?

(b) What is the ratio of production to sales workers?

Solution

(a) Total work force = 400 + 1,200 = 1,600
Sales/Total work force = 400/1,600 = 1/4

(b) Production: Sales = 1,200/400 = 3:1

An Application of Fractions

Sample Problem 10 shows an example of how fractions are used in a business problem. Study the solution and the notes that follow very carefully.

SAMPLE PROBLEM 10

The production of Silverite requires a combination of three metals: $\frac{1}{4}$ Rosine, $\frac{1}{5}$ Lomelite, and the remainder, Anderite. If 800 lbs. of Silverite is to be produced, find the number of pounds of each metal required:

Solution

Rosine: $\frac{1}{4} \times \frac{800}{1} = 200$ pounds

Lomelite: $\frac{1}{5} \times \frac{800}{1} = 160$ pounds

Anderite: $\frac{1}{4} + \frac{1}{5} = \frac{9}{20}$

$1 = \frac{20}{20} - \frac{9}{20} = \frac{11}{20}$ left

$\frac{11}{20} \times \frac{800}{1} = 440$ pounds

Check 200 + 160 + 440 = 800 pounds

Notes

(1) Each calculation was based on 800 pounds. The number 800 was expressed as $\frac{800}{1}$ and cancellation was used.

(2) To calculate the answer for Anderite, the first two results (for Rosine and Lomelite) were *not* used. The reason: Had either result been wrong, the answer for Anderite would also have been wrong.

(3) The "whole" in a series of fractions is always 1. In this problem, the whole was expressed as $\frac{20}{20}$ in order to subtract the sum of Rosine and Lomelite, $\frac{9}{20}$.

(4) A final check on accuracy was made by adding all of the parts. The result of the addition equaled the original total.

TEST YOUR ABILITY

1. Solve each of the following problems:
   (a) $2 \times \frac{1}{4}$
   (b) $5 \times 3\frac{1}{2}$
   (c) $8\frac{1}{2} \times 9\frac{1}{6}$
   (d) $5 \div \frac{2}{3}$
   (e) $\frac{3}{4} \div 2\frac{1}{2}$
   (f) $5\frac{1}{2} \div 3\frac{1}{2}$
   (g) $1\frac{1}{4} \div 6\frac{1}{8}$
2. M. Morris, a travel agency, has 80 clerical workers, and 20 researchers who travel in order to compare prices, hotels, etc.
   (a) What fraction of the staff does clerical work?
   (b) What is the ratio of researchers to clerical workers?
3. An enormous wedding cake weighing 720 pounds is made up of four ingredients: $\frac{1}{6}$, strawberries; $\frac{1}{8}$, ice cream; $\frac{2}{3}$, shortcake; the remainder, whipped cream. Find the number of pounds of each ingredient.

CHECK YOUR READING

1. In adding and subtracting fractions, why should the common denominator be the lowest possible?
2. Fractions can be raised to higher terms, or reduced to lower terms, without changing the value of the fraction. Explain why.
3. Draw a rectangle and use it to explain that $\frac{1}{2} = \frac{2}{4}$.
4. Explain what is meant by the *greatest common divisor*.
5. Add: $\frac{1}{4} + \frac{1}{8} + 1\frac{1}{6}$
6. Subtract: $\frac{7}{8} - \frac{5}{9}$
7. Multiply: $2\frac{1}{2} \times 5$
8. Divide: 16 by $3\frac{1}{3}$
9. Change the following improper fractions to mixed numbers:
   (a) $\frac{28}{3}$
   (b) $\frac{46}{7}$
   (c) $\frac{69}{9}$
10. Change the following mixed numbers to improper fractions:
    (a) $4\frac{1}{3}$
    (b) $5\frac{1}{6}$
    (c) $7\frac{2}{5}$

EXERCISES

1. South Oaks University has a staff of 840, divided as follows: $\frac{1}{4}$ are full professors; $\frac{3}{10}$ are associate professors; $\frac{7}{20}$ are assistant professors; and the remainder are instructors. Find the number of staff members in each group.
2. Reduce the following fractions to lowest terms:
   (a) $\frac{15}{65}$ (e) $\frac{162}{216}$
   (b) $\frac{12}{78}$ (f) $\frac{35}{84}$
   (c) $\frac{28}{120}$ (g) $\frac{300}{960}$
   (d) $\frac{39}{96}$ (h) $\frac{7}{210}$

3. Add. Reduce answers to lowest terms.
   (a) $\frac{2}{11} + \frac{5}{11}$ (d) $\frac{1}{3} + \frac{1}{8} + \frac{1}{10}$
   (b) $\frac{5}{9} + \frac{5}{18}$ (e) $\frac{1}{5} + \frac{1}{6} + \frac{1}{7}$
   (c) $\frac{2}{5} + \frac{4}{15}$ (f) $\frac{2}{10} + \frac{1}{9} + \frac{1}{6} + \frac{1}{5}$
4. Subtract. Reduce answers to lowest terms.
   (a) $\frac{7}{9} - \frac{2}{3}$
   (b) $\frac{11}{20} - \frac{4}{11}$
   (c) $\frac{3}{4} - \frac{7}{15}$
   (d) $\frac{7}{8} - \frac{1}{9}$
5. Multiply. Reduce answers to lowest terms.
   (a) $\frac{4}{5} \times \frac{1}{9}$
   (b) $\frac{6}{11} \times \frac{5}{6}$
   (c) $\frac{5}{12} \times \frac{2}{3}$
   (d) $9 \times \frac{1}{15}$
6. Divide. Reduce answers to lowest terms.
   (a) $\frac{5}{9} \div \frac{3}{5}$
   (b) $\frac{3}{4} \div \frac{11}{12}$
   (c) $\frac{1}{4} \div \frac{5}{8}$
   (d) $16 \div \frac{3}{4}$
7. Perform the required operations with the following mixed numbers:
   (a) $12 \div \frac{5}{6}$ (g) $2\frac{1}{2} \div 5\frac{1}{9}$
   (b) $6 \times \frac{5}{9}$ (h) $5\frac{1}{8} \times 3\frac{1}{2}$
   (c) $12\frac{1}{5} + 9\frac{3}{8}$ (i) $4\frac{2}{3} - 2\frac{4}{5}$
   (d) $9 - 4\frac{7}{8}$ (j) $7\frac{7}{8} \div 1\frac{1}{5}$
   (e) $11 \times 3\frac{1}{3}$ (k) $6\frac{5}{6} + 4\frac{4}{9}$
   (f) $\frac{4}{7} \div 3\frac{1}{5}$
8. Of Pedro Walker's employees, 80 are nonunion members and 220 are union members.
   (a) What fraction of the staff is nonunion?
   (b) What fraction of the staff is union?
   (c) What is the ratio of union to nonunion members?
9. Johnson, Lane, and King made investments in a business as follows:

Johnson	$20,000
Lane	38,000
King	12,000

   What is the ratio of each partner's investment to the total investment?
10. Of the 6,000 students at Maryvale College, 800 are business majors. What is the ratio of business majors to the total enrollment?
11. Dave Rodriguez earns $355 per week. Of this amount, Dave's employer withholds $47 in income taxes. What is Dave's ratio of taxes to earnings?
12. The Lakeside Grocery Store occupies its building as follows: $\frac{1}{5}$, produce department; $\frac{1}{7}$, meat department; and the remainder, grocery department. If the building has a total of 35,000 square feet, how many square feet are in each department?

13. On July 17, Bettye Spence purchased $65\frac{1}{8}$ acres of land. She purchased an additional $24\frac{3}{5}$ acres on August 5. On August 31, she sold $12\frac{1}{6}$ acres. How many acres did she have left?
14. At the beginning of this year, Jane Griffith purchased a car for $12,000. At the end of the year, Jane estimates that, through depreciation, the car had lost $\frac{1}{3}$ of its value. What is the value of the car at the end of the year?

CHAPTER **3**

Decimals

Like a fraction, a *decimal* is also a means of expressing less than a whole unit. A decimal represents one or more tenths, hundredths, thousandths, etc., of a whole number. Since a decimal refers to less than a whole unit, it is often called a *decimal fraction.* Since our number system, and our monetary system, is the decimal system, you need a firm foundation in the uses and applications of decimals.

Your Objectives

After studying Chapter 3, you should be able to

1. Read and interpret decimal numbers.
2. Round off decimal numbers.
3. Add, subtract, multiply, and divide decimal numbers.
4. Use decimal numbers in business applications.
5. Convert fractions, decimals, and percents to other forms.

Reading Decimal Numbers

The sign of the decimal is the *decimal point.* It is used to separate whole numbers from decimal fractions. You learned in Chapter 1 that "places" to the left of the decimal point are worth ten times the value of the previous place; and places to the right of the decimal point are worth one-tenth the value of the previous place. The following illustration shows the place value to billions (for whole numbers) and to billionths (for decimals).

Whole Numbers	*Billions*
	Hundred millions
	Millions
	Hundred thousands
	Ten thousands
	Thousands
	Hundreds
	Tens
	Units
	· Decimal point
Decimals	*Tenths*
	Hundredths
	Thousandths
	Ten thousandths
	Hundred-thousandths
	Millionths
	Ten millionths
	Hundred-millionths
	Billionths

You will recall from Chapter 1 that there is a correct way to read decimal numbers. For example, the number 55.5 is read fifty-five and five tenths, which means *five tens, five units* and *five tenths of another unit.* You will also recall from Chapter 1 that, when reading mixed numbers (mixed decimals), the *and* is pronounced to indicate a division between the whole number and the decimal fraction.

SAMPLE PROBLEM 1

Read aloud, or write, each of the following numbers:

1. .45	4. .0012	7. 19.145
2. .455	5. .0005	8. .86431
3. .205	6. 2.5005	

Solution

1. .45 = forty-five hundredths.
2. .455 = four hundred fifty-five thousandths.
3. .205 = two hundred five thousandths.
4. .0012 = twelve ten thousandths.
5. .0005 = five ten thousandths.
6. 2.5005 = two *and* five thousand five ten thousandths.
7. 19.145 = nineteen *and* one hundred forty-five thousandths.
8. .86431 = eighty-six thousand, four hundred thirty-one hundred thousandths.

Rounding Decimal Numbers

Calculation can result in answers with more places than are necessary. Since our money system, for example, uses dollars and cents, a final answer of $73.372 is unacceptable. Dollars and cents are typically rounded to the nearest hundredth—two places after the decimal point. Thus, the answer above would be rounded to $73.37. At times, a firm will round dollars to the nearest unit, disregarding cents. In filing our individual income tax returns, the government permits us to round to even dollars. Thus, if an amount is $.50 or higher, it is rounded up to the next even dollar; or, if an amount is $.49 or less, it is rounded down to the next even dollar. Accordingly, $25.67 would be rounded to $26, and $56.43 would be rounded to $56.

To round a number, drop all the digits to the right of the place to which the number is to be rounded. If the first digit to be dropped is 5 or more, increase the last digit to be kept by one. If the first digit to be dropped is less than 5, do not change the last digit that is being retained. Compare the following two examples:

6.743 rounded to tenths is 6.7
6.751 rounded to tenths is 6.8

Sample Problem 2 shows how a number can be rounded in several ways. Notice that rounding to the nearest whole number is the same as rounding to the nearest unit or to the nearest *integer*.

SAMPLE PROBLEM 2

Round the number 4,376.538 to the nearest (a) thousand; (b) hundred; (c) ten; (d) integer; (e) tenth; (f) hundredth.

Solution

(a) 4,000 (c) 4,380 (e) 4,376.5
(b) 4,400 (d) 4,377 (f) 4,376.54

A rounding variation that you may find in other studies is the *computer rule,* which can avoid errors in rounding large series of numbers. By means of this rule, some numbers may become larger; others, smaller. The rule applies *only* to situations in which the first digit to be dropped is a 5 or a 5 followed by zeroes. In either case, make the number *even.* If it is already even, your work is done; if the number is odd, add 1 to the last digit to be kept. For example:

3.750 rounded to tenths is 3.8
3.850 rounded to tenths is 3.8

The computer rule will *not* be followed in working any problems in this book.

TEST YOUR ABILITY

Round each of the following numbers as indicated:

1. 75.66666 to tenths
2. 18.0058 to thousandths
3. 97.5521 to hundredths
4. .99999 to hundredths
5. 1.24678 to thousandths
6. 55.55555 to nearest whole number

Adding Decimal Numbers

The successful addition of decimal numbers requires the alignment of decimal points. To accomplish this, put the numbers in a column by (1) lining up the decimal points and (2) filling in zeroes for any numbers with fewer decimal digits than the longest number. In Sample Problem 3, 63.5 becomes 63.500 in order to align the numbers correctly.

SAMPLE PROBLEM 3

Add the following decimal numbers. Round the answer to the nearest hundredth.

4,137.25 + 47.305 + .173 + 63.5 + 2.45

Solution

$$
\begin{array}{rcr}
4{,}137.25 & = & 4{,}137.250 \\
47.305 & = & 47.305 \\
.173 & = & .173 \\
63.5 & = & 63.500 \\
2.45 & = & 2.450 \\
\hline
 & & 4{,}250.678 = 4{,}250.68
\end{array}
$$

TEST YOUR ABILITY

Add the following decimal numbers. Round the answer to the nearest hundredth.

1. .25 + .67 + .0035 + .09
2. .005 + .456 + .0025 + .0096
3. .995 + .205 + .8 + .06 + .992
4. .30 + 1.45 + 22.5 + .8 + 15
5. 255 + .90 + .8995 + 3.5
6. 55.6 + 1,933.78 + 22.005
7. $1,245.33 + $45.90
8. $255.50 + $0.95
9. 5,678.9982 + 1.2
10. 1 + .0092

Subtracting Decimal Numbers

In decimal subtraction, as in addition, be sure to align the numbers around the decimal point. First, line up the decimal points; then, equalize the digits in each number by filling in zeroes where necessary. To subtract 258.7169 from 3,471.29, change the second number to 3,471.2900 in order to equalize the digits. If the answer is required in hundredths, round to two decimal places.

$$
\begin{array}{r}
3{,}471.2900 \\
-\ \ 258.7169 \\
\hline
3{,}212.5731 = 3{,}212.57
\end{array}
$$

TEST YOUR ABILITY

Subtract the following. Round answers to the nearest hundredth.

1. .56 from 3.45
2. .0025 from .7891
3. .0569 from 2.5
4. .069 from .8876
5. .005 from .5
6. 4.55 from 8.2
7. 25.665 from 250.28
8. $25 from $245.50
9. $12.88 from $12,278.90
10. .009876 from 2

Multiplying Decimal Numbers

When multiplying numbers containing decimals, count the number of decimal places in the multiplicand and the number of decimal places in the multiplier, and put a decimal point before the combined number of places in the product. When multiplying 5.89 x .3684, for example, the answer will have 6 places (4 + 2) to the right of the decimal point, as shown on the following page.

```
   .3684     (4 places)
 × 5.89      (2 places)
 ------
  33156
 29472
18420
--------
2.169876     (6 places)
```

When multiplying some decimal numbers, zeroes must be inserted in the product to yield the appropriate number of places. Sample Problem 4 illustrates how the product must be stated when the combined number of places in the multiplicand and the multiplier exceed the digits in the product.

SAMPLE PROBLEM 4

Multiply 1.45 × .005

Solution

```
  1.45    (2 places)
 ×.005    (3 places)
 ------
.00725    (5 places)
```

TEST YOUR ABILITY

Multiply the following. Round answers to the nearest hundredth.

```
1.  $76.90      6.  289.754     11.  3.025
    ×    9          × 12.75          × 2.2

2.  85.999      7.  .0014       12.  1,250
    ×   12          × .021           × 3.02

3.  12.905      8.  1.25        13.  12,900
    ×  .09          × .25            × 125.8

4.  .09867      9.  288.9       14.  245.5
    × .002          × 4.2            ×2.355

5.  145.8      10.  30.12       15.  .0002
    × 15.4          × .12            × .02
```

Dividing Decimal Numbers

To divide numbers containing decimals, proceed as follows:

1. Move the decimal point of the divisor all the way to the right to make it a whole number.
2. Move the decimal point of the dividend the *same number* of places to the right, attaching zeroes if necessary.
3. Put the decimal point in the quotient directly above the newly located decimal point in the dividend.

Let's now look at Sample Problem 5 in which three situations are illustrated which call for the division of decimal numbers.

SAMPLE PROBLEM 5

Divide each of the following:
(a) 87.3 by 2.26
(b) 36.6 by 6
(c) 72 by .12

Solution

```
                          38.6              6.1                600.
(a) 2.26)87.300 = 226)8730.0      (b) 6)36.6      (c) .12)72.00
                      678
                      1950
                      1808
                       1420
                       1356
                         64
```

Notice in Part (b) above that it was not necessary to move the decimal point in the divisor since it was already a whole number. Also notice that, in Part (c), it was necessary to move the decimal point in the dividend, even though it was a whole number. Remember that when the decimal point is moved in the divisor, it *must also be moved the same number of places in the dividend.*

Estimating Answers An excellent way to check if the decimal point is in the correct position in division is to estimate your answer by rounding both the dividend and the divisor to convenient numbers. In Part (a) of Sample Problem 5, for example, 87.3 is close to 90, and 2.26 is close to 2. Dividing 90 by 2 gives 45, an incorrect answer—but an accurate indication that the exact answer will have two digits to the left of the decimal point.

Rounding and estimating will speed up the process of division. You can guess at digits in the answer with a high degree of accuracy. For example, to divide 4,175 by 49, first round the two numbers: 4,000 and 50. Dividing 4,000 by 50 gives a round answer of 80, so start your division with 8.

```
       8
49)4175.00
   392
    255
```

Now prepare a revised estimate. The number 255 is about 250, and 49 is about 50. Dividing 250 by 50 gives an answer of 5. Try 5 as the second number in the quotient.

```
      85.
49)4175.00
   392
    255
    245
     100
```

Similarly, since 100 divided by 50 is 2, use 2 as your third digit. Continue the process until you have reached the appropriate place. The steps in the full solution are as follows, rounded to the nearest tenth:

```
      85.20
49)4175.00
   392
    255
    245
     100
      98
       20
```

Let's summarize the division procedure:

1. Adjust the decimal points in the divisor and the dividend so that the divisor is a whole number.
2. Be prepared to carry out division one place to the right of the last digit retained in order to round to the required place.
3. Estimate the final answer in order to check the placement of the decimal point.
4. Estimate each digit in the quotient by approximating the division of each remainder by the divisor.

Sample Problem 6 illustrates a business application in which it is necessary to divide decimal numbers.

SAMPLE PROBLEM 6

Magill Pipes produced 6,475 units during the current year at a total cost of $27,475.25. Find the cost per unit to the nearest cent.

Solution

```
      $     4.243 = $4.24 per unit
6475)$27475.250
```

TEST YOUR ABILITY

Divide the following. When applicable, round the quotient to the nearest thousandth.

1. $55\overline{)278.79}$	6. $.002\overline{)60}$	11. $.0765\overline{)63.28}$
2. $20\overline{)88.02}$	7. $65.2\overline{)368}$	12. $1.254\overline{)68.9}$
3. $75\overline{)567.9}$	8. $3.45\overline{)12.6}$	13. $48.125\overline{)96.5}$
4. $.80\overline{)5.80}$	9. $.08\overline{)64}$	14. $4.8\overline{)156}$
5. $1.2\overline{)24.845}$	10. $.125\overline{).80}$	15. $215.82\overline{)890.35}$

Converting Fractions, Decimals, Percents

As we have learned, part of a whole unit can be expressed as a fraction, or as a decimal. Parts of a whole can also be expressed as a *percent*. The ability to convert from one form to another is often required in business problems. Some calculations are easier to perform with decimals, others with fractions. Or, you may be given two sets of data, one in fraction form and the other in percent form.

Whatever the problem, be flexible in deciding which form is the easiest to work with.

Each type of conversion is explained below.

Converting a Fraction to a Decimal

To convert a fraction to a decimal, divide the numerator by the denominator and carry the division as required for rounding. To do so, insert a decimal point and add as many zeroes as necessary to the dividend. To convert $\frac{7}{8}$ to decimal form, with the answer rounded to thousandths, write 7 as 7.0000 and divide by 8.

$$8\overline{)7.0000}\quad .8750$$

Rounded to thousandths, $\frac{7}{8} = .875$. An alternate expression of the answer is $.87\frac{1}{2}$.

Converting a Decimal to a Percent

A *percent* means a part of 100; therefore, 1% is 1 part of 100 and 10% is 10 parts of 100. Expression as a *decimal* means part of 1. Thus, to convert from decimal form to percent form, multiply by 100 by moving the decimal point two places to the right (adding zeroes where necessary), and add a percent sign (%). The following examples show how this is done:

$$\begin{aligned} .836 &= 83.6\% \\ 2.13 &= 213\% \\ .07 &= 7\% \\ .004 &= .4\% \\ 5 &= 500\% \\ .635 &= 63.5\% \end{aligned}$$

Converting a Fraction to a Percent

To convert a fraction to a percent, first convert the fraction to a decimal; then convert the decimal to a percent. To convert $\frac{4}{7}$ to a percent:

$$\frac{4}{7} = 7\overline{)4.00}\quad .57\tfrac{1}{7} = 57\tfrac{1}{7}\%$$

It was assumed that the answer would be carried to two decimal places and would also include a fractional remainder.

Converting a Decimal to a Fraction

A decimal number is actually a fraction. The number .8 when read aloud is read as "eight tenths" and can be written as the fraction $\frac{8}{10}$. By the same means, .07 can be read as "seven hundredths" and written as $\frac{7}{100}$.

The conversion from decimal to fraction relies on this alternate expression, writing the decimal as a fraction and reducing it to lowest terms:

$$.006 = \frac{6}{1000} = \frac{3}{500}$$

Or, instead, follow these three steps: (1) omit the decimal point; (2) place the number over 1 followed by as many zeroes as there are digits after the decimal point; (3) reduce to lowest terms. For example, this is how .006 is converted:

(1) .006 = 6 (drop zeroes)

(2) $\frac{6}{1 + 3 \text{ zeroes}} = \frac{6}{1000}$

(3) $\frac{6}{1000} = \frac{3}{500}$

For another example:

$$.625 = \frac{625}{1000} = \frac{5}{8}$$

Converting a Percent to a Decimal

To convert a percent to a decimal, reverse the procedure used in converting a decimal to a percent. First, drop the percent sign; second, move the decimal point two places to the left, inserting zeroes if necessary.

$$70\% = .70$$
$$53.2\% = .532$$
$$46\tfrac{1}{2}\% = .46\tfrac{1}{2} \text{ or } .465$$
$$.9\% = .009$$
$$112\% = 1.12$$

Converting a Percent to a Fraction

To convert a percent to a fraction, first, convert the percent to decimal form; then, convert the decimal to fraction form.

$$25\% = .25 = \frac{25}{100} = \frac{1}{4}$$

To convert a percent containing a fraction, an additional step is needed. First, change the number to a fraction.

$$57\tfrac{1}{7}\% = .57\tfrac{1}{7} = \frac{57\frac{1}{7}}{100}$$

Working with a fraction within a fraction is difficult. We can simplify our work by taking the denominator of the upper fraction—7—and multiplying both terms of the entire fraction by it.

$$\frac{57\frac{1}{7}}{100} \times \frac{7}{7} = \frac{57\frac{1}{7} \times \frac{7}{1}}{\frac{100}{1} \times \frac{7}{1}} = \frac{\frac{400}{\not{7}_1} \times \frac{\not{7}^1}{1}}{\frac{100}{1} \times \frac{7}{1}} = \frac{400}{700}$$

Finally, reduce to lowest terms.

$$\frac{400}{700} = \frac{4}{7}$$

Conversion Table

To assist you in the conversion of the more common fractions, decimals, and percents, Table 3.1 has been prepared. For example, let's assume that you want to find the decimal equivalent of $\frac{7}{12}$. Look down the column headed "7" for the numerator and across the row headed "12" for the denominator until you find the intersection, $.58\frac{1}{3}$, which is the answer.

Table 3.1 does not contain all possible combinations, but it is usable for all fractions up to and including twelfths.

Table 3.1

Fraction, Decimal, and Percent Equivalents

	Numerator										
Denominator	**1**	**2**	**3**	**4**	**5**	**6**	**7**	**8**	**9**	**10**	**11**
2	.50 50%										
3	$.33\frac{1}{3}$ $33\frac{1}{3}\%$	$.66\frac{2}{3}$ $66\frac{2}{3}\%$									
4	.25 25%	.50 50%	.75 75%								
5	.20 20%	.40 40%	.60 60%	.80 80%							
6	$.16\frac{2}{3}$ $16\frac{2}{3}\%$	$.33\frac{1}{3}$ $33\frac{1}{3}\%$	.50 50%	$.66\frac{2}{3}$ $66\frac{2}{3}\%$	$.83\frac{1}{3}$ $83\frac{1}{3}\%$						
7	$.14\frac{2}{7}$ $14\frac{2}{7}\%$	$.28\frac{4}{7}$ $28\frac{4}{7}\%$	$.42\frac{6}{7}$ $42\frac{6}{7}\%$	$.57\frac{1}{7}$ $57\frac{1}{7}\%$	$.71\frac{3}{7}$ $71\frac{3}{7}\%$	$.85\frac{5}{7}$ $85\frac{5}{7}\%$					
8	$.12\frac{1}{2}$ $12\frac{1}{2}\%$	.25 25%	$.37\frac{1}{2}$ $37\frac{1}{2}\%$	.50 50%	$.62\frac{1}{2}$ $62\frac{1}{2}\%$	.75 75%	$.87\frac{1}{2}$ $87\frac{1}{2}\%$				
9	$.11\frac{1}{9}$ $11\frac{1}{9}\%$	$.22\frac{2}{9}$ $22\frac{2}{9}\%$	$.33\frac{1}{3}$ $33\frac{1}{3}\%$	$.44\frac{4}{9}$ $44\frac{4}{9}\%$	$.55\frac{5}{9}$ $55\frac{5}{9}\%$	$.66\frac{2}{3}$ $66\frac{2}{3}\%$	$.77\frac{7}{9}$ $77\frac{7}{9}\%$	$.88\frac{8}{9}$ $88\frac{8}{9}\%$			
10	.10 10%	.20 20%	.30 30%	.40 40%	.50 50%	.60 60%	.70 70%	.80 80%	.90 90%		
11	$.09\frac{1}{11}$ $9\frac{1}{11}\%$	$.18\frac{2}{11}$ $18\frac{2}{11}\%$	$.27\frac{3}{11}$ $27\frac{3}{11}\%$	$.36\frac{4}{11}$ $36\frac{4}{11}\%$	$.45\frac{5}{11}$ $45\frac{5}{11}\%$	$.54\frac{6}{11}$ $54\frac{6}{11}\%$	$.63\frac{7}{11}$ $63\frac{7}{11}\%$	$.72\frac{8}{11}$ $72\frac{8}{11}\%$	$.81\frac{9}{11}$ $81\frac{9}{11}\%$	$.90\frac{10}{11}$ $90\frac{10}{11}\%$	
12	$.08\frac{1}{3}$ $8\frac{1}{3}\%$	$.16\frac{2}{3}$ $16\frac{2}{3}\%$	.25 25%	$.33\frac{1}{3}$ $33\frac{1}{3}\%$	$.41\frac{2}{3}$ $41\frac{2}{3}\%$	.50 50%	$.58\frac{1}{3}$ $58\frac{1}{3}\%$	$.66\frac{2}{3}$ $66\frac{2}{3}\%$	.75 75%	$.83\frac{1}{3}$ $83\frac{1}{3}\%$	$.91\frac{2}{3}$ $91\frac{2}{3}\%$

TEST YOUR ABILITY

1. Convert each of the following fractions into (i) decimals and (ii) percents. Express uneven answers in (i) to three decimal places. Use Table 3.1 where possible.
   (a) $\frac{5}{8}$
   (b) $\frac{3}{5}$
   (c) $\frac{7}{10}$
   (d) $\frac{5}{9}$
   (e) $\frac{1}{7}$
   (f) $\frac{11}{16}$
   (g) $\frac{1}{4}$
   (h) $\frac{1}{2}$
2. Convert each of the following percents, first, to decimals and, second, to fractions; then reduce to lowest terms. Use Table 3.1 where possible.
   (a) 80%

(b) 42%
(c) 87.5%
(d) $6\frac{1}{4}$%
(e) $5\frac{1}{2}$%
(f) $85\frac{5}{7}$%
(g) $\frac{1}{5}$%
(h) 250%

CHECK YOUR READING

1. Write the following numbers in terms of place values (how many thousands, hundreds, etc.).
   (a) 5,421.083
   (b) 644.0025
   (c) 7,890.2
   (d) 25,678.004
   (e) .002508
   (f) 7.00045
2. Round the following numbers to hundredths.
   (a) 75.667 (d) .086578
   (b) 12.0034 (e) 1.258
   (c) 89.6666 (f) 17.0015
3. Round 57,271.386 to the nearest (a) hundredth; (b) whole number; (c) hundred; (d) thousand; (e) ten; (f) tenth.
4. Add: 1.2 + .0025 + 78 + .004
5. Subtract: 68.2 − 12.125
6. Multiply: 12.5 × 10.25
7. Divide: 144.2 by 12.8
8. Convert the following decimals to percents:
   (a) .09
   (b) .26
   (c) .007
   (d) 1.37
   (e) 1.816
   (f) $.05\frac{1}{2}$
   (g) .7

EXERCISES

1. Add the following. Where applicable, round answers to the nearest thousandth.
   (a) .02 + 56.799 + .002 + .0025 + 6,789.5
   (b) 15,680 + 456.7 + 89.552 + 32.0068 + 12.6
   (c) $122,899.60 + $12,000.09
   (d) $69,600 + $32.90
   (e) 187,968.28479 + 498.2855 + .09954
2. Subtract. Round as directed.

(a)	(b)	(c)
$927.54	527.341	37.4275
− 803.69	− 29.865	− 9.6481
(nearest dollar)	(nearest tenth)	(nearest hundredth)

   (d) 83.27 − .946 (nearest tenth)

(e) 546 − 39.92 (nearest tenth)
(f) .5275 − .0461 (nearest hundredth)
(g) 4.9362 − .49781 (nearest thousandth)
(h) 1 − .00125 (nearest hundredth)

3. Multiply .0125 by .00255 rounding your final answer to (a) the nearest tenth and (b) the nearest thousandth.
4. Multiply. Round all decimal answers to hundredths.

(a) $\begin{array}{r} 5.67 \\ \times 38.1 \\ \hline \end{array}$ (b) $\begin{array}{r} 397 \\ \times .0006 \\ \hline \end{array}$ (c) $\begin{array}{r} .409 \\ \times 4.35 \\ \hline \end{array}$

(d) 49 × .6001
(e) 5.6 × $3.45
(f) 5,663 × .0933
(g) 34,585.245 × 10,000

5. John McHargue earns $5.65 per hour. Last week he worked 38 hours. What are his earnings for the week?
6. A certain brand of pencils cost $0.18 each. The River Road School decides to purchase 800. What is the total cost rounded to (a) even dollars and (b) even cents (hundredths)?
7. Divide 5354 by 47 rounding to the nearest (a) whole number; (b) tenth; (c) hundreth.
8. A fraternity that runs monthly dances mails 790 announcements in April and 813 in May. If the postage rate for a single announcement is $0.22, how much was spent on stamps in each month?
9. Divide. Round all decimal answers to thousandths, where necessary.

(a) $6.6\overline{)36}$ (b) $50\overline{)525.65}$ (c) $22.45\overline{)480.25}$
(d) 75 ÷ .5
(e) $12,400 ÷ 12.5
(f) 656.2 ÷ .85
(g) .44 ÷ .044

10. Convert the following fractions, first, to decimals, and second, to percents. Round answers to thousandths, where necessary.

(a) $\frac{1}{2}$
(b) $\frac{1}{4}$
(c) $\frac{3}{8}$
(d) $\frac{9}{11}$
(e) $\frac{4}{5}$
(f) $\frac{2}{3}$

11. Convert the following percents, first, to decimals, and second, to fractions reduced to lowest terms.

(a) 48% (e) 140%
(b) 55% (f) $12\frac{1}{2}$%
(c) 31% (g) $90\frac{10}{11}$%
(d) 7.5%

12. Twelve hundred units of a certain product cost $8,900. What is the cost per unit to the nearest cent?
13. Add $15\frac{1}{2}$ and 32.75.

14. Ray Slaker purchased 4 pounds of apples at $0.59 per pound and 6 pounds of peaches at $0.69 per pound. How much was the total cost of his purchase?
15. A local wholesaler is asking $25 for 60 pounds of nails. What is the cost per pound to the nearest cent?
16. A certain car averages 22.5 miles to a gallon of gasoline. How many gallons would it take to drive 500 miles? Round answer to the nearest tenth.
17. Add. Round as directed.
    (a) 562.3 + 479.8 + 307.6 + 812.5 (nearest whole number)
    (b) $6,274.83 + $4,177.94 + $5,900.43 + $4,182.45 (nearest cent).
    (c) 6.82 + 4.7 + 9.36 + 2.19 + 4.62 (nearest tenth)
    (d) 47.83 + 18.556 + 16.1 + 19.575 (nearest hundredth)

(e)	(f)	(g)
5,721.83	$386.30	572.36
8,164.19	415.79	4.7152
2,925.11	374.02	1937.04
5,367.47	535.18	83.625
1,423.65	527.12	200.0412
(nearest tenth)	(nearest hundredth)	(nearest tenth)

18. Nancy Chin hired three workers who produced 4,920 dresses at a total cost of $62,951.12. Determine the cost per dress to the nearest cent.

CHAPTER **4**

Base, Rate, and Percentage

In the pages that follow, mathematical techniques will be used to solve specific types of business problems. Because one of these techniques keeps recurring, we will discuss it here as a fundamental procedure—that of finding the *base*, the *rate*, or the *percentage*.

Every base, rate, or percentage problem gives two of the three factors and asks you to find the third. As a start, you already know how to solve the following problem:

Find 30% of $50
Answer: $50 × .30 = $15

In this problem, the two factors that have been given are the *rate* (the percent number—30%) and the *base* (the number on which the percent is taken—$50). You are asked to find the *percentage* (the number that is *part of* the base—$15).

In every problem in which two of these factors are given, the third can be found by means of this basic formula:

$$\text{Percentage} = \text{Base} \times \text{Rate} \qquad \text{or} \qquad P = B \times R$$

Your Objectives

After finishing Chapter 4, you should be able to

1. Identify the base, the rate, and the percentage.
2. Solve for the unknown item.
3. Apply these techniques to subsequent problems in business mathematics.

Finding the Percentage

To find a percentage, multiply base × rate, or

$$P = B \times R$$

Finding the percentage is the most familiar of the problems we will study in this chapter. In Sample Problem 1, we are solving for percentage (although the problem does not say so directly).

SAMPLE PROBLEM 1

George Markus earns $22,600 in a year. He spends 8% of this amount for electricity and gas. Determine his annual electric and gas expense.

Solution

Given $B = \$22{,}600$
$R = 8\% = .08$
Find $P =$ a part of his earnings
$P = B \times R$
$P = \$22{,}600 \times .08 = \$1{,}808$

Generally the *base* is preceded by the words "part of" or "of." The *rate* is easily recognized, since it includes a percent sign. The *percentage* is the *number* that is related to the base by the rate. Note, in Sample Problem 1, that the rate is changed into decimal form before calculations are carried out.

In some cases, logic must be applied to what is given and what is being sought, as shown in Sample Problem 2.

SAMPLE PROBLEM 2

On a certain day, 11% of the 300 workers in the Haakon Fishery were absent. How many workers were present?

Solution

Given $B = 300$
$R = 11\%$ absent
Find $300 - P =$ the number present
$P = B \times R$
$P = 300 \times .11 = 33$ absent
$300 - P = 300 - 33 = 267$ present

Alternate Solution

Given $B = 300$
$R = 100\% - 11\% = 89\%$ present
Find $P =$ the number present
$P = B \times R$
$P = 300 \times .89 = 267$ present

The rate given in Sample Problem 2 refers to the number of workers in the Haakon Fishery who were *absent*, but the problem calls for the number of workers *present*. In the first solution, the number absent is the percentage; to find the number present, the percentage must be subtracted from the total staff of 300.

In the alternate solution, the rate of those present is calculated by subtracting the rate of those absent (11%) from 100%—the base amount, *which is always equal to a rate of 100%*:

A common misconception is that the percentage is always smaller than the base. This is true *only* when the rate is less than 100%:

(a) if B = \$500 and R = $7\frac{1}{2}$%, then P = \$500 × .075 = \$37.50

(b) if B = 5,500 and R = $\frac{1}{10}$%, then P = 5,500 × .001 = 5.5

If the rate exceeds 100%, however, the percentage will be larger than the base. For example:

(a) if B = \$600 and R = 120%, then P = \$600 × 1.20 = \$720

(b) if B = 400 and R = 250%, then P = 400 × 2.50 = 1,000

Sample Problem 3 summarizes our discussion of this topic so far. Note the importance of wording shown by the difference between the two parts of the problem.

SAMPLE PROBLEM 3

Sales for T. Grace importers in 1988 were \$400,000. Find the estimated 1989 sales if (a) sales for 1989 are 130% of 1988 sales; (b) sales for 1989 are 130% *more* than 1988 sales.

Solution

(a) Given B = \$400,000
R = 130% = 1.30
Find P = sales for 1989
P = B × R
P = \$400,000 × 1.30 = \$520,000 sales in 1989

(b) Given B = \$400,000
R = 130% = 1.30
Find P + B = sales for 1989
P = B × R
P = \$400,000 × 1.30 = \$520,000 *more* sales, in 1989
400,000 sales in 1988
\$920,000 sales in 1989

Alternate Solution

(b) R = 130% + 100% = 230% = 2.30
P = B × R
P = \$400,000 × 2.30 = \$920,000 sales in 1989

TEST YOUR ABILITY

1. In each example, find P (the percentage) to the nearest cent.

	B	R
(a)	\$ 500.00	17%
(b)	\$ 600.00	.6%
(c)	\$3,500.00	$6\frac{1}{2}$%
(d)	\$ 75.85	$\frac{1}{10}$%
(e)	\$ 655.00	200%
(f)	\$4,000.00	$58\frac{1}{4}$%

2. Shirley Blake earns \$17,500 in a year and spends 32% of this amount for food. Find her annual food expense.
3. The Wintergarden Race Track, employing 750 workers, finds 20% of its

workers on the picket line one morning. How many employees did not picket?

4. The district of Amherst's school budget of \$3,000,000 was divided as follows: staff salaries, 63%; administrative salaries, 8%; supplies, 9%; maintenance, 11%; transportation, 4%; and food, 5%. Find the dollar amount for each category. (Hint: This is a convenient problem to double-check by addition.)
5. The Bellows Corporation's sales in 1988 totaled 310,000 units. Find the 1989 sales if:
   (a) 1989 sales are 120% more than 1988 sales;
   (b) 1989 sales are 120% of 1988 sales;
   (c) 1989 sales are 20% less than 1988 sales.

Finding the Base

Suppose you spent \$90 on lab fees this semester. This amount represents 5% of your college costs. What are your total college costs for the semester?

In this problem, we have been given the rate (5%) and an amount that is part of another number—\$90—the *percentage*. We are asked, indirectly, to find the *base*, the number of which \$90 is 5%. In other words, \$90 is 5% of what number?

One approach to solving this problem is to use the fact that the base in any problem represents 100%. Having realized this, we can than proceed as follows:

(a) If 5% of a number is \$90, what is 1% of that number?
$1\% = \frac{1}{5}$ of 5%; and $\frac{1}{5} \times \$90 = \$18$

(b) If $\$18 = 1\%$ of a number, what is 100% of that number?
Multiply $\$18 \times 100 = \$1{,}800$

Thus, the base is \$1,800.

A more direct approach to finding the *base* is a modification of the formula $P = B \times R$. We accomplish this by dividing both sides by R.

$$\frac{P}{R} = \frac{B \times \not{R}}{\not{R}} \quad \text{or} \quad B = \frac{P}{R}$$

The equation now allows us to find the answer for B.

If $P = \$90$ and $R = 5\%$, express R in decimal form as .05 and divide.

$$B = \frac{\$90}{.05} = \frac{\$90.00}{.05} = \frac{\$9{,}000}{5} = \$1{,}800$$

SAMPLE PROBLEM 4

Valdez Electronics desires a profit of \$45,000, which will be 6% of its annual sales. How much must it sell to realize this profit? Check your answer.

Solution

Given $P = \$45{,}000$
$R = 6\% = .06$
Find $B =$ total sales

$$B = \frac{P}{R} = \frac{\$45{,}000}{.06} = \$750{,}000$$

Check

$P = B \times R$
$P = \$750{,}000 \times .06 = \$45{,}000$

The problem was checked by substituting into the basic formula.

It is essential to establish exactly what rate the percentage represents in each problem. Always read the wording of a problem carefully. As you can see in Sample Problem 5, the difference in only one word (*increased* instead of *decreased*) makes the rate for part (a) entirely wrong for part (b). Remember: *the rate of the base is always 100%.*

SAMPLE PROBLEM 5

Sales for Mountain Christmas Trees, Inc., in 1989 were $360,000. Find the 1988 sales if (a) sales have *increased* by 20% since 1988, (b) sales have *decreased* by 20% since 1988.

Solution

(a) Given P = $360,000
R = 100% (base) + 20% = 120% = 1.20
Find B = 1988 sales

$$B = \frac{P}{R} = \frac{\$360{,}000}{1.20} = \$300{,}000$$

(b) Given P = $360,000
R = 100% (base) − 20% = 80% = .80
Find B = 1988 sales

$$B = \frac{P}{R} = \frac{\$360{,}000}{.80} = \$450{,}000$$

TEST YOUR ABILITY

1. In each example, find B (the base).

	P	R
(a)	$ 60.00	50%
(b)	$180.00	120%
(c)	$ 3.50	10%
(d)	$450.00	6%
(e)	$ 72.00	.8%
(f)	$ 83.72	200%

2. Inventory shortages of $1,911.70 were reported this month by Harding Jewelry Importers. The shortage is 7% of Harding's total inventory. Find the total inventory.
3. Quan Oil Company reports layoffs of 272 persons in July, a staff reduction of $8\frac{1}{2}$%. Find the number of employees on the staff prior to the layoffs.
4. Bob Wilbert saves $20 a week, an amount that is 8% of his weekly earnings. How much does he *spend* each week?
5. A raise of 12% brings Linda Rice's salary to $17,360 a year. Find her (a) base salary; (b) raise in dollars.
6. Eastern Motors, Inc., earned $89,100 in 1989. Find 1988 earnings if (a) 1989 earnings were 10% more than 1988 earnings; (b) 1989 earnings were 10% less than 1988 earnings.
7. Wilkes Clothing Company sells its coats for $31.78 each, a gain of $13\frac{1}{2}$% over their cost. Find (a) the cost; (b) the profit in dollars.

Finding the Rate

To convert the basic formula P = B × R into a formula to find the *rate,* both sides of the equation are divided by B.

$$\frac{P}{B} = \frac{\not{B} \times R}{\not{B}} \quad \text{or} \quad R = \frac{P}{B}$$

This formula yields a fraction, which must then be converted to either a decimal or a percent, as we saw in Chapter 3.

If you earn \$36 interest on your savings account balance of \$600, what rate are you earning? Here is a problem obviously lacking the percent, that is, the *rate.* The *base* is \$600 (the word "of" is the key), and the *percentage* is \$36.

$$R = \frac{P}{B} = \frac{\$36}{\$600} = \frac{6}{100} = .06$$

To obtain the percent from the decimal, multiply by 100. Thus, the rate is 6%.

SAMPLE PROBLEM 6

An aircraft designing firm showed profits of \$7,500. Its sales were \$300,000. What percent of its sales are its profits?

Solution

Given P = \$7,500
B = \$300,000
Find R

$$R = \frac{P}{B} = \frac{\$7{,}500}{\$300{,}000} = \frac{25}{1{,}000} = .025 = 2.5\%$$

Rounding in Base, Rate, and Percentage Problems

Unless otherwise indicated, round as follows:

(1) If the answer is in dollars, round to the nearest cent. Carry the calculation to three places; then round to two places. This rule will apply when seeking the base or the percentage in dollars.

(2) If the answer is not in dollars, round to the nearest tenth. This happens when the base or the percentage is not in dollars, or when the rate is calculated. A rate is rounded to the nearest tenth of a percent. This makes it necessary to carry the division of P by B to *four places* past the decimal point. Thus, .0374 becomes 3.74% and is rounded to 3.7%.

SAMPLE PROBLEM 7

The Appliance Department of Walden's occupies 2,360 square feet of a 45,000 square foot store. What percent of Walden's area is occupied by the Appliance Department?

Solution

Given P = 2,360 square feet
B = 45,000 square feet
Find R

$$R = \frac{P}{B} = \frac{2{,}360}{45{,}000} = .0524 = 5.2\%$$

Sample Problems 8 and 9 show further calculations needed to find the rate. Study the wording of each problem *carefully*. Notice the importance of understanding exactly what values are presented by the figures given, and of determining which figure is the base and which is the percentage.

SAMPLE PROBLEM 8

The Dakota Copper Mine has union records showing 183 union members and 87 non-union members. What percent of the staff is unionized?

Solution

Given P = union members = 183
B = total staff = 183 + 87 = 270
Find R

$$R = \frac{P}{B} = \frac{183}{270} = .6777 = 67.8\%$$

SAMPLE PROBLEM 9

Maria de la Vega's grade index was 3.1 in the fall term and 3.7 in the spring of that year. What percent of the fall grade index was her spring grade index?

Solution

Given P = 3.7
B = 3.1 ("of")
Find R

$$R = \frac{P}{B} = \frac{3.7}{3.1} = 1.1935 = 119.4\%$$

TEST YOUR ABILITY

1. In each example, find R (the rate).

	P	B
(a)	$ 5.40	$ 108.00
(b)	$ 75.00	$ 60.00
(c)	$ 36.40	$ 520.00
(d)	$ 55.80	$ 900.00
(e)	$4,000.00	$2,000.00
(f)	$ 280.00	$ 240.00

2. Simpson Sports, Inc., produced 61,642 pool tables. Its capacity is 74,000 pool tables. At what percent of capacity was Simpson operating in that year?
3. A blue turtleneck sweater sells for $26.40, an amount that is $4.40 higher than its cost. (a) Find the cost. (b) Find the rate of profit based on cost.
4. Continental Orchards, a large fruit grower, employs 36,000 married workers and 24,000 single workers. What percent of the staff is single?
5. The travel distance to Ralph Martinez's former job was 32 miles. Travel distance to his current job is 20 miles. (a) What percent of the distance to his new job is the distance to his old job? (b) What percent of the distance to his old job is the distance to his new job?

Finding the Rate of Increase or Decrease

A rate of increase (or decrease) is the difference between two numbers divided by one of them. In Sample Problem 9, Maria de la Vega's index increased from 3.1 in the fall to 3.7 in the spring. What was her *rate of increase?*

First, find the *difference:* 3.7 − 3.1 = .6

Next, determine the base. In comparing data from two different time periods, the *earlier period* is always the base. In our example, the fall is the earlier period; therefore, 3.1 is the base. We find the rate of increase by adapting the rate formula in the following way:

$$\text{Rate of increase (decrease)} = \frac{\text{Difference}}{\text{Base}}$$

$$\text{Rate of increase} = \frac{.6}{3.1} = .1935 = 19.4\%$$

As Sample Problem 10 shows, a problem may combine rates of increase and decrease.

SAMPLE PROBLEM 10

Sales figures for the Pollock Corportion were as follows: 1987, $32,000; 1988, $35,200; 1989, $31,680. Find (a) the rate of increase from 1987 to 1988; (b) the rate of decrease from 1988 to 1989.

Solution

(a) Difference $35,200 − $32,000 = $3,200
Base 1987 = $32,000

$$\text{Rate of Increase} = \frac{\$3{,}200}{\$32{,}000} = 10\%$$

(b) Difference $35,200 − $31,680 = $3,520
Base 1988 = $35,200

$$\text{Rate of Decrease} = \frac{\$3{,}520}{\$35{,}200} = 10\%$$

It's one of the ironies of math: A 10% increase is followed in the next year by a 10% decrease, and the final result ($31,680) is lower than the original amount ($32,000). Why? *The base changed* as the time changed.

TEST YOUR ABILITY

1. Everett Company's production in 1988 is 37,500 units; in 1989, 44,625 units. Find the rate of increase in production.
2. Wallace Williams received a grade of 77 on his first math exam and 72 on the second. Find the rate of decrease in his test scores.
3. A radio costing $75 is sold for $120. Find the rate of gross profit (sales price − cost) based on cost.
4. Today's production output of Penelope Weaver's Inc., is 631.806 yards of material. Originally, 686 yards were put into production. To the nearest tenth of a percent, find the rate of waste, based on the original amount.
5. Dr. Benjamin Anderson's gross royalty earnings from 1987 to 1989 were as follows: 1987, $20,910; 1988, $18,819; 1989, $25,092. Find (a) the rate of decrease from 1987 to 1988; (b) the rate of increase from 1988 to 1989.

SUMMARY

Base, rate, and percentage problems are basic to business mathematics. You will find repeated use in your personal and business life in the future for the formulas presented in this chapter. Each time that you approach a problem of this type be sure that you begin by establishing what is given and what is to be found. Then, apply one of the following formulas:

To find the percentage, multiply the base × the rate:

$$P = B \times R$$

To find the base, divide the percentage by the rate:

$$B = \frac{P}{R}$$

To find the rate, divide the percentage by the base:

$$R = \frac{P}{B}$$

To find the rate of increase or decrease, divide the difference by the base:

$$\text{Rate of increase (decrease)} = \frac{\text{Difference}}{\text{Base}}$$

CHECK YOUR READING

1. Define *base, rate,* and *percentage.*
2. Identify the base, rate, and percentage in each of the following statements:
   (a) 30% of 20 is 6.
   (b) $50 is 10% of $500.
3. What word or words normally indicate the base in a problem?
4. To what rate is the base in a problem always equivalent?
5. Is a percentage always smaller than a base? Explain.
6. What rate is represented by a percentage if there has been (a) a 15% increase from the base; (b) a 7% decrease from the base?
7. What rule is followed if a problem seeking a rate of increase or decrease does not specify a base year?

EXERCISES

1. If Gregory Taimanov's salary is $250.00 before deductions and $202.50 after deductions, what percent of his salary has been deducted?
2. Production by the Constantine Apparel Company in 1989 totaled 33,000 men's suits. Production in 1988 was 24,000 units. Find the rate of increase in suit production.
3. If waste for the Constantine Apparel Company is normally $8\frac{1}{2}$% on original material placed into production, how much will be wasted on 440 yards of original material?
4. In each example, find R (the rate) to the nearest tenth of a percent.

	P	B
(a)	$ 70.00	$1,400.00
(b)	$ 10.20	$ 85.00
(c)	$ 36.50	$ 500.00
(d)	$ 1.20	$ 600.00
(e)	$126.00	$ 63.00
(f)	$ 98.00	$2,800.00

5. All-Sport industries earned $264,000 in 1989. Find the 1988 earnings if (a) 1989 earnings are 20% more than 1988 earnings; (b) 1989 earnings are 20% less than 1988 earnings.
6. Kenzi Sugihara, a real estate broker, earns a commission of 8% on a house sold for $75,000. What is the amount of his commission?
7. In each example, find P (the percentage).

	B	R
(a)	$2,000.00	13%
(b)	$ 800.00	1.5%
(c)	$ 400.00	$6\frac{1}{2}$%
(d)	$5,600.00	.2%
(e)	$ 25.00	210%
(f)	$ 600.00	$41\frac{1}{4}$%

8. After a thorough study by its systems analyst, Marcia Optical, Inc., reduced the time of an operation from 2.7 minutes to 2.3 minutes. What rate of decrease in time was achieved?
9. A bank deposit of $75 represents 8% of Juan Garcia's earnings. Find his total income.
10. If 16% of the Penswift Design Corporation's total earnings is $105,600, find the total earnings.
11. Cynthia Najdorf earned total commissions of $1,911 on sales of $27,300. What is the rate of her commission earnings?
12. A purchase of a sofa amounts to $693.75, including freight and handling charges at 11% of the original cost. Find (a) the original cost and (b) the amount of freight and handling charges.
13. In each example, find B (the base).

	P	R
(a)	$ 28.00	7%
(b)	$156.00	5.2%
(c)	$ 6.00	$7\frac{1}{2}$%
(d)	$286.00	130%
(e)	$ 16.80	.4%
(f)	$ 75.00	300%

14. Absences because of illness rose from 72 to 97 at Pratt Motors this week. Find the rate of increase.
15. The profit made by a manufacturer of boating goods is normally 17% of cost. Today's sales, including profit, amounted to $105.30. Find the cost.
16. At Skyway Transport, 54 employees out of a total staff of 720 are on vacation this week. What percent of the staff is working?
17. A roast weighing 2.4 pounds shrank to 1.8 pounds after cooking. What is the rate of decrease in weight?
18. An 8% sales tax on a lamp amounts to $3.36. How much did the lamp cost?
19. Sales of the Lermond Corporation were 75,000 units in 1988. Find 1989 sales if (a) 1989 sales are 130% of 1988 sales; (b) 1989 sales are 130% more than 1988 sales; (c) 1989 sales are 30% less than 1988 sales.

20. A firm's delivery truck logged 58,000 miles this year. Rerouting will result in only 93% of the distance traveled. How many miles will be saved by rerouting?
21. Bill Jackson's salary is $240; a week; Martha Hight's is $300 a week. (a) What percent of Bill's salary is Martha's? (b) What percent of Martha's salary is Bill's?
22. Given the following sales figures for the Peso Oil Company, find (a) the rate of increase from 1987 to 1988; (b) the rate of decrease from 1988 to 1989.

1987	$40,500.00
1988	$43,740.00
1989	$41,115.60

UNIT 1 REVIEW

There are two purposes to a unit review. First, important terms covered in the unit are listed. You should be familiar with these terms. Basic arithmetic terms, such as *sum, difference, multiplicand,* and others are not included.

Second, a series of problems is presented, with the answers listed below, in order for you to check your new skills. Use these activities as a "self-test" to determine how well you have met the major goals of the unit.

Terms You Should Know

bank reconciliation	extension
base	outstanding check
canceled check	payee
check	per C
clearing house	per M
decimal system	percentage
drawee	rate
drawer	ratio

UNIT 1 SELF-TEST

Items 1–7 review the major concepts from Chapter 1. Try them mentally. Then, check your answers with the Key at the end of the self-test.

1. The number 4,623.042 is written in words as __________.
2. The sum of 430 + 96 + 209 is __________.
3. 709 − 271 = __________.
4. 336 × 30 = __________.
5. 800 × 25 = __________.
6. 68,034 ÷ 34 = __________.
7. 17,360 ÷ 10 = __________.

Items 8–18 review the major points of Chapter 2.

8. $\frac{78}{520}$ reduced to lowest terms is __________.
9. The greatest common divisor for $\frac{90}{800}$ is __________.
10. $\frac{4}{5} = \frac{?}{30}$
11. $\frac{217}{8}$ written as a mixed number is __________.
12. $5\frac{7}{12}$ written as an improper fraction is __________.
13. The lowest common denominator for the denominators 3, 5, 6, and 10 is __________.

14. $\frac{5}{7} + \frac{4}{6} =$ __________.
15. $\frac{5}{9} - \frac{1}{4} =$ __________.
16. $\frac{3}{5} \times \frac{4}{7} =$ __________.
17. $\frac{8}{9} \div \frac{2}{3} =$ __________.
18. If B = 500 and A = 1,500, the ratio of A to B is __________.

Items 19–27 review the major points of Chapter 3.

19. The decimal 6,378.206 rounded to the nearest whole number is __________.
20. The sum of 56.3 + .081 + 7.24 = __________.
21. 572.71 − 13.078 = __________.
22. 7.36 × 1,000 = __________.
23. In multiplying 8.371 × .042, the answer will have __________ places after the decimal point.
24. 640 ÷ .32 = __________.
25. 5.30 written as a percent is __________.
26. .03% written as a decimal is __________.
27. 35% written as a fraction is __________.

Items 28–37 review short problems and concepts from Chapter 4.

28. In the sentence "8% of 70 is 5.6," the percentage is __________.
29. If B = $3,000 and R = 7%, P = __________.
30. If 8% of a firm's staff are absent, __________ are present.
31. If sales of 20,000 units have increased by 150%, total sales are now __________.
32. If P = $42 and R = 6%, B = __________.
33. If sales are $96,000 after a 20% decrease, sales were __________ before the decrease.
34. If P = 60 and B = 400, R = __________.
35. If sales are $80,000 after a $20,000 increase, the rate of increase is __________.
36. If the difference is −$40,000 based on sales of $200,000, the rate of decrease is __________.
37. The base year in a problem concerning two years is always the __________ year.

KEY TO UNIT 1 SELF-TEST

1. Four thousand six hundred twenty-three and forty-two thousandths.
2. 735
3. 438
4. 10,080
5. 20,000
6. 2,001
7. 1,736
8. $\frac{3}{20}$
9. 10
10. 24
11. $27\frac{1}{8}$
12. $\frac{67}{12}$
13. 30
14. $\frac{58}{42}$ or $1\frac{8}{21}$
15. $\frac{11}{36}$
16. $\frac{12}{35}$
17. $\frac{4}{3}$ or $1\frac{1}{3}$

18. 3:1
19. 6,378
20. 63.621
21. 559.632
22. 7,360
23. six
24. 2,000
25. 530%
26. .0003
27. $\frac{7}{20}$
28. 5.6
29. $210
30. 92%
31. 50,000 units
32. $700
33. $120,000
34. 15%
35. $33\frac{1}{3}$%
36. 20%
37. earlier

UNIT 2

The Mathematics of Trading

Business organizations can be classified in several ways. One way is to divide them into two categories: trading firms and service firms. A trading firm buys goods and then sells them at a profit. A service firm, in contrast, sells no physical product but performs functions for its customers. Any retail store (druggist, supermarket, florist, etc.) falls into the trading category. Doctors, lawyers, and plumbers are examples of service occupations.

You, as a student of business mathematics, will want to learn about the problems of both types of firms. Since the calculations involved in a trading firm are more complex, this unit emphasizes the mathematics of trading.

Chapter 5 deals with the arithmetic of buying goods. Chapter 6 discusses pricing and selling, and calculating the profit or loss that results from the trading process. Chapter 7 presents the arithmetic for calculating the value of the inventory that remains at the end of the trading process.

CHAPTER **5**

Buying

Every person, as well as every business organization, is a buyer of goods and services. Whereas consumers buy from retailers, retailers buy from wholesalers, who in turn buy from manufacturers. Even the manufacturer is a purchaser of raw materials and supplies.

In this book, we are emphasizing the activities of the retailer, but the principles apply to any level of purchasing.

Your Objectives

After completing Chapter 5, you should be able to

1. Calculate the total list price of a purchase.
2. Use aliquot parts accurately.
3. Understand the purposes of trade and cash discounts.
4. Calculate the amount of one or more trade and cash discounts.
5. Calculate the net amount paid for a purchase after subtracting all discounts.

Computing the Total List Price of a Purchase

The *list price* of an item is the price at which it is first offered for sale to the consumer. The term comes from the fact that this amount is often found in a firm's price list, or catalog.

To calculate the total list price of a purchase, *extensions* are found and added. Sample Problem 1 shows how to do this.

SAMPLE PROBLEM 1

Calculate the total list price of a purchase consisting of 40,000 nails @ \$0.13 per C and 3 dozen hammers @ \$8.80 each.

Solution

400 × \$0.13 =	\$ 52.00
36 × \$8.80 =	316.80
Total list price	\$368.80

Notes
(1) The symbol "@" stands for the word "at."
(2) The nails were quoted per C (hundred); 40,000 = 400 hundreds. "Per M" if used means per thousand.
(3) Since the hammers were quoted at a unit price, 3 dozen was multiplied by 12 to yield 36 hammers. Watch your reading for this type of calculation. If the quotation was \$8.80 per dozen, the calculation would be 3 × \$8.80.

Aliquot Parts

When a purchase involves fractional quantities, fractional prices, or both, calculation of extensions can present some difficulty if handled as a direct multiplication of quantity × price. Consider, for example, 24 items @ $\$0.41\frac{2}{3}$. In "long form"

24	
$\times \$0.41\frac{2}{3}$	
16	⟵ $\frac{2}{3} \times 24 = 16$
24	⟵ 1 × 24 = 24; do not move left.
96	⟵ 4 × 24 = 96; move left one place.
\$10.00	⟵ only 2 places; ignore $\frac{2}{3}$ in setting decimal.

A simplified procedure for solving this type of problem is to use aliquot parts. Any number that divides exactly into another number without leaving a remainder is an *aliquot part,* or *factor,* of that number.

For example, $\$0.33\frac{1}{3}$ is an aliquot part (or factor) of \$1.00 since it divides into \$1.00 exactly 3 times. In the same manner, 9 is an aliquot part of 27; $8\frac{1}{2}$, of 17; 10, of 100. An aliquot part can be expressed as a fraction of the base number; in other words, $33\frac{1}{3}$ can be expressed as $\frac{1}{3}$ of 100; 25, as $\frac{1}{4}$ of 100; 50, as $\frac{1}{2}$ of 100, and so on.

Multiples of the basic fractions can also be considered aliquot parts. Thus, 75 is an aliquot part of 100, since 75 is a multiple of 25. Since 25 is $\frac{1}{4}$ of 100, 75 (3 times 25) is $\frac{3}{4}$ of 100.

Any number can be a base for aliquot parts, but the three most common bases are \$1, 100, and 100%. For example, 40¢ is $\frac{2}{5}$ of \$1; 40 is $\frac{2}{5}$ of 100; 40% is $\frac{2}{5}$ of 100%. Table 3.1 on page 44 is, in effect, a table of aliquot parts of 100%. If you look for the fractional equivalent of $.41\frac{2}{3}$, our introductory problem in this section, you will find $\frac{5}{12}$; therefore, $\$0.41\frac{2}{3}$ is $\frac{5}{12}$ of \$1.00.

To solve the problem $24 \times \$0.41\frac{2}{3}$ by using aliquot parts, change $\$0.41\frac{2}{3}$ to $\$\frac{5}{12}$ and multiply:

$$24 \times \$\tfrac{5}{12} = \$10$$

If you do not recognize an aliquot part, or if it is not available in Table 3.1, which only goes to 12ths, use the following technique to convert cents (or percents) to aliquot parts of \$1.00 (or 100%). In effect, you are changing a decimal—or a percent—to a fraction.

$$
\begin{aligned}
1¢ &= \$0.01 = \$\tfrac{1}{100} \\
41\tfrac{2}{3}¢ &= 41\tfrac{2}{3} \times \$\tfrac{1}{100} \\
&= \tfrac{125}{3} \times \$\tfrac{1}{100} \\
&= \$\tfrac{125}{300} \\
&= \$\tfrac{5}{12}
\end{aligned}
$$

You should know the following common aliquot parts of \$1.00:

$\$\frac{1}{2}$ = 50 cents
$\$\frac{1}{3}$ = $33\frac{1}{3}$ cents; $\frac{2}{3} = 66\frac{2}{3}$ cents
$\$\frac{1}{4}$ = 25 cents; $\frac{3}{4}$ = 75 cents
$\$\frac{1}{5}$ = 20 cents; $\frac{2}{5}$ = 40 cents; $\frac{3}{5}$ = 60 cents; $\frac{4}{5}$ = 80 cents
$\$\frac{1}{6}$ = $16\frac{2}{3}$ cents; $\frac{5}{6} = 83\frac{1}{3}$ cents
$\$\frac{1}{7}$ = $14\frac{2}{7}$ cents
$\$\frac{1}{8}$ = $12\frac{1}{2}$ cents; $\frac{3}{8} = 37\frac{1}{2}$ cents; $\frac{5}{8} = 62\frac{1}{2}$ cents; $\frac{7}{8} = 87\frac{1}{2}$ cents
$\$\frac{1}{10}$ = 10 cents; $\frac{3}{10}$ = 30 cents; $\frac{7}{10}$ = 70 cents; $\frac{9}{10}$ = 90 cents
$\$\frac{1}{12}$ = $8\frac{1}{3}$ cents
$\$\frac{1}{15}$ = $6\frac{2}{3}$ cents
$\$\frac{1}{16}$ = $6\frac{1}{4}$ cents
$\$\frac{1}{20}$ = 5 cents

Sample Problem 2 illustrates the use of aliquot parts of a dollar in calculating the total list price.

SAMPLE PROBLEM 2

Calculate the total list price of 32 brown pencils @ $12\frac{1}{2}$¢ each and 48 drawing pads @ \$$1.83\frac{1}{3}$ each.

Solution

$$12\tfrac{1}{2}¢ = \tfrac{25}{2} \times \$\tfrac{1}{100} = \$\tfrac{1}{8}$$

$$\$1.83\tfrac{1}{3} = \$1 + (\tfrac{250}{3} \times \$\tfrac{1}{100}) = \$1 + \$\tfrac{250}{300} = \$1 + \$\tfrac{5}{6} = \$\tfrac{11}{6}$$

$$
\begin{aligned}
32 \times \$\tfrac{1}{8} &= \$\ \ 4.00 \\
48 \times \$\tfrac{11}{6} &= \underline{\ \ 88.00} \\
\text{Total list price} &= \underline{\underline{\$92.00}}
\end{aligned}
$$

Notes (1) The first step in the solution was to convert both list prices into aliquot parts of \$1.00.

(2) The alternative to using aliquot parts would have been the following multiplications:

$$32 \times \$0.125$$
$$48 \times \$1.83\tfrac{1}{3}$$

As you see, multiplication by the aliquot parts saves considerable time. In fact, step (1) will become automatic with practice.

(3) $\$\frac{11}{6}$ could have been handled as $\$1\frac{5}{6}$. The arithmetic would then have been

$$\begin{aligned} 48 \times \$1 &= \$48 \\ 48 \times \$\tfrac{5}{6} &= 40 \\ &\ \overline{\$88} \end{aligned}$$

The Invoice A business purchase is recorded on an *invoice*. An invoice is a form with several carbons that serves as a record of the purchase for both buyer and seller. Sample Problem 3 shows an example of an invoice and the proper method for its completion.

SAMPLE PROBLEM 3

Compute the total list price of the following purchase and arrange it on an invoice form:

On April 8, Charles Ehl bought 36 reams of typing paper @ $\$3.08\frac{1}{3}$ per ream, 75 pads of two-column journal paper @ \$0.80 each, and 6 dozen pads of ledger paper @ \$0.875 each. The seller was the ABC Office Supply Company, 27 Spruce Street, Milford, NH 03055. All items were to be delivered by truck on April 23 to 2100 River Road, Manchester, NH 03104. Sixty-day terms of payment were requested on ABC's Invoice 536.

Solution

ABC Office Supply Company
27 Spruce Street
Milford, NH 03055

Invoice No. 536

Sold to Charles Ehl — **Date** April 8, 19—

2100 River Road — **Ship via** truck

Manchester, NH 03104 — **Delivery date** April 23, 19—

Terms n/60 — **Special instructions** None

Quantity	Description	Unit Price	Extension
36	reams of typing paper	$\$3.08\frac{1}{3}$	\$111.00
75	pads of two-column journal paper	0.80	60.00
6 dozen	pads of ledger paper	0.875	63.00
	TOTAL		\$234.00

Notes (1) The term "n/60" is an abbreviated form of "net due in 60 days."

(2) The extensions, using aliquot parts, were calculated as follows:

$$36 \times \$3\tfrac{1}{12}$$
$$75 \times \$\tfrac{4}{5}$$
$$72 \times \$\tfrac{7}{8}$$

(3) The answer of \$234.00 is the total list price of the goods, not the final price to be paid. From this amount, certain discounts may be deducted, and certain charges (freight and/or sales tax) may be added. All of these adjustments are discussed in this unit.

TEST YOUR ABILITY

1. Calculate the total list price of 39 chef's knives @ \$5.65 each and 52 casseroles @ \$2.86 each.
2. Convert each of the following amounts into aliquot parts of \$1.00. Try to calculate the conversions without using Table 3.1.

   (a) $37\frac{1}{2}$¢ (b) $16\frac{2}{3}$¢ (c) $28\frac{4}{7}$¢ (d) 75¢ (e) $44\frac{4}{9}$¢
   (f) $43\frac{3}{4}$¢ (g) $8\frac{1}{3}$¢ (h) $6\frac{1}{4}$¢ (i) $58\frac{1}{3}$¢ (j) $26\frac{2}{3}$¢

3. Calculate the extension in each of the following problems:

   (a) $75 \times \$0.40$ (b) $96 \times \$0.83\frac{1}{3}$
   (c) $80 \times \$0.75$ (d) $640 \times \$0.06\frac{1}{4}$
   (e) $63 \times \$0.71\frac{3}{7}$ (f) $33\frac{1}{3} \times \$0.30$

4. Draw an invoice form similar to the one in Sample Problem 3. Record and calculate the total list price of the following purchase:

   On July 17, Ms. Kathi Spencer of 3742 Seaway Drive, San Francisco, CA 94103 purchases from Warner Associates, San Jose, CA 95001: 5 dozen boxes of stationery @ \$1.25 per dozen, 39 pads of paper @ $\$0.33\frac{1}{3}$ per pad, and 28 reams of paper @ \$2.20 per ream. Delivery by truck will be made on Warner's Invoice 87 on August 3, with 30-day payment terms.

Trade Discounts

If a firm decides to lower its list prices, it has two means of notifying its customers. One is to print a new price list or catalog, an expensive procedure. The other choice is to issue a note to buyers that a stated percent is to be deducted from every list price.

This type of reduction is called a *trade discount.* The term "trade discount" originally referred to the granting of a price reduction to other members who were in the same type of business—the same trade. However, today it has a wider meaning and refers to any sizable discount from list price. You are probably familiar with the almost daily "special sales" in stores.

For arithmetic purposes, trade discounts are equivalent to quantity discounts (given for extra-large orders) and special-sale discounts.

Sample Problems 4 and 5 illustrate the calculation of a trade discount. This calculation uses the formula you studied in Chapter 4 for finding the percentage when you have been given the base and the rate. The problems also illustrate:

(1) the subtraction of the trade discount to arrive at the *net purchase price* and

(2) the use of aliquot parts of 100%.

SAMPLE PROBLEM 4

Find the net purchase price of a color television set listed at \$375 and sold with a 20% trade discount.

Solution

$$20\% = \tfrac{1}{5} \text{ of } 100\%$$

$$\text{Discount} = \$375 \times \tfrac{1}{5} = \$75$$

$$\text{Net purchase price} = \$375 - \$75 = \$300$$

Alternate Solution

$$20\% = \tfrac{1}{5} \text{ of } 100\%$$

$$\text{Net purchase price fraction} = 1 - \tfrac{1}{5} = \tfrac{4}{5}$$

$$\text{Net purchase price} = \$375 \times \tfrac{4}{5} = \$300$$

SAMPLE PROBLEM 5

Find the net purchase price of 36 shirts @ $\$3.37\tfrac{1}{2}$ each and 72 blouses @ $\$2.08\tfrac{1}{3}$ each, with a trade discount of $33\tfrac{1}{3}\%$ from the total.

Solution

$36 \times \$3.37\tfrac{1}{2} = 36 \times \$3\tfrac{3}{8}$	$= \$121.50$
$72 \times \$2.08\tfrac{1}{3} = 72 \times \$2\tfrac{1}{12}$	$= 150.00$
Total list price	\$271.50
$\$271.50 \times 33\tfrac{1}{3}\% = \$271.50 \times \tfrac{1}{3} =$	-90.50
Net purchase price	\$181.00

Alternate Solution

$$100\% - 33\tfrac{1}{3}\% = 66\tfrac{2}{3}\%$$

$$\$271.50 \times 66\tfrac{2}{3}\% = \$271.50 \times \tfrac{2}{3} = \$181.00$$

Notes (1) The net purchase price can be found by subtracting the calculated trade discount from the list price. This procedure was shown in the first solution to each problem. Or, the net purchase price can be found by calculating the *complement* of the trade discount rate, and then by multiplying the list price by this complement. The complement of a fraction (or percent) is that number which, when added to the fraction (or percent), totals 1 (or 100%). If the complement method is used, the amount of the trade discount is not shown.

(2) In a multiple purchase, like the one shown in Sample Problem 5, it is pointless to calculate the trade discount on each extension separately. Instead, calculate it on the total list price. Or, if you prefer, use the complement method.

Two or More Discounts

After catalog prices have been reduced by one discount, additional discounts may be allowed to take into consideration the competing markets, as well as for promotional purposes. Such *trade discount series* may be given for unusually large orders as well.

Sample Problem 6 illustrates how to calculate the net purchase price when two trade discounts are given.

SAMPLE PROBLEM 6

Find the net purchase price of a pine desk listed at \$450 and sold with trade discounts of 20% and 10%.

Correct Solution

First discount = \$450 × 20% = \$90
First net purchase price = \$450 − \$90 = \$360

Second discount = \$360 × 10% = \$36
Final net purchase price = \$360 − \$36 = \$324

Wrong Solution

20% + 10% = 30%
Discount = \$450 × 30% = \$135
Net purchase price = \$450 − \$135 = \$315

The "wrong" solution to this problem is one of the most common errors in business mathematics. *Discounts of 20% and 10% are not the same as a discount of 30%.* Why? Because the second discount rate is not based on the list price. Adding the rates treats them as both being based on the list price. If you want to find the actual rate of discount, calculate by dividing the two discounts (\$90 + \$36) by \$450—percentage by base—and you arrive at *28%*, not 30%.

Therefore, two or more rates of discount should be applied separately to successive net purchase prices. This point is seen again in Sample Problem 7.

SAMPLE PROBLEM 7

Find the net purchase price of a dishwasher listed at \$600 and sold with trade discounts of 20%, 15%, and 5%.

Solution

First discount = \$600 × $\frac{1}{5}$ = \$120
First net purchase price = \$600 − \$120 = \$480
Second discount = \$480 × $\frac{3}{20}$ = \$72
Second net purchase price = \$480 − \$72 = \$408
Third discount = \$408 × $\frac{1}{20}$ = \$20.40
Final net purchase price = \$408.00 − \$20.40 = \$387.60

Equivalent Single Discounts and Net Purchase Price Equivalents

In Sample Problems 6 and 7, a series of trade discounts was applied, one at a time, to a series of purchase prices. Such a procedure is correct but time-consuming and, therefore, prone to error.

A number of techniques have been developed in order to speed the process and minimize the chance of error. What these various techniques have in common is the determination of a single rate equivalent to a series of discount rates or to the net purchase price rate. This single rate can then be applied directly to the list price.

The first method is usable with only two rates of discount. It yields an *equivalent single rate of discount.*

(1) Add the rates.

(2) Multiply the decimal equivalents of the rates.

(3) Subtract the product in step (2) from the sum in step (1) to find the equivalent single rate of discount.

For 20% and 10%,

Add: 20% + 10% = 30%
Multiply: .20 × .10 = 2%
Subtract: 30% − 2% = 28%

A second method is time-saving with two or more discounts and yields a *rate equivalent to the net purchase price.*

(1) Express each discount as an aliquot part of 100%.

(2) Find the fractional complement of each discount.

(3) Multiply the fractional complements to find the net purchase price equivalent rate.

For 20% and 10%,

$20\% = \frac{1}{5}$; $10\% = \frac{1}{10}$

Complements: $\frac{4}{5}, \frac{9}{10}$

Multiply: $\frac{4}{5} \times \frac{9}{10} = \frac{36}{50} = 72\%$

Once we have the above answer, an additional step is required to find the discount rate: subtracting the net purchase price equivalent rate from 100%. Thus, 100% − 72% = a 28% discount rate. However, using the net purchase price equivalent saves time in finding the net purchase price as illustrated in Sample Problem 8.

SAMPLE PROBLEM 8

Find the net purchase price of a dishwasher listed at $600, sold with trade discounts of 20%, 15%, and 5%. Use a net purchase price rate.

Solution

$20\% = \frac{1}{5}$; complement $= \frac{4}{5}$

$15\% = \frac{3}{20}$; complement $= \frac{17}{20}$

$5\% = \frac{1}{20}$; complement $= \frac{19}{20}$

$$\frac{\overset{1}{\cancel{4}}}{5} \times \frac{17}{\underset{5}{\cancel{20}}} \times \frac{19}{20} = \frac{323}{500} = 64.6\%$$

Net purchase price = $600 × .646 = $387.60

As you see, the discount rate is not necessary when the net purchase price is directly obtainable. Compare Sample Problems 7 and 8 to help you decide which is the easier method for you. Before reaching a final decision, however, continue reading.

Tables are often used to save making repeated calculations of net purchase price equivalents. Table 5.1 lists equivalents for two or three discounts.

The order in which the discounts are found does not matter. Locate one of your discounts at the top of a column and locate the other discount (or discounts) on one of the rows. The entry at the intersection of column and row is the net purchase price equivalent. As an example, for a multiple discount of 20–10, follow the "20" column down to the "10" row, and at the intersection, find .72 which, as you well know, is also expressed as 72%. Or, instead, you can follow the "10" column down to the "20" row and find .72.

What net purchase price rate is equivalent to rates of 10–10–$2\frac{1}{2}$? The answer is .78975. What about $12\frac{1}{2}$–10–10? The answer is .70875.

SAMPLE PROBLEM 9

Find the net purchase price of a dishwasher listed at $600, sold with trade discounts of 20%, 15%, and 5%. Use Table 5.1.

Solution

Table entry down the 15 column and across the 20–5 row = .646.

Net purchase price = $600 × .646 = $387.60

Table 5.1

Net Purchase Price Equivalents of a Series of Discounts

Rates	$2\frac{1}{2}$	5	$7\frac{1}{2}$	10	$12\frac{1}{2}$
5	.92625	.9025	.87875	.855	.83125
10	.8775	.855	.8325	.81	.7875
15	.82875	.8075	.78625	.765	.74735
20	.78	.76	.74	.72	.70
25	.73125	.7125	.69375	.675	.65625
$33\frac{1}{3}$	.65	.633333	.616667	.60	.583333
40	.585	.57	.555	.54	.525
50	.4875	.475	.4625	.45	.4375
20–10	.702	.684	.666	.648	.63
20–5	.741	.722	.703	.684	.665
10–10	.78975	.7695	.74925	.729	.70875
10–5	.833625	.81225	.790875	.7695	.748125

Rates	15	$16\frac{2}{3}$	20	25	$33\frac{1}{3}$
5	.8075	.791627	.76	.7125	.633333
10	.765	.75	.72	.675	.60
15	.7225	.708333	.68	.6375	.566667
20	.68	.666667	.64	.60	.533333
25	.6375	.625	.60	.5625	.50
$33\frac{1}{3}$	.566667	.555556	.533333	.50	.444444
40	.51	.50	.48	.45	.40
50	.425	.416667	.40	.375	.333333
20–10	.612	.60	.576	.54	.48
20–5	.646	.633333	.608	.57	.506667
10–10	.6885	.675	.648	.6075	.54
10–5	.72675	.7125	.684	.64125	.57

Use Table 5.1 whenever possible, but learn the other methods as well because there will be times when a table is not available or it may not contain the exact rate or rates required.

TEST YOUR ABILITY

1. Find the net purchase price of a bed listing for $180 and sold with a trade discount of 35%.
2. Bill Evans purchases a pocket calculator listed at $35 and an AM-FM radio listed at $79 at a special sale, both marked 30% off. Find the net purchase price.
3. Calculate the cost of a motorcycle listed at $960 and sold with discounts of $33\frac{1}{3}$%, 10%, and 5%. Use the method of successive subtraction of discounts.
4. Using the method of complements, find the net purchase price equivalent of the following discounts:
   (a) 10–10 (b) 40–15 (c) 20–5
   (d) 25–20–5 (e) 20–$16\frac{2}{3}$–10 (f) 20–15–10
5. A stereo system listed at $960 is sold with discounts of $33\frac{1}{3}$%, 10%, and 5%. Calculate the net purchase price using the method of complements.

6. Using Table 5.1, find the net purchase price equivalent rates for the following discounts.
   (a) 10–10 (b) 40–15 (c) 20–5
   (d) 25–20–5 (e) 20–16$\frac{2}{3}$–10 (f) 20–15–10
7. Using Table 5.1, find the net purchase price in each of the following problems:
   (a) $420 less 20–2$\frac{1}{2}$
   (b) $300 less 50–5
   (c) $2,000 less 20–10–7$\frac{1}{2}$
   (d) $500 less 25–10–10
   (e) $960 less 33$\frac{1}{3}$–10–5

Cash Discount

It is often important to the seller to be paid by the buyer as soon as possible. To encourage prompt payment, the seller may offer a *cash discount* of 1, 2, or 3%.

A cash discount is an advantage to both buyer and seller: The buyer saves money, while the seller has funds to use—instead of an unpaid account.

Two subjects are of importance in studying cash discount. The first is an area of interpretation: Is the buyer entitled to a cash discount? The second is an area of calculation—figuring the amount of the cash discount and determining the net amount to be paid for a purchase.

Terms of Sale

Every sale includes the establishment of payment conditions agreed on by both buyer and seller. The most common *terms of sale* are listed and defined below:

n/30	*n*et due in 30 days
n/e.o.m.	*n*et due at *e*nd *o*f *m*onth
n/10 e.o.m.	*n*et due 10 days after *e*nd *o*f *m*onth
c.o.d.	*c*ash *o*n *d*elivery
n/10 r.o.g.	*n*et due 10 days after *r*eceipt *o*f *g*oods

Each type of terms serves an industry or a purpose. Terms such as n/30 are ordinarily found in industries where the buyer has a reasonable chance of selling the goods within the 30 days and of paying for the purchase out of the profits. Where additional processing is needed on the goods purchased, sale will be delayed, so terms of n/10 e.o.m. may be appropriate.

Where shipment is over a long distance, r.o.g. terms are available since it would be unfair to impose ordinary terms. The buyer might receive the invoice days ahead of the goods; it is only proper to count payment time from the receipt of the goods.

To understand the use of these terms, assume a sale on October 12, with delivery on October 29.

If Terms Are	**Due Date Is**
n/30	November 11
n/e.o.m.	October 31
n/10 e.o.m.	November 10
c.o.d.	October 29
n/10 r.o.g.	November 8

Exact calender days are counted in determining the due date.

When a sale is made after the 15th day of a month, n/e.o.m. terms postpone payment until the end of the following month; therefore, an invoice dated July 16 is due by August 31.

When a sale is made after the 26th of a month with terms of n/10 e.o.m., payment is due by the 10th of the second month following; therefore, an invoice dated July 27 is due by September 10.

None of the above terms includes a cash discount. If the terms are n/30 and the seller wants to encourage the buyer to pay quickly, a 2% discount may be offered for payment within 10 days. The terms would then be written as "2/10, n/30." If a sale is made on March 27 with terms of 2/10, n/30, payment by April 6 will give the buyer the 2% discount; otherwise, the final date for payment is April 26.

On a July 16 sale with terms of 2/20, 1/30, n/60, payment by August 5 will allow the buyer to deduct 2%; by August 15, 1%; and by September 14, full payment must be made.

SAMPLE PROBLEM 10

A sale of appliances for $637.27 is made on July 22 with terms of 2/10 e.o.m., n/30. Find the net amount to be paid on (a) August 8; (b) August 20.

Solution

(a) Since August 8 is within 10 days after the end of month, deduct 2%.
Cash discount = $637.27 × .02 = $12.7454 = $12.75
Net amount to be paid = $637.27 − $12.75 = $624.52

(b) August 20 is beyond the discount period.
Net amount to be paid = $637.27

Notice that we first determine if the buyer is eligible for discount. If so, calculate the discount.

What should be done if a problem involves a trade discount as well as a cash discount?

SAMPLE PROBLEM 11

A glass display case listed at $600 is sold with a trade discount of 20% and is paid in time for a cash discount of 3%. Find the net amount to be paid.

Solution

Trade discount = $600 × $\frac{1}{5}$ = $120
Net purchase price = $600 − $120 = $480
Cash discount = $480 × .03 = $14.40
Net amount to be paid = $480.00 − $14.40 = $465.60

The trade discount is always deducted first, for it is certain. The cash discount is uncertain, since it depends on the time of payment.

TEST YOUR ABILITY

1. Find the date on which an invoice must be paid if it is dated April 7 and the goods are delivered on April 16 with terms of (a) n/30; (b) n/e.o.m.; (c) n/10 e.o.m.; (d) n/10 r.o.g.; (e) c.o.d.
2. Assume a sale of 1,000 gallons of oil on August 24 and delivery on August 31. Using the terms listed in Problem 1, find each due date.
3. An invoice for $730 is dated May 11. Terms of the sale are 3/10, 2/20, 1/30,

n/60. Find the amount to be paid on (a) June 7; (b) May 29; (c) July 3; (d) May 17; (e) June 1.

4. An invoice for seven radios has a list price of $600 with trade discounts of 40–20. It is paid in time for a 2% cash discount. Find the net amount to be paid.

SUMMARY Of the concepts covered in Chapter 5, Sample Problem 12 that follows displays:

(1) Calculating list price using aliquot parts.
(2) Calculating net purchase price, using Table 5.1.
(3) Determining eligibility for cash discount.
(4) Calculating net amount to be paid.

Before looking at the solution, try the problem yourself to see if you have understood these four concepts.

SAMPLE PROBLEM 12

On July 23, a purchase is made of $3\frac{1}{2}$ dozen calendars @ $\$0.66\frac{2}{3}$ each, 57 toy watches @ $0.50 each, and 3,000 assorted novelties @ $\$0.83\frac{1}{3}$ per C. Trade discounts of 20% and $7\frac{1}{2}\%$ are granted on the purchase, which is delivered on July 28 with terms of 2/10, 1/20, n/30. If payment is made on August 6, find the net amount to be paid by the buyer.

Solution

List price:

$3\frac{1}{2}$ dozen @ $\$0.66\frac{2}{3}$ each = 42 × $\$\frac{2}{3}$ =	$28.00
57 @ $0.50 each = 57 × $\$\frac{1}{2}$ =	28.50
3,000 @ $\$0.83\frac{1}{3}$ per C = 30 × $\$\frac{5}{6}$ =	25.00
Total list price	$81.50
Net purchase price equivalent:	
20%–$7\frac{1}{2}\%$, Table 5.1	×.74
Net purchase price	$60.31

Payment on August 6, 14 days after sale, gives the 1% discount.

Cash discount = $60.31 × .01 = $0.60
Net amount to be paid = $60.31 − $0.60 = $59.71

CHECK YOUR READING

1. Distinguish *net purchase price* from *list price* and *net amount to be paid.*
2. Can you define *aliquot part?* Identify two common "wholes" on which aliquot parts can be based.
3. What is the purpose of an invoice?
4. Explain the difference between (a) the size and (b) the purpose of trade discounts and cash discounts.
5. What is meant by a *series of trade discounts?* By an *equivalent single rate of discount?* By a *net purchase price equivalent?*
6. Why can't you add a series of discount rates to find an equivalent single rate of discount?
7. Find the complement of (a) 30%; (b) 90%; (c) $\frac{3}{4}$; (d) $\frac{1}{9}$; (e) .35; (f) .975.
8. Explain the following terms: e.o.m.; n/e.o.m.; n/10 e.o.m.; 2/e.o.m.; 2/10 e.o.m.
9. Why is a trade discount deducted before a cash discount?

EXERCISES

Group A

1. Using the method of complements, find the net purchase price equivalent for the following discounts:
   (a) 10–5 (b) 40–20
   (c) $40–12\frac{1}{2}$ (d) 25–20–10
   (e) 30–20–10 (f) $33\frac{1}{3}–16\frac{2}{3}–10$
2. An invoice of $830 for a used delivery truck is dated September 20. Terms of sale are 3/10, 2/20, n/30. Find the net amount to be paid on (a) October 13; (b) October 8; (c) October 1; (d) October 19; (e) September 29.
3. Find the net purchase price of each of the following purchases, using Table 5.1.
   (a) $800 less 40–10
   (b) $750 less $33\frac{1}{3}–2\frac{1}{2}$
   (c) $400 less $16\frac{2}{3}–10–10$
   (d) $1,000 less $10–5–2\frac{1}{2}$
   (e) $600 less 20–10–5
4. A freezer, listed at $900, is sold with trade discounts of $33\frac{1}{3}–20–10$. Applying the method of complements, find the net purchase price.
5. Draw an invoice form. Calculate on it the total list price of the following purchase:
   On February 13, Mark Fritz purchased 18 dozen eggs @ $66\frac{2}{3}$¢ per dozen, 142 half-gallons of milk @ 57¢ per half-gallon, and various other grocery products totaling $73.72 for the Fritz Food Store, Main Street, Newton. The terms of sale were 2/10 e.o.m., n/30. Delivery from Hunseker Wholesalers, West Street, Newton was due the following day by delivery truck. The invoice number was 837.
6. Calculate the extension for each of the following:
   (a) 416 × $0.25
   (b) 85 × $0.40
   (c) 880 × $0.875
   (d) 16 × $$0.18\frac{3}{4}$
   (e) 72 × $$1.16\frac{2}{3}$
   (f) 50 × $0.36
7. Find the date on which an invoice must be paid if a sale of a dozen tables for a restaurant takes place on August 6, is delivered on August 12, and terms are (a) n/60; (b) n/10 r.o.g.; (c) c.o.d.; (d) n/e.o.m.; (e) n/10 e.o.m.
8. Convert each of the following percents into aliquot parts of 100%:
   (a) 80% (b) 26%
   (c) 65% (d) 32%
   (e) 95% (f) $2\frac{1}{2}$%
   (g) 70% (h) 18%
   (i) $7\frac{1}{2}$% (j) $3\frac{1}{3}$%
9. Compute the net purchase price of a ten-speed bicycle listed at $130 and sold with a special-sale discount of 35%.
10. Calculate the cost of a motorcycle listed at $900 with trade discounts of $33\frac{1}{3}–20–10$. Use the method of successive subtraction of discounts.

11. Using Table 5.1, find the net purchase price equivalents, in percent form, of the following discounts:
    (a) 15–10–5
    (b) 20–$16\frac{2}{3}$–5
    (c) 10–$7\frac{1}{2}$–5
    (d) 10–10–10
    (e) 25–10–5
12. Find the date on which an invoice must be paid if a sale on January 28 is delivered on February 2 and the terms are (a) n/30; (b) n/e.o.m.; (c) n/10 r.o.g.; (d) n/10 e.o.m.; (e) c.o.d.
13. Eileen Weisner purchases 16 yards of satin @ $3.10 a yard and 24 yards of cotton @ $1.87 a yard. Calculate the total list price.
14. James Reilly buys 6 pounds of lamb @ $3.90 a pound and $4\frac{1}{2}$ pounds of beef @ $3.70 a pound. He is granted a $12\frac{1}{2}$% trade discount on this purchase. Find the net purchase price.
15. An invoice listed at $1,000 is sold with trade discounts of $33\frac{1}{3}$–10 and is paid in time for a 2% cash discount. Find the net amount to be paid.

Group B

1. Using the alternate solution shown in Sample Problems 4 and 5, calculate the net purchase price of 55 wrenches @ $4.25 each and 75 pliers @ 0.83\frac{1}{3}$ each, with a trade discount of 25% from the total.
2. Trying the "add-multiply-subtract" approach, find the equivalent single rate of discount to the following:
    (a) 20–5
    (b) $33\frac{1}{3}$–10
    (c) 25–20
    (d) 50–10
    (e) $16\frac{2}{3}$–10
3. An invoice totaling $1,450 for decorating a hotel is dated July 29 and is paid on September 1. Find the net amount to be paid under each of the following terms:
    (a) n/45
    (b) 2/10, n/30
    (c) n/10 e.o.m.
    (d) 3/20, 2/30, 1/60, n/90
    (e) 2/10 e.o.m.
4. Calculate the net amount to be paid on the following purchases made on March 18, with terms of 2/10, 1/20, n/30, paid on April 9, and with trade discounts of 20% and 15%.

    $5\frac{1}{2}$ dozen brooches @ $1.22 each

    78 buttons @ 0.08\frac{1}{3}$ each

    51 scissors @ 2.33\frac{1}{3}$ each
5. An invoice shows a net purchase price of $331.74 *after* subtraction of discounts of 20%, 10%, and 5%. What was the list price?
6. After receiving a 2% discount, Kate Parker, who lives in Florida, pays $682.57 in settlement of an invoice for boat repairs. What was the full amount of the invoice?

CHAPTER **6**

Pricing and Selling

After purchasing a supply of goods, the retailer assigns prices to the merchandise in order to sell it at a profit to the consumer. In this chapter, we examine the mechanics and terminology of this operation. In other words, you'll see why you pay a certain price for a product, and why that product will cost more in some stores than in others.

Your Objectives

After studying Chapter 6, you will be able to

1. Understand the terminology of pricing.
2. Calculate markups, markdowns, and related amounts in order to arrive at a retail price.
3. Interchange between two different bases for markups.
4. Calculate gross and net profits.
5. Apply the terminology of freight charges.
6. Calculate amounts of sales tax.

Setting a Price

Price-setting is a two-way street. One approach is to start from the cost of a product and add its expenses and the *net profit* (the real profit). For example, if you purchase an article for $10, expect expenses of $5, and want a net profit of $3, then the selling price must be $18.

Sales price = Cost + Expenses + Net profit

SAMPLE PROBLEM 1

Calculate the selling price of a watch costing $57.50, with anticipated expenses of 20% of cost, and a desired net profit of $12.50 on the sale.

Solution

Cost	$57.50
Expenses (20% × $57.50)	11.50
Net profit	12.50
Sales price	$81.50

Markup The proper term for the expression "expenses + net profit" is the word *markup* or *markon*. *Markup* is the amount that is added to the cost of an item to arrive at the sales price, or *original retail price*.

$$\text{Original retail price} = \text{Cost} + \text{Markup}$$

Markup can be expressed in dollars (in Sample Problem 1, a $24.00 markup, $81.50 − $57.50, was added to the cost), or as a percent. In the earlier example, in which the article cost $10 and had an $8 markup, the markup could be expressed as a rate of 80% of cost ($8 ÷ $10). Generally, the percent form is used so that a firm can apply the same rate of markup to all of its products, as shown in Sample Problem 2.

SAMPLE PROBLEM 2

The Willougby Company applies a uniform markup rate of 35% to the cost of all its products. Find the original retail price of (a) a toaster costing $18.00 and (b) an iron costing $7.95.

Solution

$$\text{Original retail price} = \text{Cost} + \text{Markup}$$

	(a)	(b)
Cost	$18.00	$ 7.95
Markup (35% × cost)	6.30	2.78
Original retail price	$24.30	$10.73

The iron might be priced at a round figure, such as $10.75.

The second approach to the two-way street of price-setting goes in the opposite direction. For instance, a firm knows its markup rate, but to meet competition, the sale price of an item has been decided in advance. The firm must then calculate the highest cost it can afford to pay for the item in order to be able to sell the merchandise at that price.

Suppose that a firm has to sell a fishing reel for $17.50 and wants a $5.00 markup. What is the maximum price for which the item can be purchased? We subtract $5.00 from $17.50, and $12.50 becomes the cost; therefore, the formula for the other direction is:

$$\text{Cost} = \text{Original retail price} - \text{Markup}$$

Again, markup is more likely expressed as a percent. If a sales price, or original retail price, is to be \$17.50 with a markup of 25% on cost, then sales price = cost + 25% of cost. A base, rate, and percentage problem now appears in which B (cost) is unknown, R = 125%, and P = \$17.50. Solving for B, we have:

$$B = \frac{P}{R} = \frac{\$17.50}{1.25} = \frac{\$1750.}{125.} = \$14.00$$

The firm can pay no more than \$14 for the item.

SAMPLE PROBLEM 3

A tea set is to retail for \$27.30 after a 30% markup on cost. What is the maximum price that the seller can afford to pay for the item?

Solution

Given: P = sales price = \$27.30
R = markup = cost + 30% = 130%
Find: B = cost

$$B = \frac{P}{R} = \frac{\$27.30}{1.30} = \$21.00$$

Check 30% (markup) × \$21.00 (cost) = \$6.30
\$21.00 + \$6.30 = \$27.30

Regardless of the approach that a firm uses to set its prices, it may be necessary because of increased competition or difficulty in selling to lower its original retail prices—causing a *markdown.*

Markdown

A markdown is based on the original retail price. As we have already learned, a markup can be based on cost; we will learn later that it can also be based on the original retail price. In any problem involving both a markup and a markdown, *the markup must be computed first.*

Final retail price = Original retail price − Markdown

SAMPLE PROBLEM 4

Find the final retail price of a table costing \$50, marked up by 20% of cost, and then marked down by $33\frac{1}{3}$%.

Solution

Cost	\$50.00
Markup ($\frac{1}{5}$ × \$50)	10.00
Original retail price	\$60.00
Markdown ($\frac{1}{3}$ × \$60)	−20.00
Final retail price	\$40.00

The markup was based on cost because the problem said so. *The markdown is always based on the original retail price.*

Prices, as you have probably noticed, move up and down—although the current trend is upward. Watch a single item over a period of a month. The price will vary as time passes. Any rise in price after a markdown, up to its original retail price, is termed a *markdown cancellation.* The difference between markdowns and cancellations is termed *net markdown,* which is expressed in the following formula:

Net markdown = Markdown − Markdown cancellation

Finally, we have a revised formula:

Final retail price = Original retail price − Net markdown

SAMPLE PROBLEM 5

Find the final retail price of a piston with an original retail price of $75, reduced by 30%, and later raised by 15%. What is the net markdown?

Solution

Original retail price	$75.00
Markdown (30% × $75)	−22.50
	$52.50
Markdown cancellation (15% × $52.50)	7.88
Final retail price	$60.38

Net markdown = $22.50 − $7.88 = $14.62 (or $75.00 − $60.38 = $14.62).

Summary of Pricing Terms

Cost is the net amount the retailer pays for an item.

Markup, or *markon,* is the expenses and net profit added to the cost to arrive at the *original retail price.*

Markdown is a reduction from the original retail price.

Markdown cancellation is a reduction of the markdown, or an increase in price after a markdown, up to the original retail price.

Net markdown is the final difference between markdowns and markdown cancellations.

Final retail price is the price at which an article is sold after all markups and markdowns.

TEST YOUR ABILITY

1. Find the original retail price of a couch costing $250, marked up by 40% of cost.
2. The Baxter Cloth Company marks up uniformly all goods at a rate of $28\frac{4}{7}$% of cost. Find the original retail price of items costing (a) $42.00; (b) $99.40.
3. A one-speed bicycle is to retail for $68.90 after a 30% markup based on cost. What is the maximum amount that can be paid for the bicycle in order to cover anticipated expenses and the desired profit?
4. What is the final retail price of a tire costing $25, marked up by 30% on cost, and later reduced by 20%?
5. A stereo costing $64.50 is marked up by $16\frac{2}{3}$% of cost and later reduced by 20%. Find (a) the original retail price and (b) the final retail price.

6. A desk costing \$43.75 is marked up by \$15.00, reduced by \$12.00, and later raised by \$4.50. Find (a) the original retail price; (b) the final retail price; (c) the markup; (d) the markdown; (e) the markdown cancellation; (f) the net markdown.

Interchanging Bases in Markup Problems

So far, we have considered markup as a percent of cost and have answered the following questions:

(1) If C (Cost) = \$60 while M (Markup) = 20% of C, find M and S (Sales price).

Answer: M = \$60 × .20 = \$12
S = \$60 + \$12 = \$72

(2) If S (Sales price) = \$90 while M (Markup) = 50% of C (Cost), find M and C.

Answer: S = 150% of C
C = \$90 ÷ 150% = \$60
M = \$90 − \$60 = \$30

However, it is a much more common practice to express the markup rate as a percent of a different base: the original retail price, or sales price—that is, base a markup on *retail*. In other words, we are interchanging bases. Given a markup rate at 30% of S, and given S = \$40, we can proceed with little difficulty, for we know both the rate and the number on which it is based; therefore, M = \$40 × .30 = \$12. What is C? C = \$40 − \$12 = \$28.

But consider the following situation: C = \$40; M = 20% of S. The obvious \$40 × .20 is wrong since the markup is not based on cost. In this case, we do not know both the rate and the number on which it is to be based as we did previously. There are two ways to solve this type of problem. Let's label them simply Method 1 and Method 2.

Method 1

We know that whether expressed in dollars or percents the following formula is always true:

C (Cost) + M (Markup) = S (Sales price)

We also know that if M is based on C, then C is the base, or the 100% factor. Or, if M is based on S, then S is the base, or the 100% factor.

Using the figures C = \$40 and M = 20% of S, the data can be expressed in both dollars and percents in the following form:

Given:	\$	%
C	40	
M		20
S		100

Why is S the 100% factor? Because it is the base for M in this problem.

Percent terms: C + M = S; C + 20% = 100%; and C = 80%.
C (\$40) is 80% of S (the base). B = P ÷ R, so
S = \$40 ÷ .80 = \$50.

Dollar terms: C + M = S; \$40 + M = \$50; and M = \$10.

Check

	$	%
C	40	80
M	10	20
S	50	100

Is M ($10) 20% of S ($50)? Yes. The answer is correct.

The key to solving this type of problem is to first find either C or S in both dollar and percentage terms; then, divide the dollar amount by the percent in order to find the base.

Let's compare Sample Problems 6, 7, and 8 in terms of what is given and see how each of them is solved.

SAMPLE PROBLEM 6

If C = $55 and M = 10% of C, find M and S.

Solution

M = $55 × 10% = $5.50
S = $55 + $5.50 = $60.50

SAMPLE PROBLEM 7

If S = $80 and M = $37\frac{1}{2}$% of S, find M and C.

Solution

$37\frac{1}{2}\% = \frac{3}{8}$
M = $80 × $\frac{3}{8}$ = $30
C = $80 − $30 = $50

SAMPLE PROBLEM 8

If C = $72 while M = 40% of S, find M and S.

Solution

Given:

	$	%
C	72	
M		40
S		100

S = $72 ÷ 60% = $120
M = $120 − $72 = $48

Check

	$	%
C	72	60
M	48	40
S	120	100

Is M ($48) 40% of S ($120)? Yes.

Can you see the difference among the three problems? The solution is found by applying what is given in each case (although it is sometimes easy to be fooled by what is and is *not* given).

Study Sample Problem 9, which is another solution to Sample Problem 3.

SAMPLE PROBLEM 9

A tea set is to retail for $27.30 after a 30% markup on cost. Find the cost and the markup.

Solution

Given:

	$	%
C		100
M		30
S	27.30	

S = 130%
C = $27.30 ÷ 130% = $21.00
M = $27.30 − $21.00 = $6.30

Check

	$	%
C	21.00	100
M	6.30	30
S	27.30	130

Is M ($6.30) 30% of C ($21.00)? Yes, it is.

The tabular form is always useful when you know either C or S but not the markup stated on the base that is known. And always remember that in both *dollars* and *percents:*

$$C + M = S$$

Method 2

The second method uses a "translating" device for situations in which we know C while M is a % of S and vice versa. Suppose we have the following situation: C = $20, M = $10, and S = $30. What fraction of C is M? What fraction of S is M?

$$M \div C = \$10 \div \$20 = \tfrac{1}{2}$$
$$M \div S = \$10 \div \$30 = \tfrac{1}{3}$$

Try another: C = $40; M = $10; and S = $50. M is $\frac{1}{4}$ of C; $\frac{1}{5}$ of S. These sequences ($\frac{1}{2}$—$\frac{1}{3}$; $\frac{1}{4}$—$\frac{1}{5}$) are not accidental. There is a regular pattern to the relationship among the three numbers when they are expressed as fractions.

If the difference between the higher number (S) and the lower number (C) is 1/n of the lower number, it is 1/n + 1 of the higher number. If the difference is 2/n of the lower number, it is 2/n + 2 of the higher number. And, if it is x/n of the lower number, it is x/n + x of the higher number. So if M is $\frac{1}{2}$ of C, it is $\frac{1}{2+1}$ or $\frac{1}{3}$ of S; if $\frac{2}{5}$ of C, it is $\frac{2}{5+2}$ or $\frac{2}{7}$ of S.

For example, if C = $50, M = $30, and S = $80, then M = $\frac{3}{5}$ of C and $\frac{3}{5+3}$ or $\frac{3}{8}$ of S. Notice that $\frac{3}{5}$ becomes $\frac{3}{8}$ by adding the numerator to the denominator. If M is $\frac{5}{21}$ of C, what fraction is it of S? The correct answer is $\frac{5}{21+5} = \frac{5}{26}$. Sample Problem 10 shows an application of this translating device in solving Sample Problem 9.

SAMPLE PROBLEM 10

If S = \$27.30 while M = 30% of C, find M and C.

Solution

$M = 30\% = \frac{3}{10}$ of C, so $M = \frac{3}{13}$ of S
$M = \$27.30 \times \frac{3}{13} = \$6.30$
$C = \$27.30 - \$6.30 = \$21.00$

The rate of markup must be expressed in fractional form to use Method 2. With a rate that is not an aliquot part of 100%, such as 37%, use Method 1.

Conversely, if a difference is expressed initially as 1/n of the higher number, it is $1/n - 1$ of the lower number. If M is $\frac{1}{4}$ of S, it is $\frac{1}{4-1}$ or $\frac{1}{3}$ of C. If the difference is x/n of the higher number, it is $x/n - x$ of the lower. If M is $\frac{5}{12}$ of S, what fraction is M of C? It is $\frac{5}{12-5}$ or $\frac{5}{7}$ of C. Sample Problem 11 is another solution of Sample Problem 8.

SAMPLE PROBLEM 11

If C = \$72 while M = 40% of S, find M and S.

Solution

$M = 40\%$ or $\frac{2}{5}$ of S, so $M = \frac{2}{3}$ of C
$M = \$72 \times \frac{2}{3} = \$48$
$S = \$72 + \$48 = \$120$

Practice both Methods 1 and 2 and choose the one that you prefer.

TEST YOUR ABILITY

1. Copy and complete the following table, using the formula C + M = S.

	C	M	S
(a)	\$30	\$20	
(b)	\$55		\$95
(c)		\$47	\$86
(d)	30%		100%
(e)	100%	40%	
(f)		70%	100%

2. If S = \$112 and M = $37\frac{1}{2}$% of S, find M and C.
3. If C = \$288 while M = 28% of S, find S and M, using Method 1.
4. A lamp is to retail for \$12.65 after a 15% markup on its cost. Find the cost and the markup, using Method 1.
5. In each of the following problems, markup (M) is given as a fraction or a percent of cost (C). Translate each figure into a fraction or percent of sales price (S).
   (a) $\frac{1}{7}$C $= ? \times S$
   (b) $\frac{2}{9}$C $= ? \times S$
   (c) $\frac{4}{11}$C $= ? \times S$
   (d) 25% of C $= ? \times S$

(e) $66\frac{2}{3}\%$ of C = ? × S

(f) 100% of C = ? × S

6. In each of the following problems, markup (M) is given as a fraction or percent of sales price (S). Translate each into a fraction or percent of cost (C).

(a) $\frac{1}{10}$S = ? × C

(b) $\frac{3}{11}$S = ? × C

(c) $\frac{11}{25}$S = ? × C

(d) 20% of S = ? × C

(e) $16\frac{2}{3}\%$ of S = ? × C

(f) 50% of S = ? × C

Computing Profit

Profit, a goal of business, is a word that has many definitions. One meaning of profit is commonly used in everyday life. Suppose you bought a house for $45,000 and sold it for $75,000, making a $30,000 profit. The business terminology to express these three facts is as follows:

Sales price	$75,000
Cost	−45,000
Gross profit	$30,000

Gross profit is the excess of the sales price over the cost of an item. Gross profit can be calculated for a firm's total sales, or item by item. On a single product sold at its original retail price, gross profit will equal markup. If a product is sold below cost, a *gross loss* will be the result.

SAMPLE PROBLEM 12

Find the gross profit on a toaster oven costing $40, marked up $33\frac{1}{3}\%$ of sales price, reduced by 20%, and then sold.

Solution

Cost	$40.00
Markup ($\frac{1}{3}S = \frac{1}{2}C$)	20.00
Original retail price	$60.00
Markdown ($\frac{1}{5} \times \$60$)	−12.00
Final retail price	$48.00
Cost	−40.00
Gross profit	$ 8.00

The calculation of gross profit on a firm's total sales will be discussed in detail in Chapter 10. For the present, let's look at the following formula:

Gross profit = Sales price − Cost

This formula deals with only part of the profit picture. The reason: a person in business must consider operating expenses—everyday costs of running a business—including rent, salaries, insurance, and so forth. In Unit 3, you will find a full discussion of these expenses. So a firm might sell its goods at a $2,000 gross profit and have operating expenses of $1,700. The firm will therefore show a net profit of only $300. Stated as a formula,

$$\text{Net profit} = \text{Gross profit} - \text{Operating expenses}$$

Net profit is the excess of gross profit over operating expenses. Should operating expenses exceed gross profit, the firm will then have a *net loss*.

SAMPLE PROBLEM 13

If a firm sells a trailer costing $775 for $1,025 and has expenses of $310 in connection with the sale, find the gross profit and the net profit or loss.

Solution

Sales price	$1,025
Cost	−775
Gross profit	$ 250
Operating expenses	−310
Net loss	($ 60)

Notes Either parentheses or red ink can be used to indicate a negative figure.

TEST YOUR ABILITY

1. Find the gross profit or loss on a suit costing $75.00 and sold at a markup of 25% on cost.
2. A paint set costing $37.50 is marked up $16\frac{2}{3}$% of sales price, reduced by 20%, and then sold. Find the gross profit or loss.
3. A large wicker basket costing $29.50 is sold for $41.25. Find the net profit or loss if operating expenses are (a) $5.75; (b) $14.50.

Other Selling Factors

As we have seen, a seller's gross profits are reduced by operating expenses. Some expense items, however, are often—or always—passed directly on to the buyer. Let's look at two of the items in this category.

Freight Charges

Most business purchases are shipped by the seller to the buyer. The seller can absorb these delivery expenses by increasing the markup or by reducing the net profit—or, instead, the seller can directly charge the buyer for them. The choice often depends on the shipping terms of the sale.

The most common shipping, or freight, term is *FOB*, which means *free on board*. Every shipment has an FOB point—the point at which title (ownership) passes from the seller to the buyer. Terms are either FOB shipping point or FOB destination.

When the terms are *FOB shipping point*, title to the goods passes to the buyer at the moment the goods are delivered by the seller to the *carrier* (truck, plane, train, etc.). During the entire period of transportation, the buyer owns the goods and must pay the freight charge. A sale from Chicago to Los Angeles with terms of "FOB Chicago" is an FOB shipping point sale.

When the terms are *FOB destination*, title to the goods remains with the seller during shipment; the title passes when the goods are transferred from the carrier to the buyer. The freight charge is then paid by the seller. In the Chicago–Los Angeles sale, terms of "FOB Los Angeles" are FOB destination terms.

Notice that the seller bills the buyer for freight in only one instance—when terms are FOB shipping point and the seller, in order to ship the goods, must *prepay* the freight charge for the buyer. The freight charge is added to the buyer's bill. However, any discounts granted to the buyer do not apply to the freight charge.

SAMPLE PROBLEM 14

The Allard Company of Macon, Georgia, sells machine parts at a price of $950 to Andrew McKeever of Charleston, South Carolina. The terms of sale are 2/10, n/30, FOB Macon. Allard prepays freight charges of $45 for McKeever. What must McKeever pay within the discount period?

Solution

Sales price	$950
Freight charges	45
	$995
Cash discount (2% × $950)	−19
Amount due	$976

If terms were FOB Charleston, no freight charge would have been made to McKeever.

Sales Tax

Most states raise money by levying a sales tax on a variety of goods and services. It is the responsibility of the seller to pay the tax to the government. Thus, in most instances, the seller will collect sales tax from the buyer.

Rates vary from state to state—and even within a state, since counties and cities are allowed to levy taxes of their own. The items taxed also vary. In our problems here assume that the rate is 7% and that all sales are taxable.

To calculate sales tax, figure 7% on whole dollar amounts and use Table 6.1 below for amounts under $1.00.

Table 6.1

7% Sales Tax Table

Amount of Sale	Tax
$0.00–0.7	0¢
.08–.20	1¢
.21–.33	2¢
.34–.47	3¢
.48–.62	4¢
.63–.76	5¢
.77–.91	6¢
.92–.99	7¢

SAMPLE PROBLEM 15

Find the amount of sales tax on a sale of (a) $10.00; (b) $7.36; (c) $137.65.

Solution

(a)	7% × $10 =	$0.70
(b)	7% × $7 =	$0.49
	Table 6.1 on $0.36 =	0.03
	Total tax	$0.52
(c)	7% × $137 =	$9.59
	Table 6.1 on $0.65 =	0.05
	Total tax	$9.64

On any sale, the tax rate that applies depends on the *place of delivery*. If goods are delivered in a 7% tax rate area, the 7% rate applies. If goods are delivered in a tax-free area, no tax is charged, even if the goods are delivered from an area with a tax. However, if a purchaser takes goods from a state with a 7% sales tax to a state with no sales tax, the 7% sales tax must be paid—the goods were "delivered" in the 7% area.

SAMPLE PROBLEM 16

A sale of a sweater for $42.50 is made from one state to another. The terms of sale are FOB shipping point, and charges of $2.75 are prepaid by the seller. The sales tax rate in the seller's state is 5%; in the buyer's, 7%. Find the total amount to be billed to the buyer.

Solution

Sales price	$42.50
Freight charges	2.75
Sales tax (7% on $42.50)	2.98
Total	$48.23

Notes (1) The sales tax rate is 7%, the rate in the state of delivery.
(2) No sales tax is charged on the freight.

TEST YOUR ABILITY

1. A sale of a grandfather clock is made from New York to a buyer in San Francisco. The grandfather clock is priced at $325. Freight charges are $25. What amount will the buyer pay the seller under each of the following terms:
   (a) FOB New York and the buyer pays the carrier
   (b) FOB New York and the seller prepays the freight
   (c) FOB San Francisco
2. An air conditioner is sold for $410 with a 20% trade discount and prepaid freight charges of $30. What amount will be billed to the buyer?
3. Find the sales tax on each of the following amounts, using Table 6.1:
   (a) $ 4.55
   (b) $ 11.37
   (c) $ 16.91
   (d) $210.83
   (e) $453.05

4. A sale of lumber is made from a 5% sales tax state to a 6% sales tax state. The lumber retails for $300, with a prepaid freight charge of $30. Find the amount to be billed to the buyer.

SUMMARY

The major concepts presented in Chapter 6—markup, markdown, changing bases, sales tax, freight charges, gross and net profits and losses–are summarized in Sample Problem 17. Use this problem to check your mastery of these key ideas.

SAMPLE PROBLEM 17

A washing machine, purchased for $210, is marked up by $14\frac{2}{7}\%$ of its sales price. It is then marked down by 10% and sold with a 7% sales tax and prepaid freight charges of $18.75. The seller's operating expenses are $25. Find (a) total price billed to the buyer; (b) gross profit or loss; (c) net profit or loss.

Solution

(a)	Cost	$210.00
	Markup ($14\frac{2}{7}\% = \frac{1}{7}S = \frac{1}{6}C$; $\frac{1}{6} \times \$210$)	35.00
	Original retail price	$245.00
	Markdown (10% × $245)	−24.50
	Final retail price	$220.50
	Sales tax (7% × $220 = $15.40 + $0.04)	15.44
	Freight charges	18.75
	Total billed price	$254.69

(b)	Final retail price	$220.50
	Cost	−210.00
	Gross profit	$ 10.50

(c)	Gross profit	$ 10.50
	Operating expenses	−25.00
	Net loss	($ 14.50)

Notice that neither the sales tax nor the freight charges affects the profit calculation. The seller is being reimbursed by the buyer for both of these charges.

CHECK YOUR READING

1. Define the terms *markup, markdown, markdown cancellation,* and *net markdown.*
2. On what bases can markup be calculated?
3. Distinguish between *markup* and *gross profit.* When are these two arithmetically the same?
4. Explain how you would know where to place the 100% factor in a problem of C, M, and S.
5. Can you distinguish between *gross profit* and *net profit?*
6. Under what circumstances could there exist a gross profit and a net loss? A gross loss and a net loss? A gross loss and a net profit?

7. Differentiate between *FOB shipping point* and *FOB destination* by indicating who is responsible for the freight charges in each instance.
8. What is the circumstance in which the seller will charge the buyer for freight charges?
9. State A has a 5% sales tax rate; State B, 7%. What rate is payable if (a) goods are delivered to State A; (b) goods are taken from State A by the buyer, a resident of State B; (c) goods are delivered to State B?

EXERCISES *Group A*

1. Dental supplies with a final retail price of $273.00 are shipped FOB destination from an area with a 6% sales tax to one with a 7% sales tax. If freight charges are $19.50, find the amount billed to the buyer.
2. A recliner costing $98.75 is marked up by $12.75, reduced by $6.30, raised by $3.50, and then sold. Find the (a) original retail price; (b) final retail price; (c) markdown; (d) markup; (e) markdown cancellation; (f) net markdown.
3. An invoice for camping equipment purchased at $730 shows an additional prepaid freight charge of $66 and is paid in time for a 3% cash discount. What amount will be paid?
4. Find the gross profit on the sale of a fish tank marked up 30% on its cost of $55.00.
5. A shirt costing $14.30 is sold for $21.10. Find the net profit or loss if operating expenses were (a) $5.35; (b) $9.65.
6. A filing cabinet costing $84.00 is marked up $16\frac{2}{3}$% on its cost, reduced by 10%, and sold after a price rise of $1.80. Find the (a) original retail price; (b) final retail price.
7. A bed is to retail for $204.75 after a 17% markup on its cost. Find (a) the cost and (b) the markup.
8. A sale of an unfinished pine desk is made by a seller in Peoria, Illinois, to a buyer in Dover, New Hampshire. The desk retails for $125.00. Freight charges on the shipment are $19.50. Find the amount the buyer will owe the seller under the following terms:
   (a) FOB Peoria, buyer pays the carrier
   (b) FOB Dover
   (c) FOB Peoria, seller prepays the freight
9. If C = $165.60 while M = 31% of S, find S and M.
10. Find the gross profit on a telescope costing $60.00, marked up by 20% of sales price, and then reduced for sale by 12%.
11. In each of the following problems, markup (M) is given as a fraction or a percent of cost (C). Translate each into a fraction or a percent of sales price (S).
    (a) 40% of C $= ? \times S$
    (b) $\frac{5}{11}C$ $= ? \times S$
    (c) $16\frac{2}{3}$% of C $= ? \times S$
    (d) $\frac{10}{17}C$ $= ? \times S$
    (e) $\frac{6}{5}C$ $= ? \times S$
    (f) $9\frac{1}{11}$% of C $= ? \times S$
12. Using Table 6.1 for the cent amounts, find the 7% sales tax on each of the following:
    (a) $1.90

(b) $5.45
(c) $11.21
(d) $635.07
(e) $81.37

13. Copy and complete the following table, using the formula C + M = S:

	C	M	S
(a)		$20	$93
(b)	$37		$75
(c)	$46	$27	
(d)		27.2%	100%
(e)	54.6%		100%
(f)	100%		135%

14. A pair of boots is to retail for $52.36 after a 36% markup on cost. What is the maximum amount that can be paid to purchase the item?
15. Find the final retail price of a bench costing $59.95, marked up by 30% of cost, and later reduced by 10%.
16. Find the original retail price of a woolen blanket marked up by $62\frac{1}{2}$% of its cost of $72.80.
17. Wexler's Sporting Goods Company marks up all goods at a uniform rate of $22\frac{2}{9}$% on cost to cover expenses and net profit. Find the original retail price of items costing (a) $186.30; (b) $72.45.
18. In each of the following, markup (M) is expressed as a fraction or percent of sales price (S). Translate each figure into a fraction or percent of cost (C).
    (a) 10% of S = ? × C
    (b) $\frac{1}{12}$S = ? × C
    (c) $\frac{2}{3}$S = ? × C
    (d) $6\frac{1}{4}$% of S = ? × C
    (e) $\frac{7}{19}$S = ? × C
    (f) $\frac{2}{17}$S = ? × C
19. If S = $350.00 and M = 27% of S, find M and C.

Group B

1. A bar costing $1,950.00 is marked up 30% on cost, reduced by 25%, and later raised by 10%. Find the final retail price and label all markup, markdown, and cancellation factors.
2. Complete the following tabular presentation by supplying figures for the blanks:

	Cost	Markup % on Cost	Markup in $	Original Retail Price	Markdown in $	Final Retail Price
(a)	$100	20	___	___	$15	___
(b)	___	___	$35	$105	___	$90
(c)	$ 60	___	$18	___	___	$72
(d)	___	30	___	$260	$40	___
(e)	$ 72	___	___	___	$25	$95

3. In each of the following, find the missing factors, using the translating method presented in this chapter:

	S	M Based on C	M Based on S	M	C
(a)	$100	$\frac{1}{3}$	___	___	___
(b)	$120	$\frac{1}{5}$	___	___	___
(c)	$ 80	$\frac{2}{3}$	___	___	___
(d)	$ 84	75%	___	___	___
(e)	$ 77	$16\frac{2}{3}$%	___	___	___
(f)	___	___	$\frac{1}{9}$	___	$ 80
(g)	___	___	$\frac{1}{11}$	___	$ 45
(h)	___	___	$\frac{3}{8}$	___	$ 60
(i)	___	___	$28\frac{4}{7}$%	___	$ 90
(j)	___	___	$66\frac{2}{3}$%	___	$100

4. A chair costing $75.00 is marked up 25% on retail price, reduced by 20%, and later sold after an increase of 10%. Find the gross profit on the sale of the chair for its final retail price.
5. Ten fireplace accessory kits are purchased by the Ashmont Company @ $240.00 each. The kits are marked up by $33\frac{1}{3}$% on cost, reduced by 10%, and sold to Arthur Barnes, who lives in a 5% sales tax state. Freight charges of $275.00 are prepaid on the shipment. Operating expenses are estimated at $40.00 per kit. Find for all ten kits: (a) final retail price; (b) total price billed to Barnes; (c) gross profit; (d) net profit.

CHAPTER 7

Inventories

Merchandise inventory refers to the value of the goods a firm holds for resale to customers in the normal course of business. For many businesses, the amount of merchandise inventory represents a significant investment. Consequently, the purchase, control, and valuation of the inventory should be carefully monitored by management. In this chapter, we will discuss how the quantity of merchandise on hand is determined, and methods of assigning a value to inventory.

Your Objectives

After completing Chapter 7, you should be able to

1. Differentiate between a periodic inventory system and a perpetual inventory system.
2. Determine the dollar value of a merchandise inventory using each of the following methods: specific identification; weighted average; first-in, first-out (FIFO); and last-in, first-out (LIFO).
3. Price inventories according to the cost or market rule.
4. Estimate the value of inventory using the retail method and the gross profit method.

Determining the Quantity of Inventory on Hand

There are two systems used to determine the quantity of goods remaining on hand: one system is the *periodic inventory system,* and the other is the *perpetual inventory system.*

Under the *periodic system,* goods are physically counted at regular intervals of time, such as every month or every year, and a value is assigned to the goods remaining on hand. The actual counting of goods is often done on weekends or after operating hours.

Under the *perpetual system,* written records are maintained which show a continuous or "running" inventory balance throughout the period. When new items are received, they are added to the inventory records; when items are sold, they are subtracted from the inventory records. Even with the use of the perpetual records, a periodic physical count is still necessary to verify their accuracy.

The next step after the quantity of goods on hand has been determined is to assign a dollar value to the goods. Inventory must be valued for a variety of reasons, such as: insurance, taxes, and determining profit or loss. Methods of assigning a value to inventory are discussed next.

Inventory Valuation Methods

Assigning a dollar value to goods on hand on a specific date is referred to as *inventory valuation.* There are several acceptable methods available for assigning a value to inventory. The most common are: *specific identification; weighted average; first-in, first-out (FIFO);* and *last-in, first-out (LIFO).*

Specific Identification

Specific identification refers to a method of valuing inventory in which each item of merchandise on hand is specifically identified with its cost. Under this method, incoming items are often tagged to indicate their cost and the specific category of inventory to which they belong. Determining the dollar value of the inventory can be an easy process: count what is there and record each unit's cost. Then add all of the individual costs to arrive at the total value of the inventory. Sample Problem I illustrates this process.

SAMPLE PROBLEM I

Harry Sachs, an auto dealer in a small town, counts seven cars in his lot at the end of May. Specific costs of each vehicle are as follows:

Vehicle Number	Cost
76	$ 7,400
105	6,280
119	5,495
204	7,210
207	3,860
219	11,240
255	9,130

Calculate the value of Harry's ending inventory.

Solution

$7,400 + $6,280 + $5,495 + $7,210 + $3,860 + $11,240 + $9,130 = $50,615

Harry's inventory calculation was a simple one because he knew *exactly* which autos were unsold. Thus, he was able to "specifically identify" the units on hand with their actual cost. Sample Problem 2 illustrates this method once again, but uses a different type of data. The information to be used for Sample

Problem 2 concerns Barbara Butler's purchases during the month of July for her jeans store. The data are as follows:

Date	Quantity (pairs)	Unit Cost
7/3	20	$7.60
7/16	40	7.80
7/24	10	8.00
7/31	50	8.20

SAMPLE PROBLEM 2

Barbara Butler counts her July 31 inventory and finds that 25 pair of jeans are on hand. Barbara can tell by codes she entered on the labels at the time of receipt that 18 of the pairs are from the July 31 purchase and the remaining 7 are from July 16. Calculate the value of her July 31 inventory using the specific identification method.

Solution

$$\begin{aligned} 18 \times \$8.20 &= \$147.60 \\ 7 \times \$7.80 &= 54.60 \\ & \ \underline{\underline{\$202.20}} \end{aligned}$$

The specific identification method of inventory works quite well when a business's merchandise consists of just a small number of easily identified items. However, if you picture a large department store, a variety store, or a food store, this method is far from practical.

The Weighted Average Method

Other methods have been created by business people and by accountants to deal with large and varied inventories. In one of these methods, the *weighted average* method, an average cost per unit is calculated for all units acquired during the period. The value of the ending inventory is then determined by applying the average cost per unit to the number of units on hand at the end of the period. Sample Problem 3 illustrates this method.

SAMPLE PROBLEM 3

On July 31 the Computer Shoppe showed the following data for one of its best selling video games, "The Inner Connection."

Date	Description	Quantity	Unit Cost
July 1	Beginning inventory	30	$10.00
8	Purchase	40	10.50
15	Purchase	40	10.75
26	Purchase	25	11.00

On July 31, a physical count revealed that 29 units remained on hand. Calculate the value of the ending inventory using the weighted average method.

Solution

Since the weighted cost method uses an average cost of all units, it is best to work in steps, as follows:

Step 1. Determine the cost of goods available for sale:

		Units	×	Unit Cost	=	Total Cost
July 1	Beginning inventory	30	×	$10.00	=	$ 300
8	Purchase	40	×	10.50	=	420
15	Purchase	40	×	10.75	=	430
26	Purchase	25	×	11.00	=	275
Units available for sale		135				$1,425

Step 2. Determine the average cost per unit.

$$\frac{\$1,425}{135} = \$10.56 - \text{average cost per unit}$$

Step 3. Apply the average cost per unit to the number of units remaining on hand:

$$\$10.56 \times 29 \text{ units} = \$306.24 - \text{Inventory value}$$

It should be stressed that the $10.56 average cost per unit is a *weighted* average, that is, an average in which the *quantity* purchased at each of the prices is included in the calculation, rather than a *simple average* in which the unit prices are simply totaled and the total divided by the number of prices.

First-In, First-Out (FIFO) Method

Two other inventory valuation methods remain to be learned. Each assumes a *flow of goods,* that is, a sequence in which goods are expected to be sold. For example, if you had perishable goods (milk, eggs, salad greens) in your inventory, you would most likely attempt to sell the older items first. The *first-in, first-out (FIFO)* method of inventory valuation assumes this flow. Thus, under FIFO, the first units acquired are assumed to be the first units sold, so what is left on hand are the most recent purchases—the latest ones bought. Sample Problem 4 shows The Computer Shoppe's calculation of inventory value using the FIFO method.

SAMPLE PROBLEM 4

Calculate the value of the Computer Shoppe's 29-unit inventory of "The Inner Connection" using the FIFO method.

Solution

Description	No. of Units		Unit Cost		Total Cost
From July 26 purchase:	25	×	$11.00	=	$275
From July 15 purchase:	4	×	10.75	=	43
Ending inventory:	29				$318

Under the FIFO method, the 25 units purchased on July 26 are assumed to remain on hand, and 4 of the units purchased on July 15 are assumed to remain on hand. All other units are assumed to be sold. Since most businesses attempt to sell old merchandise before new merchandise, FIFO is generally in harmony with the physical movement of goods in a business.

Last-In, First-Out (LIFO) Method

Under the *LIFO* method, the last units received are assumed to be the first units sold. Thus, the units on hand are considered to be a part of the earliest units acquired. Sample Problem 5 illustrates the LIFO method.

SAMPLE PROBLEM 5

Calculate the Computer Shoppe's 29-unit inventory of the "Inner Connection" using the LIFO method.

Solution

Since the use of the LIFO method assumes that inventory is composed of the earliest costs, the 29 units on hand on July 31 are assigned a value based on the value of the July 1 inventory.

29 units × $10.00 = $290

In most businesses, the LIFO method does not parallel the physical movement of goods. Remember, however, that this is a method of assigning a value to inventory—not a method of determining the quantity of goods on hand. Consequently, the use of LIFO is legally permissible, even in those businesses in which LIFO is opposite the actual movement of goods.

Comparison of Inventory Valuation Methods

Sample Problem 6 illustrates a single problem which includes all four inventory valuation methods.

SAMPLE PROBLEM 6

Bobby Gaumont's toy store shows the following data for 1988:

Beginning inventory, Jan. 1	300 units @ $7.00
Purchase, March 1	700 units @ 7.20
Purchase, June 1	1,200 units @ 7.25
Purchase, September 1	1,600 units @ 7.30
Purchase, October 1	200 units @ 7.40

On December 31 a physical count revealed that 400 units remained on hand. Calculate the value of the 400 units under each of the following methods:

(a) Specific identification, assuming that 100 of the units were from the January 1 inventory, 100 were from the June 1 purchase, and 200 were from the October 1 purchase.
(b) Weighted average
(c) FIFO
(d) LIFO

Solution

(a) Specific identification

100 × $7.00 =	$ 700
100 × 7.25 =	725
200 × 7.40 =	1,480
	$2,905

(b) Weighted average

No. of Units	×	Unit Cost	=	Total Cost
300	×	$7.00	=	$ 2,100
700	×	7.20	=	5,040
1,200	×	7.25	=	8,700
1,600	×	7.30	=	11,680
200	×	7.40	=	1,480
4,000 units		TOTALS		$29,000

$29,000 ÷ 4,000 units = $7.25 average cost per unit
400 units × $7.25 = $2,900

(c) FIFO

200 units × $7.40 = $1,480
200 units × 7.30 = 1,460
$2,940

(d) LIFO

300 units × $7.00 = $2,100
100 units × 7.20 = 720
$2,820

Notes (1) The beginning inventory balance is always included in the weighted average calculation.
(2) The beginning inventory—not the first purchase—is the first item considered in the LIFO method. This always holds true when there is a beginning inventory.

A Final Word on Inventory Valuation Methods

Since the use of the specific identification method to assign a value to inventory is very time consuming, many businesses look for a more efficient method that will yield acceptable results. Consequently, most businesses use an "assumed cost flow" method, that is, a method that assumes that goods move in a certain direction, and they value their ending inventory accordingly. The inventory methods we have worked with in this chapter are all acceptable for tax purposes and various other reports used by businesses. As a general rule, a business is free to choose the inventory method it desires to use. Once a method has been selected, however, it must be used from period to period unless the business can show a valid reason for changing. In most cases, written permission from the Internal Revenue Service is required before a business can make a change from one inventory method to another.

TEST YOUR ABILITY

1. Debbie Steven's inventory of electric guitars consists of six items valued as follows: #472, $625.00; #495, $837.50; #510, $415.00; #711, $360.00; #712, $522.50; #803, $1,472.75. Calculate the value of her inventory of guitars.

2. Calculate the weighted average unit price in each of the following cases.
   (a) 40 units @ $6
   20 units @ $9
   (b) 100 units @ $7.10
   200 units @ 7.20
   200 units @ 7.30
   (c) 50 units @ $9.00
   170 units @ 9.30
   210 units @ 9.60
   70 units @ 9.90
3. Art Welch began the month with 30 auto batteries valued at $40 each. He then purchased 20 more batteries during the month at a cost of $45 each. At the end of the month, 10 units remained in his stockroom. Calculate the ending inventory value by the weighted average method.
4. Phyllis Hanley purchased 37 coats on July 1 at $29.95 each, and 46 more on July 22 at $28.40 each. If 15 coats remain on hand at the end of the month of July, what is their value using the FIFO method?
5. Gene Fiset began the month of March with a supply of 19 notebooks at $0.22 each. During March, he made two purchases: 106 at $0.23 and 32 at $0.24. If 45 remain at the end of March, determine their value using the FIFO method.
6. The Fairgrounds School Store uses the LIFO inventory method. In January, it began with 175 posters priced at $0.71 each and then purchased 380 more on the 15th at $0.74 each. What is the value of the 105 that remain on hand when the January 31 inventory is taken?
7. Margaret Jacobelli's supply of pocket calculators on February 1 is 46 at $9.45. During February she purchased, in sequence, 103 @ $9.50 and 109 @ $9.95. Using LIFO costing, calculate the value of the 96–unit ending inventory.
8. Ivan Davis's inventory data show the following:

Beginning inventory, Jan. 1	700 units @ $4.60
Purchase, March 1	2,300 units @ 4.64
Purchase, July 1	1,800 units @ 4.70
Purchase, October 1	1,200 units @ 4.75
Balance on hand, December 31	1,400

   Calculate the dollar value of the 1,400 units remaining, assuming each of the following valuation methods:
   (a) Specific identification, if 200 units are from the beginning inventory on January 1; 300 are from the March 1 purchase; and the remainder are from the October 1 purchase.
   (b) Weighted average, rounding the unit average cost to the nearest tenth of a cent.
   (c) FIFO
   (d) LIFO

The Rule of Lower of Cost or Market

In each of the inventory methods discussed so far, the final inventory value was stated at cost. An acceptable alternative is to compare the cost assigned to the inventory with the market price of the inventory and use the lower of the two. This is an application of the *rule of cost or market, whichever is lower.*

In the phrase "lower of cost or market," *market* means the cost to replace the merchandise on the date of the inventory. In determining the market or replacement cost, merchandise prices should be based on quantities typically purchased from the usual source of supply.

In practice, it is possible to apply the lower of cost or market rule to (1) each item in inventory, (2) major categories of the inventory (by departments or merchandise lines), and (3) the inventory as a whole. The usual practice is to apply the rule to each item in inventory. This procedure is illustrated in Sample Problem 7.

SAMPLE PROBLEM 7

Cost and market data of items in the inventory of the O'Malley Company are shown below. Determine the value of the inventory using the lower of cost or market rule.

		Unit Price	
Description	**No. of Units**	**Cost**	**Market**
Item A	200	$20	$22
Item B	160	18	16
Item C	80	35	24
Item D	225	40	42

Solution

Item	No. of Units	×	Lower of cost or market	=	Value
A	200	×	$20 (cost)	=	$ 4,000
B	160	×	16 (market)	=	2,560
C	80	×	24 (market)	=	1,920
D	225	×	40 (cost)	=	9,000
Value of inventory					$17,480

TEST YOUR ABILITY

1. Using the lower of cost or market rule, determine the value of the following inventory:

		Unit Price	
Description	**No. of Units**	**Cost**	**Market**
A11G	600	$2.20	$2.25
B14	200	2.90	2.80
D611	900	1.25	1.30
E17	50	4.77	4.60

2. Determine (a) the value of the ending inventory by applying the lower of cost or market rule to each item in inventory; (b) What is the cost of the inventory as a whole? (c) What is the market value of the inventory as a whole?

(d) which answer—(b) or (c)—would be used in valuing the inventory as a whole under lower of cost or market?

Description	No. of Units	Unit Price Cost	Unit Price Market
Item C	120	$12	$15
Item F	300	16	14
Item G	230	30	28
Item I	600	40	34
Item L	290	60	62

3. Buracker Company maintains a regular stock of Item 11J. On June 30 a physical count revealed that 25 items were still on hand. The current replacement cost is $30 per unit. Using the data shown below, determine the value of the inventory by FIFO cost, or market, whichever is lower.

Beginning inventory, June 1	20 units @ $28
Purchase, June 12	18 units @ 28
Purchase, June 18	15 units @ 29
Purchase, June 24	21 units @ 31

Estimating the Value of Inventory

The inventory of many businesses includes a great variety of merchandise, much of which consists of items with low unit costs. Consequently, taking a physical inventory can be a very time consuming and costly process. Many businesses take an inventory only after the end of their accounting year. There are times during the year, however, when an inventory is needed. Many businesses prepare monthly financial reports, and monthly inventories are needed for this. Additionally, monthly inventories are often needed for planning and expanding a product line. Also, an inventory could be needed if a part of the merchandise were to be destroyed by fire or flood.

What avenues are open to a business that needs a short-term inventory when a physical inventory is not possible? One answer has already been discussed: maintain perpetual inventory records that indicate a running balance of items on hand. But perpetual records are also very expensive to maintain, too expensive, in fact, for many retail stores of the variety nature. Additionally, perpetual records indicate what should be on hand. There is no allowance for lost, damaged, or stolen items. What, then, is left? Estimate the inventory. There are two methods that are commonly used for estimating the value of an inventory: the *retail method* and the *gross profit method.*

The Retail Method

As the name implies, the *retail method* of inventory valuation is widely used by retail businesses, particularly variety and department stores. It is accomplished by determining the percentage relationship between the cost of goods available for sale at retail, and the cost of goods available for sale at cost; and applying this percentage to the retail price of units left on hand. Cost and retail figures are maintained for the beginning inventory and all units purchased during the period. The amount of goods available for sale is then determined by adding together the beginning inventory and the purchases.

The retail method is best accomplished by following these four steps:

1. Determine the amount of goods available for sale at cost and at retail (add beginning inventory and purchases).
2. Determine the *cost ratio* by dividing the amount of goods available for sale at cost by the amount of goods available for sale at retail.
3. Subtract the amount of sales for the period from the amount of goods available for sale at retail to get the ending inventory at retail prices.
4. Determine the ending inventory at cost by multiplying the ending inventory at retail by the cost ratio.

The retail method is illustrated in Sample Problem 8.

SAMPLE PROBLEM 8

The following data pertain to the inventory of The Twin City Variety Shop:

	Cost	Retail
Beginning inventory	$ 60,000	$ 90,000
Purchases during period	120,000	160,000
Sales for the period		202,000

Determine the value of the ending inventory by the retail method.

Solution

Step 1: Determine goods available for sale:

	Cost	Retail
Beginning inventory	$ 60,000	$ 90,000
Purchases	120,000	160,000
Goods available for sale	$180,000	$250,000

Step 2: Determine the cost ratio:

$$\frac{\$180,000}{\$250,000} = 72\%$$

Step 3: Determine the ending inventory at retail:

Goods available at retail	$250,000
Less: Sales	202,000
Ending inventory at retail	$ 48,000

Step 4: Convert the ending inventory at retail to cost:

$$\$48,000 \times .72 = \$34,560$$

The Gross Profit Method

The value of the inventory can also be approximated by the *gross profit method,* which is based on the calculation of the gross profit rate for the period. When the amount of merchandise purchased during a period is added to the inventory of merchandise at the beginning of that period, the total shows the value of *goods available for sale* during the period. If we subtract the estimated cost of goods that were sold during the period from the goods available for sale, we obtain the estimated ending inventory. Sample Problem 9 illustrates this method.

SAMPLE PROBLEM 9

The following data pertain to the inventory of Rhonda's Variety Shop:

Beginning inventory	$ 60,000
Purchases during period	90,000
Sales during period	170,000
Estimated gross profit rate	40%

Estimate the value of the ending inventory using the gross profit method:

Solution

Beginning inventory		$ 60,000
Add: Purchases during the period		90,000
Goods available for sale		$150,000
Sales	$170,000	
Less estimated gross profit ($170,000 × .40)	68,000	
Estimated cost of goods sold		102,000
Estimated ending inventory		$ 48,000

TEST YOUR ABILITY

1. The Stossel Company had the following data on December 31:

	Cost	Retail
Beginning inventory	$ 40,000	$ 72,000
Purchases	100,000	145,000
Sales for the period		190,000

   Determine the value of the ending inventory using the retail method.
2. The following data relate to Chuck's Computer Supplies:

Beginning inventory	$ 30,000
Purchases	70,000
Estimated gross profit rate	30%
Sales for the period	$120,000

   Estimate the value of the ending inventory using the gross profit method.

CHECK YOUR READING

1. What is the difference between the periodic inventory system and the perpetual inventory system?
2. Explain why the *specific identification* inventory method is not widely used in actual practice.
3. In the phrase "first-in, first-out," what is meant by "first in?"
4. In the *last-in, first-out* method of inventory valuation, are the last goods acquired assumed to make up the inventory? Explain.
5. Explain what is meant by "a flow of costs."
6. In the phrase "lower of cost or market," what is meant by "market?"
7. Why are there occasions when a business would need to estimate the ending inventory?
8. What prices are required in the retail method but not in the gross profit method?

EXERCISES *Group A*

1. Tabthip Amoradhat sells basketballs. She begins the year with a supply of 334 balls at a cost of $8.77 each and makes two purchases during the year of 710 @ $8.82 and 920 @ $8.81. Calculate the value of an ending inventory of 480 basketballs, using LIFO.
2. Billy Denya's supply of large and small yachts is running low. As of today, his entire stock consists of eight boats, as follows:

   #14, $4,130; #17, $5,210; #19, $6,390; #22, $17,605;
   #37, $24,380; #42, $9,520; #46, $8,360; #48, $11,260.

   Calculate the value of his inventory.
3. The following data pertain to the inventory of Fred T. Chamberlain, The Jeweler:

		Unit Price	
Description	**No. of Units**	**Cost**	**Market**
Item C-3	50	$10	$12
Item D-9	75	9	8
Item B-2	10	90	85
Item J-14	18	95	99
Item P-11	40	16	18

   Determine the value of the ending inventory by applying the lower of cost or market rule to each item in inventory.
4. Nishma Duffy's inventory data show the following:

Balance, January 1	60 units @ $60
Purchase, March 1	80 units @ $61
Purchase, May 1	90 units @ $62
Purchase, July 1	40 units @ $63
Purchase, October 1	130 units @ $64
Balance, December 31	45 units

   Calculate the value of her ending inventory using each of the following methods:
   (a) Specific identification, assuming that 30 of the units are from January; 10 are from March; and 5 are from October.
   (b) Weighted average
   (c) FIFO
   (d) LIFO
5. Cost data for the inventory of Dave Rodriguez's pet supplies follow:

Beginning inventory	$30,000
Purchases during period	80,000
Normal markup rate	40%
Sales for the period	70,000

   Estimate the value of the ending inventory using the gross profit method.
6. McNab Company maintains a stock of Item 15K. On August 31, a physical count shows 47 units on hand. The current replacement cost is $45 a unit.

Using the data that follow, determine the value of the ending inventory by FIFO cost or market, whichever is lower.

Beginning inventory, August 1	35 units @ $41
Purchase, August 10	16 units @ 43
Purchase, August 19	21 units @ 44
Purchase, August 30	34 units @ 46

7. Joan DiSantos purchased 36 lawn mowers on May 1 at a cost of $119.95 each and 62 more on the 13th at $121.95 each. If 22 mowers remain unsold at the end of May, calculate the value of the inventory using the FIFO method.
8. Angie Nagorski sells childrens's shirts and uses the LIFO method of inventory costing. On January 1 of this year, she has on hand 42 shirts @ $7.10. During January, she buys 106 @ $7.20. Calculate the value of 31 shirts remaining on hand on January 31.
9. Cost and retail data for O'Malley Company are shown below:

	Cost	Retail
Inventory, January 1	$10,000	$16,000
Purchases during January	53,000	74,000
Sales		86,000

Determine the value of the ending inventory, using the retail method.

10. The following data relate to the ending inventory of Lori Purdue's Beauty Supply Company.

Item	Cost	Market
A	$10	$12
B	8	7
C	11	10
D	15	18
E	30	28
F	2	3
G	14	15

(a) Apply the lower of cost or market rule to each item and determine the value of the ending inventory.
(b) Apply the lower of cost or market rule to the total inventory and determine the value of the ending inventory.

11. Phil Bagley began this month with 36 sets of china in stock at a cost of $950 each. During the month, he purchased 64 additional sets at a cost of $975 each. At month's end, only 62 units remain in the store. Calculate the cost of this inventory, using the weighted average method.
12. Eduardo Sanchez's opening balance of barbeque grills on June 1 is 43 at a cost of $28.10 each. During June, he purchases 61 @ $25.90 and 22 at $27.60. If 57 remain on June 30, calculate their value under FIFO costing.
13. Calculate the weighted average unit price to the nearest cent in each problem:

(a) 50 units @ $9.00
40 units @ $9.90

(b) 110 units @ $2.25
150 units @ $2.50
140 units @ $2.75

(c) 400 units @ $0.14
1,000 units @ $0.15
1,500 units @ $0.16
1,100 units @ $0.17

Group B

1. Wills Company ends the month of March with an inventory at FIFO cost of 43 units @ $7.50 and 61 units @ $7.35. During April purchases are made in the following order: 125 @ $7.55, 140 @ $7.60, and 75 @ $7.65. Using FIFO, calculate the value of an April 30 inventory of 106 units.
2. Work question 1 again, but assume LIFO costing instead.
3. A firm using the retail method shows the following data:

Purchases at cost	$60,000
Purchases at retail	80,000
Ending inventory at cost	17,280
Beginning inventory at retail	45,000
Beginning inventory at cost	30,000

Find sales at retail.

4. A firm using the gross profit method shows the following data:

Sales at retail	$85,000
Purchases at cost	45,000
Ending inventory at cost	24,000
Beginning inventory at cost	30,000

Find the rate of gross profit.

UNIT 2 REVIEW

Terms You Should Know

The following alphabetic list presents the business terms covered in this unit. Check back in the text for any that are unfamiliar.

aliquot part
c.o.d.
carrier
cash discount
complement
cost
e.o.m.
equivalent single rate of discount
FIFO
FOB
FOB destination
FOB shipping point
final retail price
gross profit method
gross profit or loss
inventory valuation
invoice
LIFO
list price
markdown
markdown cancellation
market
markup
merchandise inventory
net markdown
net profit or loss
net purchase price
net purchase price equivalent
original retail price
periodic inventory
perpetual inventory
retail method
r.o.g.
sales tax
specific identification
terms of sale
trade discount
trade discount series
weighted average

UNIT 2 SELF-TEST

Items 1–10 below review the highlights of Chapter 5. Try doing them mentally; use Table 5.1 where needed.

1. The list price of 75 items @ $0.20 is ________.
2. 0.08\frac{1}{3}$ is ________ of a dollar.
3. A 16$\frac{2}{3}$% discount on $1,200 amounts to ________.
4. If a trade discount is 40% of the list price, then the net purchase price is ________ of the list price.
5. If a net purchase price is .632 of the list price, then the trade discount is ________ of the list price.
6. The single rate of discount equivalent to discounts of 20% and 5% is ________.
7. The net purchase price equivalent to 20% and 5% discounts is ________.

8. An invoice listed at $1,000 less 10–10–5 will have a net purchase price of __________.
9. An invoice dated October 17, with terms of 2/10, 1/20, n/30, must be paid by __________ to receive any discount.
10. An invoice dated March 28, with terms of 2/10 e.o.m., must be paid by __________ to receive a discount.

Items 11–24 review major concepts from Chapter 6.

11. An art book costing $40 and marked up 30% on cost will retail for __________.
12. If a tennis racket is priced to sell at $50 after a 25% markup on cost, its maximum cost is __________.
13. An easel costing $25 is marked up by 20% of cost and then reduced by 10%. Its final retail price is __________.
14. If the easel in Item 13 is sold at its final retail price, the gross profit will be __________.
15. The original retail price of a record album is $35. If the markdown is 20% and the markdown cancellation is 10%, the net markdown is __________.
16. If C = $80 while M = 20% of S, M = __________.
17. If S = $75 while M = 25% of C, M = __________.
18. If M = $\frac{1}{10}$ of S, it is __________ of C.
19. If M = $\frac{1}{12}$ of C, it is __________ of S.
20. If Sylvia's Plants and Things shows a gross profit of $2,500 and a net loss of $400, the shop's operating expenses are __________.
21. On an FOB __________ sale, the seller is responsible for the freight charges.
22. On a sale from Des Moines to Pittsburgh, terms FOB Des Moines, the freight charges are the ultimate responsibility of the __________.
23. The 7% sales tax on a $2.50 item is __________.
24. On a sale of woolen mufflers for $600, prepaid freight charges are $50 and the sales tax is 7%. The amount billed to the buyer is __________.

Item 25 is a comprehensive problem concerning all of the major topics from Chapters 5 and 6. Round to the nearest cent in all calculations.

25. A purchase is made of 360 pens @ 0.33\frac{1}{3}$ each. Discounts of 20% and 10% are granted on this FOB destination transaction with a 7% sales tax. Freight charges are $5.50. The invoice, dated October 22, includes terms of 2/10, n/30 and is paid on October 31.

    The amount paid is the cost to the buyer, who marks up the goods 30% on cost, marks them down by 20%, raises the price by 10%, and finally sells the pens—charging a 7% sales tax and prepaying freight charges of $4.95. Neither the sales tax nor the freight affects the profit on the sale; however, operating expenses of $7.10 are connected with the sale.

    Find: (a) the list price; (b) the trade discount; (c) the net purchase price; (d) the total of the purchase invoice; (e) the amount paid on October 31; (f) the markup; (g) the original retail price; (h) the markdown; (i) the markdown cancellation; (j) the net markdown; (k) the final retail price; (l) the sales tax; (m) the amount billed on the sales invoice; (n) the gross profit or loss; (o) the net profit or loss.

Items 26–32 review concepts from Chapter 7.

26. Purchases are made as follows: 2 @ $5, 3 @ $6, and 5 @ $7. Total goods purchased amount to ________.
27. If 3 are on hand from the purchases in item 26, all of which are from the second purchase, the inventory value by specific identification is ________.
28. An inventory of 3 units from item 26 valued by the weighted average method is worth ________.
29. Using FIFO, the 3 units are worth ________.
30. Using LIFO, the 3 units are worth ________.
31. In the ________ method, the ratio of cost to retail is used.
32. The inventory estimation method that does not require all retail values is the ________ method.

KEY TO UNIT 2 SELF-TEST

1. $15.00
2. $\frac{1}{12}$
3. $200.00
4. 60%
5. .368
6. 24%
7. 76%
8. $769.50
9. November 6
10. May 10
11. $52.00
12. $40.00
13. $27.00
14. $2.00
15. $4.20
16. $20.00
17. $15.00
18. $\frac{1}{9}$
19. $\frac{1}{13}$
20. $2,900.00
21. destination
22. buyer
23. $0.18
24. $692.00
25. (a) $120.00
    (b) $33.60
    (c) $86.40
    (d) $92.45
    (e) $90.72
    (f) $27.22
    (g) $117.94
    (h) $23.59
    (i) $9.44
    (j) $14.15
    (k) $103.79
    (l) $7.27
    (m) $116.01
    (n) $13.07 gross profit
    (o) $5.97 net profit
26. $63
27. $18
28. $18.90
29. $21
30. $16
31. retail
32. gross profit

Note In 25(e), the discount is based on $86.40.

UNIT 3

The Mathematics of Operating

The ultimate aim of any business is to operate at a profit. Gross profit was defined in Unit 2 as the difference between sales price and the cost of the goods sold. Net profit was defined as the difference between gross profit and operating expenses.

It is these operating expenses on which Unit 3 is focused. The basic arithmetic processes that you have learned earlier will be applied to typical business expenses.

Chapter 8 is a discussion of payroll computation and related expenses—a very handy chapter, especially if you start your own company.

Chapter 9 illustrates the types of business insurance and how to calculate their cost to the firm.

Chapter 10 emphasizes the arithmetic of depreciation as well as other business expenses. It concludes with a review of profit calculation for both service and trading firms.

CHAPTER 8

Payroll

For many business firms, the largest operating expense is its payroll, including wages, salaries, related taxes, and fringe benefits. As you see, a firm's payroll costs include not only what it pays directly to its employees, but also what is paid to the government and other agencies on behalf of its employees.

The discussion in this chapter emphasizes three areas: computing an employee's gross earnings; determining the deductions to be made from an employee's gross earnings in arriving at net pay; and calculating the payroll costs of the employer.

Study this chapter from both the employer's viewpoint ("What do my employees cost me?") and the employee's ("How is my take-home pay calculated?").

Your Objectives

Upon completing Chapter 8, you will be able to

1. Calculate gross wages by several methods.
2. Determine amounts of payroll deductions and arrive at take-home pay.
3. Compute the cost of payroll and related expenses to the employer.

Bases for Gross Wages

A *salary* is a payment based on a certain time period—such as a week, month, or year—regardless of how much an employee produces.

A *wage* is a payment based on the employee's actual working time (so much money per hour) or actual production (so much money per unit produced).

Most executives and supervisory personnel receive a salary, while most factory and clerical workers receive a wage. In addition, *commission*, a payment based on an employee's sales, generally forms all or part of a sales representative's earnings.

Gross wages are total earnings for a specific period, whether the base for earnings is a salary, a wage, or a commission.

Flat Salary

Many employees receive a stated amount of pay either weekly, monthly, or annually, A worker's flat salary may be, for example, $250 a week, $900 a month, or $13,000 a year.

An employee who earns an annual salary will still be paid on a regular basis, such as weekly, biweekly, semimonthly, or monthly. Sample Problem 1 illustrates how the gross amount of each paycheck is determined.

SAMPLE PROBLEM 1

Pedro Rojas, an office manager, earns $15,000 a year. Find the gross amount of each paycheck if salaries are paid (a) weekly; (b) biweekly (every two weeks); (c) semimonthly (twice a month); (d) monthly.

Solution

(a) Weekly payment means 52 checks a year.
$15,000 ÷ 52 = $288.46 per check.
(b) Biweekly payment means 26 checks a year.
$15,000 ÷ 26 = $576.92 per check.
(c) Semimonthly payment means 24 checks a year.
$15,000 ÷ 24 = $625.00 per check.
(d) Monthly payment means 12 checks a year.
$15,000 ÷ 12 = $1,250.00 per check.

Hourly Wages

The basic computation for determining gross earnings based on an hourly wage is:

$$\text{Gross wages} = \text{Hours worked} \times \text{Rate per hour}$$

An employee working 40 hours a week at a rate of $6 per hour earns gross wages of $240 a week.

Timekeeping

An hourly wage system must rely on an accurate record of hours worked—*timekeeping*. Some firms use a sign-in sheet on which every worker writes in the times of arrival at and departure from work each day. Other firms use a time-card system, in which every worker will "punch" in and out each day. At the end of the week, the timekeeper (or the computer) adds up the hours worked. Another clerk (or the computer) will then double-check the addition and compute gross wages.

Figure 8.1 illustrates a time card that has been filled in.

Figure 8.1

Time Card

EMPLOYEE NAME T. Rhodes CARD # 76

Date	In	Out	In	Out	Total Hours
2/7	8:00	12:00	1:00	5:00	8
2/8	7:53	12:00	12:57	5:01	8
2/9	7:58	12:03	1:00	5:04	8
2/10	7:54	12:06	12:55	5:03	8
2/11	7:59	12:01	12:59	5:02	8
Total hours					40
Rate per hour					$5.00
Gross wages					$200.00

An efficient timekeeping system must set rules for determining exactly what is "on time" and what is "late." An employee may lose a full quarter-hour's pay for being three or more minutes late in one firm; in another, only the actual number of minutes late may be lost. The common factor in all timekeeping systems is that gross pay is determined from the hours shown on the time card or sign-in sheet.

Sample Problem 2 illustrates the calculation of gross pay based on hours worked.

SAMPLE PROBLEM 2

Nancy Burke's time card shows the following hours worked for the past week:

Monday	8 hours
Tuesday	$7\frac{3}{4}$ hours
Wednesday	$7\frac{1}{2}$ hours
Thursday	8 hours
Friday	$7\frac{1}{4}$ hours

If her hourly wage rate is $8.40, find her gross wages.

Solution

$$8 + 7\tfrac{3}{4} + 7\tfrac{1}{2} + 8 + 7\tfrac{1}{4} = 38\tfrac{1}{2} \text{ hours}$$

$$\times \$8.40$$

Gross wages $323.40

Overtime

Hours worked can be grouped into *regular*, or *normal*, hours and *overtime* hours—those in excess of regular. What determines regular hours and overtime is a combination of company and union agreements, subject to limits set by federal law.

Today, "regular" hours range from seven to eight hours a day, totaling between 35 and 40 hours a week. Overtime will be any time worked beyond these hours. Let's suppose an employee works the following hours in one week: 8, 9, 7, 7, 10. If the firm's regular hours are defined as 40 per week, then the total of 41 worked consists of 40 regular hours and one overtime hour. If, however, overtime is defined as any time over eight hours a day, then the employee has one overtime hour on the second day and two on the last day. The 41 hours are then divided into 38 regular and three overtime.

In the problems in this book, unless otherwise specified, we will assume that regular hours means 40 hours a week.

Overtime wages can be calculated basically in three ways. The concept of *time-and-a-half* gives the worker credit for $1\frac{1}{2}$ hours for each overtime hour worked. Thus, for a worker who works 4 hours overtime and earns an hourly rate of $6.00, the overtime pay amounts to $36.00, calculated as follows:

$$4 \text{ hours} \times 1\tfrac{1}{2} \times \$6.00 = 6 \text{ hours} \times \$6.00 = \$36.00$$

To the $36.00 will be added the worker's regular earnings of 40 hours @ $6.00 = $240.00 for total earnings of $276.00.

A second approach to calculating overtime pay can be termed the *bonus method*. For each overtime hour worked, a $\frac{1}{2}$ hour bonus is added. The following tabulation shows this idea:

If a Person Works	Regular Hours Are	Overtime Hours Are	Bonus Hours Are	Total Paid Hours Are
38 hours	38	0	0	38
40 hours	40	0	0	40
41 hours	40	1	$\frac{1}{2}$	$41\frac{1}{2}$
44 hours	40	4	2	46
$48\frac{1}{2}$ hours	40	$8\frac{1}{2}$	$4\frac{1}{4}$	$52\frac{3}{4}$
$49\frac{1}{2}$ hours	40	$9\frac{1}{2}$	$4\frac{3}{4}$	$54\frac{1}{4}$

The worker on the fourth line has worked 46 hours. At a rate of $6.00 an hour, the total earnings are 46 × $6.00 = $276.00.

SAMPLE PROBLEM 3

Arthur Fonzerelli worked $45\frac{1}{2}$ hours in the current week. If his hourly rate is $5.10, find his (a) regular hours; (b) overtime hours; (c) bonus hours; (d) paid hours; (e) gross wages.

Solution

(a) Regular hours	=	40
(b) Overtime hours	=	$5\frac{1}{2}$
(c) Bonus hours	=	$2\frac{3}{4}$
(d) Paid hours	=	$48\frac{1}{4}$
Hourly rate	=	× $5.10
(e) Gross wages	=	$246.08

The third method, called *rate-and-a-half,* pays the worker $1\frac{1}{2}$ times the regular hourly rate for overtime hours. An individual who earns $6.00 for a regular hour will earn $1\frac{1}{2} \times \$6.00$, or $9.00 for an overtime hour. An individual who works 44 hours at a $6.00 regular rate will have gross pay computed as follows:

40 regular hours × $6.00 =	$240.00
4 overtime hours × $9.00 =	36.00
Gross pay	$276.00

Sample Problem 4 repeats the data in Sample Problem 3, but uses the rate-and-a-half approach.

SAMPLE PROBLEM 4

Arthur Fonzerelli worked $45\frac{1}{2}$ hours in the current week. If his hourly rate is $5.10, find his gross wages, using the rate-and-a-half method.

Solution

Regular rate is $5.10; overtime rate is $1\frac{1}{2} \times \$5.10 = \$7.65$

Regular wages: 40 hours × $5.10 =	$204.00
Overtime wages: $5\frac{1}{2}$ hours × $7.65 =	42.08
Gross wages	$246.08

TEST YOUR ABILITY

1. Cathy Morash earns an annual salary of $31,200. Find the gross amount of each paycheck if the pay periods are (a) weekly; (b) bimonthly; (c) monthly.
2. Stanley Kowalski works at an hourly rate of $5.20 and clocks $7\frac{1}{2}$, 8, $8\frac{1}{4}$, 7, and $6\frac{1}{2}$ hours for the five days of the current week. Assuming time-and-a-half for hours over 40 a week, find his gross wages.
3. Alice Clark's time card shows hours worked from Monday to Friday as 8, 9, 7, $9\frac{1}{2}$, and 10. She earns $7.20 an hour, with time-and-a-half for overtime. Find her (a) regular hours; (b) overtime hours; (c) bonus hours; (d) total paid hours; (e) gross wages.
4. Solve Problem 3 using the rate-and-a-half method.

Piecerate Plans

An employee receiving an hourly wage is paid strictly for the number of hours worked and/or the actual output. The hourly rate often reflects what is expected from the worker in terms of output or responsibility. With this in mind, a firm may offer an *incentive wage* in order to motivate its workers. The incentive wage is calculated by means of a piecerate plan—so much money per each unit produced. Thus, an employee's pay will depend on that person's initiative.

The purpose of an incentive plan is to reward the better worker and encourage the poor worker to become a better one. However, so as not to penalize the slower worker, a minimum wage is usually guaranteed. Calculation of gross wages is expressed as:

$$\text{Gross wages} = \text{Units produced} \times \text{Rate per unit}$$

A wide variety of piecerate plans is found in the business world. We will consider two of them. *Straight piecework* follows the formula just seen: so much per unit with a guaranteed minimum wage. Sample Problem 5 illustrates this plan.

SAMPLE PROBLEM 5

Bill Allen and Ray Valliere are two co-workers who receive $0.38 per unit produced. A weekly wage of $300 is guaranteed. Their production statistics for this week are:

	M	T	W	T	F
Allen	226	219	232	205	199
Valliere	135	141	138	124	107

Calculate the gross earnings of each for the week.

Solution

	Allen	Valliere
Total units	1,081	645
Rate per unit	× $0.38	× $0.38
Gross wages	$410.78	~~$245.10~~ $300.00 minimum

Valliere's wages were adjusted to the minimum of $300.

A logical reaction at this point is to be very sympathetic to Allen, who produced over 400 more units than did Valliere but earned only $110.78 more. What is Allen's incentive for high production? Frederick W. Taylor, a pioneer in scientific management, developed the idea of a *differential piecerate plan:* (1) Establish what the average worker can do—a standard; (2) for production up to and including the standard, pay a given rate per unit; (3) for production over standard, pay a higher rate for *all* units produced. Assume, for example, that a worker is paid 10¢ a unit; however, the rate becomes 15¢ if more than 10 units are produced. The worker producing 10 units receives 10 × 10¢ = $1.00; the worker producing 11 units receives 11 × 15¢ = $1.65. In other words, that employee has an incentive. Sample Problem 6 reviews the case of Allen and Valliere under a differential piecerate system.

SAMPLE PROBLEM 6

Bill Allen and Ray Valliere are now being paid under a differential piecerate system as follows: $0.38 per unit up to and including a standard of 1,025 units; $0.48 per unit if production exceeds standard. A $300 weekly wage is guaranteed. Find the gross wages of each man, using the production figures from Sample Problem 5.

Solution

Allen: 1,081 units × $0.48 = $518.88
Valliere: 645 units × $0.38 = $245.10 = $300 minimum

Allen's incentive is better rewarded under a differential piecerate plan.

Commission

Over and above any salary earned, a person engaged in selling a firm's products or services is usually paid a *commission*—a percentage of sales or a fixed amount per unit sold. The gross wages of a person paid on a *straight commission* depend strictly on the amount of sales and the rate of commission.

SAMPLE PROBLEM 7

Pearl Poisson, a sales representative, is paid a straight commission of $33\frac{1}{3}\%$ on all sales. If her March sales are $1,375.50, find her gross earnings.

Solution

$33\frac{1}{3}\% = \frac{1}{3}$
Gross wages = $1,375.50 × $\frac{1}{3}$ = $458.50

A monthly gross of $458.50 is not very high. Because of the lack of security in the sales profession, many firms pay *salary plus commission* to sales representatives.

SAMPLE PROBLEM 8

Richard McGuire, a sales representative, is paid a $37\frac{1}{2}\%$ commission on all sales, in addition to a weekly salary of $45.00. If this week's sales are $760.00, find his gross earnings.

Solution

$37\frac{1}{2}\% = \frac{3}{8}$

Commission ($\frac{3}{8}$ of $760.00)	$285.00
Salary	45.00
Gross wages	$330.00

Another way to pay a commission is by using a *quota* system—in return for a set salary, no commission is paid until the sales representative sells more than a certain amount. This system is illustrated in Sample Problem 9.

SAMPLE PROBLEM 9

Ann Dougherty has a weekly gross wage of $280 plus a 15% commission on total sales over $400. Given the following sales figures for the four weeks of February, find her gross wages for the month:

Week of February	1	$400
	8	650
	15	375
	22	740

Solution

Week	1	2	3	4
Sales over quota	$ –0–	$250.00	$ –0–	$340.00
Rate		×.15		×.15
Commission earnings		$ 37.50		$ 51.00
Salary	280.00	280.00	280.00	280.00
Gross wages	$280.00	$317.50	$280.00	$331.00

Total gross wages for the month = $1,208.50

Another commission plan is called a *graduated commission scale.* To encourage higher sales, a higher rate of commission is offered as sales increase. Unlike the differential piecerate plan, the higher rates do not apply to all sales, but just to sales over the stated limits. The scale may be arranged as follows:

10% on the first $1,000 of sales
15% on the second $1,000 of sales
20% on all sales over $2,000

SAMPLE PROBLEM 10

Julie Hearn and Phil Atkinson work under the graduated commission scale listed above. Find their gross wages if Hearn's sales are $2,600 and Atkinson's are $1,700.

Solution

	Hearn	Atkinson
10% × $1,000	$100	$100
15% × $1,000	150	
15% × $700		105
20% × $600	120	
Gross wages	$370	$205

Because there are so many different combinations of quotas, salaries, and graduated commission scales in business, be sure that you know which one you are using in each problem.

TEST YOUR ABILITY

1. Fred Carter's output is 420 units this week under a straight piecework plan; Adele Walsh's is 605 units. The piece rate is $0.56, with a guaranteed wage of $240. Find each person's gross wages.
2. Employees at the Geneva Glassworks Company are lucky enough to work under the Taylor differential piecerate plan. They earn $0.46 per unit if production is at standard (700) or below, but earn $0.60 per unit if production exceeds standard. A minimum wage of $300 is guaranteed. Determine the gross wages of an individual producing (a) 600 units; (b) 700 units; (c) 800 units.
3. Andrea Smith earns a commission of $16\frac{2}{3}$% of all sales. If she sells 30 beds this month at $210 each, find her gross wages.
4. Bill Shikibu earns a $70 weekly salary plus a $14\frac{2}{7}$% commission on all sales. If his sales are $847 this week, find his gross earnings.
5. Iris Haines earns a 25% commission on all sales over $600 a month, plus a $125 monthly salary. If her sales this month are $2,170, find her gross earnings.
6. A graduated commission scale for the Dove Missile Company reads as follows: 10% on the first $1,000 of sales; 15% on the next $1,000; 20% on all sales over $2,000. On sales totaling $4,200, what commission will be earned?

Payroll Deductions

Now that we have studied the various ways of earning gross wages, let's turn to the calculation of those amounts that are subtracted from gross wages—*payroll deductions*—to obtain *net wages*. It has been said that of every $1 of gross wages, 65 to 70 cents is taken home, the other 30 or 35 cents taking the form of various deductions.

Payroll deductions fall into two categories: those required by law (social security tax and income tax) and those of a voluntary nature (health and life insurance, union dues, bonds, and others). The two required deductions will be explained in detail in the following pages.

FICA Tax

With certain technical exceptions, everyone who works in the United States is subject to the FICA (Federal Insurance Contributions Act) tax, more commonly known as social security tax. The law requires that a certain percentage of each employee's wages be deducted by the employer and sent to the government. (See also "The Employer's Expenses," covered later in this chapter.)

The purposes of the tax when it began in 1937 were to (1) provide protection for a worker who becomes unable to work (disability benefits); (2) provide a worker with a pension after retirement (old-age benefits); and (3) provide an income for the family of a deceased worker (survivors' benefits). In 1966, protection against illness (Medicare) was added as a fourth benefit.

The original rate of 1% on the first \$3,000 of earnings has gradually increased to its 1986 level of 7.15% on the first \$42,000 of earnings; the 1986 maximum tax is $42{,}000 \times .0715 = \$3{,}003$ a year. The planned 1987 rate as of this writing is the same—7.15%—but the base will increase in accordance with the average wage level.

Computing FICA Tax

The standard way to compute FICA tax is to multiply gross wages by 7.15% (which, as we well know, is .0715).

SAMPLE PROBLEM 11

Everett Sims's gross wages for the week are \$300. Calculate his FICA tax deduction.

Solution

$\$300 \times .0715 = \$21.45$

The work is easy: We are given a base and a rate and are asked to find a percentage. However, it's also too easy to make a math error, so a calculator memory key set at a constant (see the Appendix) helps immensely.

The Taxable Limit

Finally, we must understand the "limit" concept in FICA taxation. Earnings are taxed only up to a certain amount, which is \$42,000 as of 1986. As soon as this limit is reached, no additional tax is deducted during that year. This is another reason why a firm must keep accurate earnings records. The "limit" is shown in Sample Problem 12.

SAMPLE PROBLEM 12

Audrey Brown has already earned \$36,300, and this week she earns \$800. Lara Bugakov's earnings have reached \$42,100, to which \$900 is added this week. Andrew Black's figures are now \$41,850 and \$850 for the week. Find the amount of FICA tax to be deducted from each person's earnings this week.

Solution

Brown (\$36,300 + \$800 = \$37,100)7.15% × \$800 = \$57.20
Bugakov (already over limit)
Black (\$41,850 + \$150 = \$42,000 limit; other \$650 not taxed)
7.15% × \$150 = \$10.73

Of the three, only Brown will have additional tax deducted this year; however, in several more weeks, she will also clear the limit.

Withholding Tax

The other principal deduction required in the United States is the federal income tax. All wage earners are too familiar with income tax—by April 15 of each year, we must all go through the taxing procedure of filing a return for the previous year.

It would be unfortunate for most of us to pay our income tax in one lump sum on or before April 15. In order to avoid this problem and insure a steady flow of money to the government, almost every employer is required by federal law to "withhold" a certain amount from every paycheck. The employer in turn must send these deductions to the government at regular intervals.

When April 15 comes, the worker will compare the income tax bill with the withholding statement and pay whatever additional tax is due, or, in the event too much was withheld, receive a refund, which is always a pleasant surprise.

Withholding Tax Computations

The size of the payroll deduction for withholding tax depends on three factors: gross wages, marital status, and the number of exemptions claimed by the employee.

The facts of life are that the higher a worker's gross wages, the higher the tax. For a single worker, the tax is higher than for a married worker.

An *exemption* is an allowance claimed for each person whom you support. Exemptions are granted for yourself, your spouse, your children, other dependents who meet certain legal tests of support, and old age or medical conditions. The higher the number of exemptions, the *lower* the tax withheld.

Individuals must report the exact number of exemptions allowable when filing tax returns. For purposes of payroll deductions, however, individuals may claim fewer exemptions than allowed—even none at all. In such a case, more tax will be deducted from each paycheck than is necessary, but a tax refund after April 15 is very likely.

The amount of tax to be withheld can be determined in two ways. One is a percentage method in which stated rates are applied to gross earnings. The second and more common way is by referring to tables such as 8.1 and 8.2. Each of these tables is for *weekly* earnings only; other tables are used for biweekly, semimonthly, and monthly pay periods. Table 8.1 for *single* persons; 8.2 for *married* persons. To use the tables, look at the following two examples:

(1) A single person claims one exemption and earns \$300 a week. In Table 8.1, read the "at least" \$300 line across to the "1" column and find \$40. (Single with *one* exemption is often abbreviated as *S-1*, a form used in our problems.)

(2) A *married* person with *four* exemptions (*M-4*) earns \$400 a week. The tax withheld, according to Table 8.2 is \$38.

SAMPLE PROBLEM 13

Jerry Briggs earns $300 this week, fully subject to FICA tax. His withholding status is M-3.

(a) Find his FICA tax.
(b) Find his federal withholding tax.
(c) Assuming no other deductions, find his net wages.

Solution

(a) FICA tax (.0715 × $300)	= $21.45
(b) Table 8.2, withholding tax on $300.00, status M-3 =	25.00
Total deductions	46.45

(c) Net wages = $300.00 − $46.45 = $253.55

Other Deductions

Payroll deductions are made for other reasons besides the two required taxes just discussed. A number of states and cities also withhold income tax. In some states, a disability insurance amount is withheld—usually ½ of 1% (.005), or so. Firms may also deduct for pension plans, union dues, bonds, insurance, etc.

To summarize our discussion, consider the case of Joanne Watts, who begins with a weekly gross of $300.

SAMPLE PROBLEM 14

Joanne Watts earns $300 a week gross. Her status is S-0. She has the following deductions, in addition to FICA and federal income taxes: $3 for union dues, $2.75 for life insurance, ½ of 1% for disability insurance, $2 for savings bonds, 5% for her pension, 3% for state income tax, and 1% for city income tax. Assuming that her previous earnings are $11,300, calculate her net pay.

Solution

Gross wages		$300.00
Deductions:		
FICA tax ($300 × .0715)	$21.45	
Federal withholding tax		
(Table 8.1, S-0)	45.00	
Union dues	3.00	
Life insurance	2.75	
Disability insurance		
(½ of 1% = .005 × $300)	1.50	
Savings bonds	2.00	
Pension		
(5% × $300)	15.00	
State withholding tax		
(3% × $300)	9.00	
City withholding tax		
(1% × $300)	3.00	
Total deductions		102.70
Net pay		$197.30

From $300 down to $197.30! Let's hope Ms. Watts' rent isn't too high.

Table 8.1

Single Persons—Weekly Payroll Period
(For Wages Paid After December 1985)

And the wages are—		And the number of withholding allowances claimed is—										
		0	1	2	3	4	5	6	7	8	9	10
At least	But less than	The amount of income tax to be withheld shall be—										
$ 0	$ 32	$ 0	$ 0	$ 0	$ 0	$ 0	$ 0	$ 0	$ 0	$ 0	$ 0	$ 0
32	34	1	0	0	0	0	0	0	0	0	0	0
34	36	1	0	0	0	0	0	0	0	0	0	0
36	38	1	0	0	0	0	0	0	0	0	0	0
38	40	1	0	0	0	0	0	0	0	0	0	0
40	42	2	0	0	0	0	0	0	0	0	0	0
42	44	2	0	0	0	0	0	0	0	0	0	0
44	46	2	0	0	0	0	0	0	0	0	0	0
46	48	2	0	0	0	0	0	0	0	0	0	0
48	50	2	0	0	0	0	0	0	0	0	0	0
50	52	3	0	0	0	0	0	0	0	0	0	0
52	54	3	1	0	0	0	0	0	0	0	0	0
54	56	3	1	0	0	0	0	0	0	0	0	0
56	58	3	1	0	0	0	0	0	0	0	0	0
58	60	4	1	0	0	0	0	0	0	0	0	0
60	62	4	1	0	0	0	0	0	0	0	0	0
62	64	4	2	0	0	0	0	0	0	0	0	0
64	66	4	2	0	0	0	0	0	0	0	0	0
66	68	5	2	0	0	0	0	0	0	0	0	0
68	70	5	2	0	0	0	0	0	0	0	0	0
70	72	5	3	0	0	0	0	0	0	0	0	0
72	74	5	3	0	0	0	0	0	0	0	0	0
74	76	6	3	1	0	0	0	0	0	0	0	0
76	78	6	3	1	0	0	0	0	0	0	0	0
78	80	6	4	1	0	0	0	0	0	0	0	0
80	82	6	4	1	0	0	0	0	0	0	0	0
82	84	7	4	2	0	0	0	0	0	0	0	0
84	86	7	4	2	0	0	0	0	0	0	0	0
86	88	7	4	2	0	0	0	0	0	0	0	0
88	90	8	5	2	0	0	0	0	0	0	0	0
90	92	8	5	2	0	0	0	0	0	0	0	0
92	94	8	5	3	0	0	0	0	0	0	0	0
94	96	8	6	3	1	0	0	0	0	0	0	0
96	98	9	6	3	1	0	0	0	0	0	0	0
98	100	9	6	3	1	0	0	0	0	0	0	0
100	105	9	7	4	1	0	0	0	0	0	0	0
105	110	10	7	4	2	0	0	0	0	0	0	0
110	115	11	8	5	3	0	0	0	0	0	0	0
115	120	12	9	6	3	1	0	0	0	0	0	0
120	125	12	9	6	4	1	0	0	0	0	0	0
125	130	13	10	7	4	2	0	0	0	0	0	0
130	135	14	11	8	5	2	0	0	0	0	0	0
135	140	15	11	9	6	3	1	0	0	0	0	0
140	145	15	12	9	6	4	1	0	0	0	0	0
145	150	16	13	10	7	4	2	0	0	0	0	0
150	160	17	14	11	8	5	3	0	0	0	0	0
160	170	19	16	13	10	7	4	1	0	0	0	0
170	180	20	17	14	11	8	5	3	0	0	0	0
180	190	22	19	16	12	9	6	4	1	0	0	0
190	200	24	20	17	14	11	8	5	2	0	0	0
200	210	25	22	19	15	12	9	6	4	1	0	0
210	220	27	24	20	17	14	11	8	5	2	0	0
220	230	29	25	22	18	15	12	9	6	4	1	0
230	240	31	27	23	20	17	14	11	8	5	2	0
240	250	32	29	25	22	18	15	12	9	6	4	1
250	260	34	31	27	23	20	17	14	10	8	5	2
260	270	36	32	29	25	22	18	15	12	9	6	3
270	280	38	34	30	27	23	20	17	13	10	7	5
280	290	40	36	32	28	25	21	18	15	12	9	6
290	300	43	38	34	30	27	23	20	16	13	10	7
300	310	45	40	36	32	28	25	21	18	15	12	9
310	320	47	42	38	34	30	26	23	20	16	13	10
320	330	49	45	40	36	32	28	24	21	18	15	12
330	340	52	47	42	38	34	30	26	23	19	16	13
340	350	54	49	45	40	36	32	28	24	21	18	15
350	360	56	52	47	42	38	34	30	26	23	19	16
360	370	59	54	49	44	40	36	32	28	24	21	18
370	380	62	56	51	47	42	38	34	30	26	23	19
380	390	64	59	54	49	44	40	36	31	28	24	21

Table 8.2

Married Persons—Weekly Payroll Period
(For Wages Paid After December 1985)

And the wages are—		And the number of withholding allowances claimed is—										
At least	But less than	0	1	2	3	4	5	6	7	8	9	10
		The amount of income tax to be withheld shall be—										
$ 0	$ 54	$ 0	$ 0	$ 0	$ 0	$ 0	$ 0	$ 0	$ 0	$ 0	$ 0	$ 0
54	56	1	0	0	0	0	0	0	0	0	0	0
56	58	1	0	0	0	0	0	0	0	0	0	0
58	60	1	0	0	0	0	0	0	0	0	0	0
60	62	1	0	0	0	0	0	0	0	0	0	0
62	64	1	0	0	0	0	0	0	0	0	0	0
64	66	2	0	0	0	0	0	0	0	0	0	0
66	68	2	0	0	0	0	0	0	0	0	0	0
68	70	2	0	0	0	0	0	0	0	0	0	0
70	72	2	0	0	0	0	0	0	0	0	0	0
72	74	3	0	0	0	0	0	0	0	0	0	0
74	76	3	0	0	0	0	0	0	0	0	0	0
76	78	3	1	0	0	0	0	0	0	0	0	0
78	80	3	1	0	0	0	0	0	0	0	0	0
80	82	3	1	0	0	0	0	0	0	0	0	0
82	84	4	1	0	0	0	0	0	0	0	0	0
84	86	4	2	0	0	0	0	0	0	0	0	0
86	88	4	2	0	0	0	0	0	0	0	0	0
88	90	4	2	0	0	0	0	0	0	0	0	0
90	92	5	2	0	0	0	0	0	0	0	0	0
92	94	5	2	0	0	0	0	0	0	0	0	0
94	96	5	3	0	0	0	0	0	0	0	0	0
96	98	5	3	1	0	0	0	0	0	0	0	0
98	100	5	3	1	0	0	0	0	0	0	0	0
100	105	6	4	1	0	0	0	0	0	0	0	0
105	110	6	4	2	0	0	0	0	0	0	0	0
110	115	7	5	2	0	0	0	0	0	0	0	0
115	120	8	5	3	1	0	0	0	0	0	0	0
120	125	8	6	3	1	0	0	0	0	0	0	0
125	130	9	6	4	2	0	0	0	0	0	0	0
130	135	9	7	5	2	0	0	0	0	0	0	0
135	140	10	8	5	3	1	0	0	0	0	0	0
140	145	11	8	6	3	1	0	0	0	0	0	0
145	150	11	9	6	4	2	0	0	0	0	0	0
150	160	13	10	7	5	2	0	0	0	0	0	0
160	170	14	11	8	6	4	1	0	0	0	0	0
170	180	15	12	10	7	5	2	0	0	0	0	0
180	190	17	14	11	8	6	3	1	0	0	0	0
190	200	18	15	12	10	7	5	2	0	0	0	0
200	210	20	17	14	11	8	6	3	1	0	0	0
210	220	21	18	15	12	9	7	4	2	0	0	0
220	230	22	19	17	14	11	8	6	3	1	0	0
230	240	24	21	18	15	12	9	7	4	2	0	0
240	250	26	22	19	16	14	11	8	6	3	1	0
250	260	27	24	21	18	15	12	9	7	4	2	0
260	270	29	25	22	19	16	13	11	8	5	3	1
270	280	30	27	24	21	18	15	12	9	7	4	2
280	290	32	29	25	22	19	16	13	10	8	5	3
290	300	34	30	27	24	21	18	15	12	9	7	4
300	310	35	32	28	25	22	19	16	13	10	8	5
310	320	37	33	30	27	23	20	18	15	12	9	6
320	330	39	35	32	28	25	22	19	16	13	10	8
330	340	40	37	33	30	27	23	20	17	14	12	9
340	350	42	38	35	32	28	25	22	19	16	13	10
350	360	44	40	37	33	30	27	23	20	17	14	11
360	370	46	42	38	35	31	28	25	22	19	16	13
370	380	48	44	40	36	33	30	26	23	20	17	14
380	390	49	46	42	38	35	31	28	25	21	19	16
390	400	51	47	44	40	36	33	30	26	23	20	17
400	410	53	49	46	42	38	35	31	28	25	21	18
410	420	55	51	47	44	40	36	33	29	26	23	20
420	430	58	53	49	45	42	38	34	31	28	24	21
430	440	60	55	51	47	43	40	36	33	29	26	23
440	450	62	57	53	49	45	42	38	34	31	28	24
450	460	64	60	55	51	47	43	40	36	33	29	26
460	470	66	62	57	53	49	45	41	38	34	31	27
470	480	69	64	60	55	51	47	43	39	36	32	29
480	490	71	66	62	57	53	49	45	41	37	34	31
490	500	73	68	64	59	55	51	47	43	39	36	32

TEST YOUR ABILITY

1. Find the FICA tax on each of the following salaries:
   (a) $197.45 (b) $400.00
   (c) $226.83 (d) $233.08
2. Bill O'Neill has earned $41,837.83 previous to this week's $900.00 gross earnings. Find his FICA tax deduction for the week.
3. Find the federal withholding tax in each situation:

	Weekly Gross	Status
(a)	$385.00	S-1
(b)	$380.00	M-4
(c)	$155.00	M-7
(d)	$274.50	S-0

4. Mel Katz, status S-1, earns $245 gross a week. Find his net pay, assuming only the required deductions.
5. Jeanne Price, S-2, earns $350 gross a week. In addition to the required deductions, her employer deducts $1.50 for union dues, 5% for her pension, and $\frac{1}{2}$% for disability insurance. Find her weekly net pay.

The Payroll Register

The calculation of Joanne Watts' net pay in Sample Problem 14 was accurate, but lengthy. This procedure is too awkward for a firm with many employees. In order to save time and trouble, forms have been designed for recording payroll data.

A summary of all payroll data for a specific time period is called a *payroll register.* Sample Problem 15 displays this form and its completion for a firm with ten workers who have a 40-hour week, with time-and-a-half for overtime. All the workers are well below the $42,000 FICA limit.

Notice that the form lends itself to a series of simple math checks for all totals. The checking procedure is shown at the end of the solution.

SAMPLE PROBLEM 15

Prepare a payroll register for the Mitchell Granite Quarry for the week ended May 11, 19—. The pertinent information is summarized below. Use the bonus method with time-and-a-half for hours over 40.

Employee No.	Employee Name	Exemption	Hours Worked M	T	W	T	F	Rate per Hour	Other Deductions
1	*R. Boulette*	S-1	8	8	8	8	8	$4.30	$5.00
2	*S. Atkins*	S-0	8	7	8	10	8	4.50	–0–
3	*W. Sousa*	M-3	8	9	9	9	9	4.70	2.50
4	*A. Munz*	M-2	$8\frac{1}{2}$	9	9	$8\frac{1}{2}$	10	4.40	7.50
5	*C. O'Dowd*	M-4	9	$8\frac{1}{4}$	$8\frac{3}{4}$	$7\frac{1}{2}$	9	5.00	–0–
6	*L. Gagne*	S-0	8	7	7	7	—	5.60	3.00
7	*N. Morgan*	S-1	10	10	$8\frac{1}{2}$	8	11	5.00	7.00
8	*P. Reed*	S-1	8	8	8	8	$8\frac{1}{2}$	5.80	2.50
9	*S. Frazier*	M-6	$7\frac{1}{2}$	$8\frac{3}{4}$	$7\frac{1}{4}$	9	8	5.80	1.00
10	*C. Stickland*	S-1	8	9	$7\frac{1}{2}$	9	10	5.20	8.50

Solution for Problem 15

Mitchell Granite Quarry Payroll

For the Week Ended May 11, 19–

Employee No.	Employee Name	Exemption	Hours Worked M	T	W	T	F	Total Hours	Reg. Hours	OT Hours	OT Bonus	Paid Hours	Rate	Gross Wages
1	*R. Boulette*	S-1	8	8	8	8	8	40	40	0	0	40	$4.30	$ 172.00
2	*S. Atkins*	S-0	8	7	8	10	8	41	40	1	$\frac{1}{2}$	$41\frac{1}{2}$	4.50	186.75
3	*W. Sousa*	M-3	8	9	9	9	9	44	40	4	2	46	4.70	216.20
4	*A. Munz*	M-2	$8\frac{1}{2}$	9	9	$8\frac{1}{2}$	10	45	40	5	$2\frac{1}{2}$	$47\frac{1}{2}$	4.40	209.00
5	*C. O'Dowd*	M-4	9	$8\frac{1}{4}$	$8\frac{3}{4}$	$7\frac{1}{2}$	9	$42\frac{1}{2}$	40	$2\frac{1}{2}$	$1\frac{1}{4}$	$43\frac{3}{4}$	5.00	218.75
6	*L. Gagne*	S-0	8	7	7	7	—	29	29	0	0	29	5.60	162.40
7	*N. Morgan*	S-1	10	10	$8\frac{1}{2}$	8	11	$47\frac{1}{2}$	40	$7\frac{1}{2}$	$3\frac{3}{4}$	$51\frac{1}{4}$	5.00	256.25
8	*P. Reed*	S-1	8	8	8	8	$8\frac{1}{2}$	$40\frac{1}{2}$	40	$\frac{1}{2}$	$\frac{1}{4}$	$40\frac{3}{4}$	5.80	236.35
9	*S. Frazier*	M-6	$7\frac{1}{2}$	$8\frac{3}{4}$	$7\frac{1}{4}$	9	8	$41\frac{1}{2}$	40	$1\frac{1}{2}$	$\frac{3}{4}$	$42\frac{1}{4}$	5.80	245.05
10	*C. Stickland*	S-1	8	9	$7\frac{1}{2}$	9	10	$43\frac{1}{2}$	40	$3\frac{1}{2}$	$1\frac{3}{4}$	$45\frac{1}{4}$	5.20	235.30
TOTALS			83	84	81	84	$82\frac{1}{2}$	$414\frac{1}{2}$	389	$25\frac{1}{2}$	$12\frac{3}{4}$	$427\frac{1}{4}$		$2,138.05

continued

Solution for Problem 15 (continued)

Employee No.	Employee Name	FICA Tax	Federal Withholding Tax	Other Deductions	Total Deductions	Net Wages
1	*R. Boulette*	$ 12.30	$ 17.00	$ 5.00	$ 34.30	$ 137.70
2	*S. Atkins*	13.35	22.00	–0–	35.35	151.40
3	*W. Sousa*	15.46	12.00	2.50	29.96	186.24
4	*A. Munz*	14.94	14.00	7.50	36.44	172.56
5	*C. O'Dowd*	15.64	9.00	–0–	24.64	194.11
6	*L. Gagne*	11.61	19.00	3.00	33.61	128.79
7	*N. Morgan*	18.32	31.00	7.00	56.32	199.93
8	*P. Reed*	16.90	27.00	2.50	46.40	189.95
9	*S. Frazier*	17.52	8.00	1.00	26.52	218.53
10	*C. Stickland*	16.82	27.00	8.50	52.32	182.98
TOTALS		$152.86	$186.00	$37.00	$375.86	$1,762.19

Check of Totals:

1. Hours worked

M	83
T	84
W	81
T	84
F	$82\frac{1}{2}$
Total	$414\frac{1}{2}$

2.

Regular hours	389
OT hours	$25\frac{1}{2}$
Total hours	$414\frac{1}{2}$

3.

OT hours	$25\frac{1}{2}$
×	$\frac{1}{2}$
OT bonus	$12\frac{3}{4}$

4.

Regular hours	389
OT hours	$25\frac{1}{2}$
OT bonus	$12\frac{3}{4}$
Paid hours	$427\frac{1}{4}$

5.

Gross wages	$2,138.05
×	.0715
FICA tax	$ 952.87 (rounding difference)

6.

FICA tax	$152.86
Fed. W.T.	186.00
Other ded.	37.00
Total ded.	$375.86

7.

Gross wages	$2,138.05
Total ded.	−375.86
Net Wages	$1,762.19

The Employer's Expenses

Up to now, we have considered payroll calculations directly affecting the employee. However, the employer also has payroll expenses: specific payroll taxes and the cost of fringe benefits.

FICA Tax

An employer—as well as an employee—is required by law to contribute FICA tax. The employer's amount equals that deducted from the employees' pay.

In practice, the employer remits three items to the District Director of Internal Revenue every three months (or monthly if a certain dollar amount is exceeded): the federal income tax withheld, the FICA tax deducted, and the employer's own share of FICA tax. For the payroll in Sample Problem 15, the Mitchell Investigating Company is responsible for:

Federal income tax withheld	\$186.00
FICA tax deducted	152.86
Employer's share of FICA tax	152.86
Total tax owed for week	\$491.72

Unemployment Tax

Workers who lose their jobs through no fault of their own and are unable to find other suitable employment can usually claim federal and state employment insurance benefits. The money for such payments comes from federal and state taxes on *employers.* As of 1986, the basic federal tax was 6.2% of the first \$7,000 of each employee's gross annual wage. Of this rate, a credit of 5.4% is allowed the employer if the state rate is paid in full and on time. Thus, if an employer meets the state's obligations, the federal tax is actually only .8%. Let's assume in our problems that the employer's state rate is paid in full and on time.

Each state sets its own unemployment insurance tax rate, which may be more than, less than, or equal to 5.4%. Usually, a business with a high unemployment rate will pay a higher rate of tax than will a firm with little unemployment.

If the ABC Company has a state rate of 5.7%, its total rate will be 5.7% + 0.8% = 6.5% for that year. If the XYZ Company has a better record, its state rate may be only 2.1%. In that case, its total tax rate is only 2.1% + 0.8% = 2.9%.

SAMPLE PROBLEM 16

Stan Samuels earned \$6,900 prior to this week, during which he has earned an additional \$250. His employer's total unemployment rate is 6.8%. Find his employer's (a) state unemployment tax and (b) federal unemployment tax on his salary for the week.

Solution

Eligible for tax	\$7,000 − \$6,900 = \$100
(a) State tax	(.068 − .008) = .06 × \$100 = \$6.00
(b) Federal tax	.008 × \$100 = \$0.80

SAMPLE PROBLEM 17

The Bryant Company pays a 1.4% state unemployment tax rate. Its weekly gross is \$2,750 for the current week, \$800 of which is subject to unemployment tax. Find the firm's state and federal unemployment tax expenses for the week.

Solution

State: .014 × \$800 = \$11.20
Federal: .008 × \$800 = \$ 6.40

(Read this type of problem *carefully* to determine which rate is initially stated.)

Fringe Benefits

Fringe benefits apply to any money paid by an employer to employees beyond their basic wages; examples are paid holidays, sick days, pensions, insurance, uniforms, education, and others. It has been estimated that for every \$1 of gross wages another 50 to 80 cents goes to fringe benefits.

Sample Problem 18 shows two fringe benefits.

SAMPLE PROBLEM 18

The Auerback Ice Cream Company has 350 employees and a weekly payroll of \$40,000. The firm pays weekly benefits of 8% of gross wages for a pension plan and \$2.00 per person for life insurance premiums. Calculate (a) the weekly cost and (b) the annual cost of these two fringe benefits.

Solution

(a)	Pension cost (8% × \$40,000)	\$ 3,200.00
	Insurance cost (\$2.00 × 350)	700.00
	Weekly cost	\$3,900.00
(b)	Weeks in year	× 52
	Annual cost	\$202,800.00

TEST YOUR ABILITY

1. Prepare a payroll register for the week ended August 25 at the Crosby Repair Company. All employees receive time-and-a-half for hours over 40 a week. None will have reached the FICA limit with this week's earnings.

Employee No.	Employee Name	Exemption	Hours Worked M	T	W	T	F	Rate per Hour	Other Deductions
1	*M. Holt*	M-1	8	$7\frac{1}{2}$	$7\frac{1}{2}$	8	8	\$6.00	\$ 5.00
2	*P. Murphy*	M-0	$8\frac{1}{4}$	8	8	$8\frac{3}{4}$	8	5.10	8.00
3	*R. Comley*	S-1	9	9	8	$8\frac{1}{2}$	7	5.20	-0-
4	*L. Howard*	S-1	10	$7\frac{1}{4}$	9	8	$9\frac{1}{4}$	5.60	2.50
5	*W. Pinard*	M-3	10	8	8	$8\frac{1}{2}$	$7\frac{1}{2}$	4.40	7.00
6	*B. Wilson*	S-0	8	8	$9\frac{1}{2}$	$8\frac{3}{4}$	$8\frac{3}{4}$	4.80	-0-
7	*A. Karanas*	M-5	8	9	$9\frac{1}{2}$	$9\frac{1}{2}$	11	4.90	10.00
8	*F. Mai*	M-7	10	10	8	$7\frac{1}{4}$	$8\frac{3}{4}$	4.60	4.50

2. Wilma Castle has earned \$6,920 prior to this week, during which she earns \$185. Her employer's total unemployment tax rate is 6.3%. Determine the (a) state and (b) federal unemployment tax expense for the week.
3. Wages subject to unemployment tax total \$5,900 this week for the Purple Cab Corporation. If the state rate is 4.1%, find the state and federal unemployment taxes.
4. Vanity Wool, Inc., has 510 employees and a monthly payroll of \$250,000. The firm pays monthly an 8% pension contribution and \$5.00 per person for

insurance. Find the monthly and yearly costs of (a) the actual payroll; (b) the fringe benefits; (c) the total payroll and fringe benefits, excluding taxes.

CHECK YOUR READING

1. Distinguish between a *wage* and a *salary.*
2. What are *normal* hours? *Overtime* hours?
3. Fred Pierce works 8, 7, 9, 9, and 10 hours, respectively, from Monday to Friday of a given week. What are his overtime hours if overtime is based on (a) 8 hours a day; (b) 40 hours a week?
4. Distinguish between *straight piecework* and a *differential piecerate* system.
5. What is the main difference between a *straight commission* plan and a *quota* plan?
6. Explain the operation of a graduated commission plan.
7. List the four purposes of FICA tax.
8. Explain the meaning of the limit on FICA taxable earnings.
9. What is the purpose of withholding tax?
10. What three factors affect the amount of withholding tax?
11. On a gross weekly salary of $200, how much more is withheld for a married person with three exemptions than for one who is married and has four exemptions?
12. If a weekly gross wage is $150, how many exemptions must a married person claim to have no tax withheld?
13. What is the purpose of a payroll register?
14. If a state unemployment tax rate is 4.8%, what is the total federal and state rate?

EXERCISES

Group A

1. Marcia Cozzens, a sales representative, receives 10% on her first $2,000 of sales, 15% on her second $2,000, and 20% on all sales over $4,000 in one month. If June sales are $6,300, find her gross earnings.
2. Samson Cleaners has a state unemployment insurance rate of 4.6%. If taxable wages are $8,400, find the expense for (a) state unemployment tax and (b) federal unemployment tax.
3. Find the federal income tax withheld in each case:

	Weekly Gross	Status
(a)	$300.00	S-0
(b)	$450.00	M-4
(c)	$125.00	S-1
(d)	$198.75	M-6

4. Ted Zayer produces 910 units this week under a straight piecework plan, while Carl Harris produces 640 units. The piecerate is $0.42, with a $280 guaranteed wage. Find each man's gross earnings.
5. Patricia Clark earns $280 a week plus a commission at $7\frac{1}{2}$% of sales. If her sales are $1,240, find her gross earnings.
6. Mary Oliverio's hourly rate is $5.60, with overtime for hours over 40 a week. Her hours this week are 9, $8\frac{1}{2}$, $9\frac{1}{2}$, 7, and $9\frac{1}{2}$. Find her gross wages.
7. Harry Bernanos, earning $199.25, M-2, has weekly deductions of $2.25 for union dues, 6% for pension, and $1.20 for life insurance in addition to the

required deductions. Find his net wages, assuming that he has earned less than the taxable limit.

8. Marge Owens, a lawyer, earns $20,800 a year. Find her gross amount per paycheck if pay periods are (a) monthly; (b) biweekly; (c) weekly.
9. Find the FICA tax on gross salaries of (a) $217.46; (b) $144.27; (c) $163.85; (d) $709.34.
10. Rita Anderson sells 46 lamps @ $19.50 this week, on which she earns a $33\frac{1}{3}$% commission. Find her gross wages.
11. Deducting for FICA tax and federal withholding tax, find the net pay of an M-0 individual who grosses $440 this week and is not over the FICA taxable limit.
12. Izaak Walpole's time card shows $51\frac{1}{2}$ hours worked this week at an hourly rate of $4.00. Find his (a) regular hours; (b) overtime hours; (c) bonus hours; (d) total paid hours; (e) gross wages, assuming 40 hours to be a normal work load.
13. Solve Exercise 13, using the rate-and-a-half method.
14. The Buxton Pool Corporation has 175 employees and an average monthly payroll of $102,000. The firm pays monthly an additional 7% of gross wages to a pension plan and $37.00 per worker for hospitalization. Find the monthly and yearly costs for (a) the two fringe benefits; (b) the actual payroll; (c) the total payroll and fringe benefits, excluding taxes.
15. Larry Yamata earns a $390 base salary, plus a $12\frac{1}{2}$% commission on all sales over $300 per week. If this week's sales are $960, find his gross earnings.
16. Employees at the New Castle Brush Corporation work under a differential piecerate plan, getting $0.42 per unit for standard (500) or less, $0.60 per unit if production exceeds standard, or a guaranteed wage of $280 a week. Find the gross earnings of a worker who produces (a) 480 units; (b) 501 units; (c) 700 units.
17. Shirley Mackin's earnings before this week's $896.45 are $41,845.20. Find this week's FICA tax deduction.
18. Fletcher Film Associates pays a state unemployment insurance rate of 5.1%. It has three employees whose prior earnings are $7,010, $6,940, and $6,850. Their respective earnings for this week are $150, $210, and $140. Determine the employer's tax liability this week for (a) state unemployment tax; (b) federal unemployment tax.
19. Payroll data for the Youngstown Coal Company are listed below. Prepare a payroll register for the week ending October 9, assuming the standard overtime rules. No workers exceed the FICA limit.

Employee No.	Employee Name	Exemption	Hours Worked					Rate per Hour	Other Deductions
			M	T	W	T	F		
1	*L. Brown*	S-1	8	9	9	9	9	$5.00	$15.50
2	*F. Stein*	S-0	7	$8\frac{1}{2}$	7	$7\frac{1}{2}$	$7\frac{1}{2}$	4.40	12.50
3	*C. Collins*	M-1	8	$8\frac{1}{2}$	8	$9\frac{1}{2}$	8	5.20	13.00
4	*D. Callahan*	S-0	10	8	9	8	11	4.60	–0–
5	*W. Helfter*	S-1	9	11	8	9	8	5.10	12.50
6	*R. Childs*	M-2	$8\frac{1}{4}$	$8\frac{1}{4}$	$8\frac{1}{2}$	8	$9\frac{1}{4}$	5.00	17.50
7	*N. Blake*	M-4	8	$8\frac{1}{2}$	11	8	8	4.40	–0–
8	*P. Hunt*	M-3	10	8	9	9	$9\frac{1}{2}$	4.80	14.00

Group B

1. Eric Volsung, a sales manager with a yearly gross salary of $17,200, receives an 8% raise. By how much will his gross weekly paycheck increase?
2. Steve Habib's time card appears below. He earns time-and-a-half for all hours over 35 a week. Normal working hours are 9 to 12 a.m. and 1 to 5 p.m. Lateness of over five minutes in any quarter hour loses the quarter hour, and time worked beyond normal is counted in full quarter hours. Calculate his gross wages.

Employee Name Steve Habib **Card #** 406

Date	In	Out	In	Out	Total Hours
5/3	8:31	12:04	1:02	5:01	
5/4	9:00	12:01	12:58	5:03	
5/5	9:03	12:00	12:57	5:17	
5/6	8:57	12:02	1:09	7:20	
5/7	9:06	11:59	1:02	6:04	
Total hours					
Rate per hour					$8.00
Gross wages					

3. A three-differential piecerate plan is set up as follows:
   $0.155 per piece if production is up to or at 460;
   $0.265 per piece if production is up to or at 540;
   $0.375 per piece if production is over 540.
   Find the gross wages for the production of (a) 540 units; (b) 690 units.
4. Richard Peters receives a commission of 15% on all sales he makes in addition to his $75 a week. In order to earn a total of $300 this week, what must his sales be?
5. Angela Ricci's monthly earnings during 1987 are shown below. Remembering the $42,000 limit, find the 7.15% FICA tax deduction from each of her monthly paychecks.

January	$3,600	July	$3,900
February	3,450	August	3,500
March	3,500	September	3,200
April	3,750	October	3,800
May	3,350	November	3,650
June	3,400	December	3,650

6. Norma Burke has another child and goes from M-3 to M-4 while continuing to earn $180 a week. How much additional take-home pay will she have in a year?

CHAPTER **9**

Insurance

Businesses, like individuals, run the risk of financial loss from unexpected events, such as fire, car accidents, and the death of key personnel. And just like individuals, businesses can buy protection against these losses: *insurance*.

The types of insurance most commonly purchased by businesses are fire insurance; burglary insurance; liability insurance; car insurance; group life and health insurance; inland marine insurance; flood insurance; life insurance; use and occupancy insurance. In Chapter 9, we discuss three of the types in detail: fire insurance, car insurance, and life insurance. The other types will be discussed briefly in the section called "Other Business Insurance" near the end of this chapter.

Life insurance is also used by some people as a form of investment. This use of insurance will be discussed in Chapter 15.

Your Objectives

After studying Chapter 9, you will be able to

1. Calculate premiums for fire, auto, and life insurance policies.
2. Interpret a wide variety of insurance tables.
3. Determine the amounts recovered from losses due to fires and car accidents.
4. Describe the purposes of all the types of business insurance discussed in this chapter.

Terminology

Insurance is generally purchased from an *insurance company,* though many firms set aside money for a self-insurance program. The amount of insurance purchased is the *face* of the policy. The payments made by the *insured* to the insurance company are called the *premiums.*

The amount of a premium payment depends on several factors, which will be discussed as each type of insurance is presented.

Fire Insurance

Fire insurance provides a payment, or *indemnity,* to the policyholder in cases of damage and loss due to fire, lightning, or other natural causes. *Extended coverage* policies pay against other possibilities, such as damage from smoke and water, or from putting out a fire.

Computing Fire Insurance Premiums

Fire insurance premium rates are generally expressed as so many cents per \$100 of insurance desired. To find an *annual premium,* do the following:

$$\text{Annual premium} = \text{No. of hundreds} \times \text{Rate per \$100}$$

The annual premium is rounded, in practice, to the *nearest whole dollar,* a procedure that we will follow.

SAMPLE PROBLEM 1

Calculate the annual fire insurance premium for a \$22,000 policy with a rate of \$0.295 per \$100.

Solution

$\$22{,}000 \div \$100 = 220$ hundreds
$220 \times \$0.295 = \$64.90 = \$65.00$ (rounded)

Premium rates are found in table form, like Table 9.1.

Table 9.1

Annual Fire Insurance Premiums for Buildings
(Rates per \$100)

	Class or Territory					
Construction	**A**	**B**	**C**	**D**	**E**	**F**
Brick	.115	.275	.32	.335	.38	.625
Frame	.285	.32	.415	.44	.475	.655

Notes (1) The same rate is used for protection of the contents.
(2) For two-year policies, multiply the premium by 1.85.
(3) For three-year policies, multiply the premium by 2.7.

As you notice in Table 9.1, rates vary according to construction: wooden (frame) buildings have higher rates than brick buildings.

Another factor causing differing rates is the area (class or *territory*) where the property is located. The risk of fire is greater in an industrial area than in a rural one. Factors such as population density, availability of fire protection, and distance from a hydrant are also considered in determining the class rating for a building. Thus, in the table shown, Territory F is considered a greater fire risk than Territory A.

The rates for a building can also be used for the contents of the building; however, in some areas, the rates for contents are different. Sample Problem 2 uses Table 9.1 to calculate the annual fire insurance premium for a building and its contents. Rounding is done after all other calculations.

SAMPLE PROBLEM 2

A frame building for Warding T. Thomas, an architectural firm, is to be insured for $85,000 and its contents for $8,500; it is located in Territory D. Calculate the annual premium.

Solution

Building (850 hundreds × $0.44) = $374.00
Contents (85 hundreds × $0.44) = 37.40
$411.40 = $411.00

Check Since the building and contents rates are the same,

850 + 85 = 935 hundreds × $0.44 = $411.40 = $411.00

Premiums for Other than One Year

Properties can be insured for time preriods longer or shorter than one year. Rates for longer periods are multiples of the annual premium. The rate for a two-year policy, however, is not simply twice the one-year rate. The savings to the insurance company in administrative expenses keeps the two-year premium down to 1.85 times the one-year premium. Similarly, a three-year premium costs only 2.7 times the one-year premium. Both of these factors are noted at the bottom of Table 9.1.

A three-year policy is generally the longest offered to business firms, although certain public buildings are offered four- and five-year policies.

SAMPLE PROBLEM 3

Determine the three-year fire insurance premium on a brick building insured for $98,000 in Territory C.

Solution

$98,000 = 980 hundreds
980 × $0.32 = $313.60 annual premium
$313.60 × 2.7 = $846.72 3-year premium
= $847.00 (rounded)

Note that rounding is done *after* multiplication by 2.7.

While bargain rates are offered for policies lasting more than a year, the opposite is true for shorter periods. Rates for periods of less than a year are proportionately higher than annual rates. For example, a six-month period is half a year, but a six-month premium can be 60% of the annual premium. Table 9.2 is an abridged version of a *short-rate table*. A full short-rate table would contain a rate for every day of the whole year. What percent of the annual premium is paid for a 75-day policy? The answer is 31%.

Table 9.2

Short-Rate Fire Insurance Table for One Year

Days in Force	Percent of Basic Rate	Days in Force	Percent of Basic Rate
5	8%	75	31%
10	10	80	32
15	13	85	34
20	15	90 (3 months)	35
25	17	120 (4 months)	44
30 (1 month)	19	150 (5 months)	52
35	20	180 (6 months)	60
40	21	210 (7 months)	67
45	23	240 (8 months)	74
50	24	270 (9 months)	80
55	26	300 (10 months)	87
60 (2 months)	27	330 (11 months)	94
65	28	360 (12 months)	100
70	30		

Exact calendar days are counted in determining time periods, as shown in Sample Problem 4.

SAMPLE PROBLEM 4

The Boland Toy Company insures for $200,000 its brick building, located in Territory F, from October 9 to November 3 of the current year. Find its fire insurance premium.

Solution

Find the annual rate in Table 9.1, compute the annual premium, and then use Table 9.2 to find the short-rate premium.

Annual premium = 2,000 × $0.625 = $1,250.00
Time (October 9–November 3) = 25 days
Short rate = 17%
Premium = $1,250.00 × .17 = $212.50 = $213.00

Cancellations of Policies

Suppose that a company decides to take a risk and its insurance policy is canceled? How much of the premium will be refunded to the insured? The amount to be refunded depends on who is responsible for the cancellation.

If, as above, the policy is canceled by the *insured,* the short-rate table is used to determine what part of the premium is kept by the insurance company, and the balance is refunded to the insured. If, after 40 days, a one-year policy is canceled by the insured, Table 9.2 tells us that 21% of the premium is kept by the insurance company; therefore, 79% is returned to the policyholder.

SAMPLE PROBLEM 5

Frank Hansen pays a $90 premium on a one-year fire insurance policy on July 11. He cancels the policy on September 19. Find the amounts of premium (a) retained by the insurance company and (b) refunded to Hansen.

Solution

Time (July 11–September 19) = 70 days
Short rate = 30%

(a) Amount retained by insurance company = 30% × $90.00 = $27.00
(b) Amount returned to Hansen = $90.00 – $27.00 = $63.00

Check 70% × $90.00 = $63.00

On the other hand, if the *insurance company* cancels the policy, a different set of rules applies. The total premium is *prorated* (divided equally) over the entire insurance period. If a one-year policy is canceled by the insurance company after one-fourth of the year, one-fourth of the premium is kept by the company and three-fourths are refunded to the insured. Exact calendar days and a 365-day year are used. The short-rate table is *not* used.

SAMPLE PROBLEM 6

Leslie Adams pays a $75 premium on a one-year fire insurance policy on March 3. On October 8, the insurance company cancels the policy. Determine the premiums (a) retained by the insurance company and (b) refunded to Ms. Adams.

Solution

Time in effect (March 3–October 8) = 219 days

(a) Amount retained by insurance company = $\frac{219}{365}$ × $75.00 = $\frac{3}{5}$ × $75.00 = $45.00
(b) Amount refunded to Ms. Adams = $\frac{146}{365}$ × $75.00 = $\frac{2}{5}$ × $75.00 = $30.00

Had the cancellation been by Ms. Adams, the short-rate of approximately 67% (for 210 days) would have applied. In this case, however, only $\frac{3}{5}$, or 60%, of the premium was retained by the insurance company.

TEST YOUR ABILITY

1. Find the annual fire insurance premiums for each of the following policies:

	Face of Policy	Rate per $100
(a)	$7,000	$0.45
(b)	$2,500	$0.175
(c)	$ 750	$0.41

2. Using Table 9.1 find the annual premium for a fire insurance policy on each of the following buildings:

	Face of Policy	Construction	Territory
(a)	$100,000	Frame	E
(b)	$ 85,000	Frame	B
(c)	$200,000	Brick	A

3. A brick two-family house and its contents are to be insured in Territory E for $90,000 and $30,000, respectively. Calculate the annual fire insurance premium.

4. Find the total fire insurance premium for each of the following buildings:

	Face of Policy	Construction	Territory	Time
(a)	$40,000	Frame	A	2 years
(b)	$92,000	Brick	B	3 years
(c)	$90,000	Brick	D	3 years

5. Using Table 9.2, find the fire insurance premiums for each of the following short-rate policies:

	Face of Policy	Time	Annual Premium Rate
(a)	$ 25,000	55 days	$0.20
(b)	$140,000	10 months	$0.40
(c)	$ 1,650	10 days	$0.35

6. The Hasbury Silverware Company insures its $70,000 frame building in Territory F from March 15 to April 19. Use Tables 9.1 and 9.2 to find the premium for fire insurance.
7. A fire insurance policy is taken out on July 10 and canceled on July 25 by the insured. If the initial premium paid was $146, find (a) the amount retained by the insurance company and (b) the refund to the insured.
8. If, in Problem 7, cancellation had been made by the insurance company, how much would have been refunded?

Compensation for Loss

Up to now, we have looked at the paying and refunding of premiums. Now let's see what actually happens when a fire causes a loss.

Should a fire destroy all or part of an insured property, the insurance company will pay the insured. Under a normal policy, the insured recovers either the face of the policy or the loss, whichever is *lower.*

If a policy for $50,000 is taken out on property worth $70,000, and a $30,000 loss occurs, then:

Face = $50,000
Loss = $30,000
Recovery = $30,000, the lower

Should the property be completely destroyed, then:

Face = $50,000
Loss = $70,000
Recovery = $50,000, the lower

In the second case, the owner was not completely insured and must absorb the $20,000 loss beyond the face of the policy. Why did the loss occur? Perhaps the owner was not willing to pay the rates on a $70,000 policy. Or perhaps the property value increased but the owner never extended the coverage.

Two or More Insurers

Insurance companies will sometimes limit the amount of insurance they are willing to grant. A firm may desire a $100,000 policy on a certain property, but Insurance Company A will only issue $50,000 on the property. The firm must then buy insurance from other companies in order to reach the $100,000. Then, if a loss occurs, each insurance company will share the payment. The share paid is determined by (1) calculating the fraction of the total insurance provided by each insurance company and (2) applying this fraction to the loss.

SAMPLE PROBLEM 7

The Alexander Manufacturing Corporation has insured its property as follows: Company A, $100,000; Company B, $60,000; Company C, $80,000. A loss of $48,000 occurs. How much will Alexander Manufacturing recover from each insurance company?

Solution

Company A	$100,000
Company B	60,000
Company C	80,000
Total	$240,000

A has $\frac{100{,}000}{240{,}000} = \frac{5}{12}$ of the coverage, so A pays $\frac{5}{12} \times \$48{,}000 = \$20{,}000$

B has $\frac{60{,}000}{240{,}000} = \frac{1}{4}$ of the coverage, so B pays $\frac{1}{4} \times \$48{,}000 = 12{,}000$

C has $\frac{80{,}000}{240{,}000} = \frac{1}{3}$ of the coverage, so C pays $\frac{1}{3} \times \$48{,}000 = 16{,}000$

Total $48,000

Suppose, in the situation just presented, the loss was $240,000. What would each company pay? What if the loss were $300,000? Both answers are the same: the face of each policy. In the second instance, Alexander Manufacturing would lose $60,000 ($300,000 – $240,000).

Coinsurance

The chance of a property being completely destroyed by fire is rare; more often, fire results in partial destruction. This is the reason why most firms carry policies for less than the full value of their property. Insurance companies, on the other hand, cannot afford to pay losses in full—if they did, their rates would become so high that no one would want to be insured. To assist both sides, a practice known as *coinsurance* has evolved.

In coinsurance, the insured agrees to carry a policy for a stated percentage of the property's value—usually 80%. For example, with an 80% coinsurance clause, property valued at $200,000 would be insured for $160,000. If carrying at least $160,000, the owner will receive in full all losses, *up to the face of the policy.* However, should the policy be less than $160,000, the insured will receive only a part of the losses; this is determined by the fraction the actual insurance is of $160,000. If a policy for $120,000 is carried, then 120,000 / 160,000 or $\frac{3}{4}$ of all losses will be collected. The other $\frac{1}{4}$ is not collected; by being underinsured, the policyholder shares the loss and is termed a *coinsurer.*

$$\text{Recovery} = \frac{\text{Insurance carried}}{\text{\% Coinsurance} \times \text{Value of property}} \times \text{Loss}$$

SAMPLE PROBLEM 8

Marvin Fitzgerald has a policy for $70,000 carried on property valued at $100,000 under an 80% coinsurance clause. A loss of $20,000 occurs. How much will he recover?

Solution

Insurance carried = \$70,000
Insurance required = 80% × \$100,000 = \$80,000

$$\frac{70{,}000}{80{,}000} = \frac{7}{8}$$

Mr. Fitzgerald will recover $\frac{7}{8}$ × \$20,000 = \$17,500

Although the loss is less than the face of the policy, the loss is not paid in full because Mr. Fitzgerald carried less than what was required. Had the face of the policy been \$80,000, he would have had full payment.

Coinsurance problems are tricky. Here are some other examples:

(1) Face = \$90,000 Value = \$100,000
Loss = \$95,000 90% coinsurance

$$\text{Recovery} = \frac{\$90{,}000}{90\% \times \$100{,}000} \times \text{Loss} = \frac{\$90{,}000}{\$90{,}000} \times \$95{,}000 = \$95{,}000$$

It seems that there's going to be extra money, but recovery cannot exceed face of policy, so *\$90,000* is recovered.

(2) Face = \$60,000 Value = \$80,000
Loss = \$12,000 60% coinsurance

$$\text{Recovery} = \frac{\$60{,}000}{60\% \times \$80{,}000} \times \text{Loss} = \frac{\$60{,}000}{\$48{,}000} \times \$12{,}000 = \$15{,}000$$

Once again, there's extra money in the air. But remember that recovery cannot exceed the loss; so *\$12,000* is recovered.

TEST YOUR ABILITY

1. The James Candle Corporation is insured with Company A, Company B, and Company C for \$30,000, \$60,000, and \$40,000, respectively. If a loss of \$39,000 occurs, how much will be recovered from each company?
2. Solve each of the following problems in coinsurance by determining the amount recovered by the insured:

	Face of Policy	Coinsurance	Value of Property	Amount of Loss
(a)	\$ 80,000	80%	\$200,000	\$40,000
(b)	\$ 72,000	80%	\$120,000	\$48,000
(c)	\$ 50,000	70%	\$100,000	\$70,000
(d)	\$ 20,000	90%	\$ 20,000	\$10,800
(e)	\$ 40,000	80%	\$ 50,000	\$50,000
(f)	\$150,000	60%	\$250,000	\$15,000
(g)	\$ 25,000	70%	\$ 30,000	\$21,000
(h)	\$ 63,000	70%	\$120,000	\$96,000

Automobile Insurance

Since an automobile can be a source of major injury to life and/or property, almost all states require businesses and people to insure their autos. Even in the absence of law, it is handy to carry automobile insurance. The following discussion covers the various types of coverage available and/or required by different states.

Driver Classification A key factor in determining car insurance premiums is the driver's classification, which is made up of factors such as age, sex, use of car for business or pleasure, distance to work, marital status, experience with driver training, etc. Table 9.3 shows a summary of classifications used in one state. Since car insurance laws are regulated by state, other states may have either more or fewer categories.

Table 9.3

Driver Classifications

Class	Class Definitions
10	Individually owned, no operator under age 25, no business use, not used to or from work over 10 miles one way.
12	Same as Class 10, except used to or from work over 10 miles one way.
15	Same as Class 10, except principal operator over 65.
20	Individually owned, with male operator under 25 not owner or principal operator.
22	Individually owned, with male operator under 25 owner or principal operator.
24	Individually owned, with female operator under 25.
26	Female operator under 25 with driver training.
40	Same as Class 20, but with driver training certificate.
42	Same as Class 22, but with driver training certificate.
50	Male under 25, married.
30	Used in business and not qualifying for any other private passenger classification.
31	Vehicles owned by a corporation, partnership, or unincorporated association.

Liability Insurance *Liability insurance* on cars covers damage to persons other than yourself and to property other than your own. The "10/20/5" formula is typical in liability insurance, representing $10,000, $20,000, and $5,000. The first figure, $10,000, is the maximum payment to any one person for injuries caused in a single accident; $20,000 is the maximum payment to everyone injured in a single accident, other than the driver at fault; and $5,000 is the maximum payment for property damaged in an accident. Sample Problem 9 shows how the 10/20/5 works.

SAMPLE PROBLEM 9

Bill Carson runs into Andrea Ellis's car causing Ellis and her husband injuries. Carson carries 10/20/5 insurance. Ellis is awarded $11,500 for her injuries and $8,000 for her husband's. Damage to Ellis's car is determined to be $4,000. How will Carson and his insurance company pay for (a) personal injury and (b) property damage?

Solution

(a) Personal injury: \$11,500 + \$8,000 = \$19,500, but \$10,000 is the maximum to any one person.
Insurance company pays \$10,000 + \$8,000 = \$18,000
Carson pays \$19,500 − \$18,000 = \$1,500

(b) Property damage: \$4,000
Insurance company pays in full.

Carson could have purchased higher liability insurance. The difference in cost is small enough—certainly less than the \$1,500 that it cost Carson for his insufficient coverage.

Table 9.4 shows typical premiums for liability insurance. An urban area would have higher premiums, just as a rural area would have lower ones; therefore, your own premiums may not agree with the various tables given in this chapter.

Table 9.4

Automobile Liability Basic Insurance Premiums

Driver Class	Bodily Injury				Property Damage		
	10/20	**20/40**	**50/100**	**100/300**	**\$5,000**	**\$25,000**	**\$100,000**
10	\$ 62	\$ 81	\$103	\$118	\$ 43	\$ 46	\$ 51
12	65	84	108	123	45	49	53
15	59	77	99	113	41	44	48
20	165	215	275	315	129	139	152
22	178	232	297	340	140	151	165
24	95	124	158	181	71	77	84
26	81	105	134	164	58	63	68
30	73	95	121	139	73	79	86
31	78	115	147	169	77	83	91
40	114	148	189	217	86	93	101
42	152	198	253	290	118	127	139
50	100	131	167	192	75	81	89

Automobile insurance policies are issued for only one year at a time. Sample Problem 10 shows how to calculate liability premiums.

SAMPLE PROBLEM 10

Calculate the premiums for 50/100/25 liability coverage for each of the following cases:

(1) Joe Sears, 22, single, owner, with driver training.
(2) Mary Walsh, 21, single, owner, no driver training.
(3) A vehicle owned by a corporation.

Solution

(1) Class 42:

50/100	\$253.00
25	127.00
Total	\$380.00

(2) Class 24:	50/100	$158.00
	25	77.00
	Total	$235.00
(3) Class 31:	50/100	$147.00
	25	83.00
	Total	$230.00

Medical Payments

Liability insurance covers the driver's responsibility for medical expenses of other people. *Medical payments coverage* protects the *driver* who has been injured. Table 9.5 shows typical premiums for such protection. Driver classification generally has no effect on this coverage though it is considered in some states.

Table 9.5

Automobile Insurance Medical Payments

Limit	Premium
$ 5,000	$ 6
10,000	10
15,000	13
20,000	15
25,000	17

Collision Insurance

Liability insurance protects against damage to the property of *others.* To protect your own car in an accident, *collision insurance* is necessary. It protects the driver's auto against damages when the driver is at fault—or if the other driver is at fault but uninsured (the other driver may not be able to pay for the damages). Table 9.6 presents a set of collision premiums and requires careful examination. Collision rates depend on (1) driver class; (2) car size; (3) car age; and (4) deductible clause.

Driver class has been discussed. *Car size* is determined by referring to a master table, not available here, in which every automobile is listed and given a size symbol. A small car might be "size 2" and a large car might be "size 7"; the larger a car, the higher the premium. *Car age* is defined in the footnotes to Table 9.6. The newer an automobile, the higher the premium.

It seems that the deductible clause almost slipped by, but it is too important to ignore. The *deductible clause* refers to the amount that the owner has to pay in case of a collision. If a policy states "$200 deductible," the owner pays the first $200 while the insurance company pays the remainder of the repairs. Table 9.6 shows premiums for $200 deductible policies, with a footnote to add 25% for $100 deductible policies. A non-deductible policy would cost too much. Sample Problem 11 displays all of these points.

SAMPLE PROBLEM 11

Alex Boland was 45 years old this past Tuesday. He drives only for pleasure and owns a size 5 car that he bought last year.

(a) Find his collision premium on a $200 deductible policy.

Table 9.6

Automobile Collision Insurance Premiums—$200 Deductible[1]

Size	Age Group[2]	Class 10–12	15	20	22	24	26	30–31	40–50	42
1,2	1	$119	$107	$238	$327	$155	$142	$137	$190	$297
	2,3	89	80	178	245	116	107	103	142	223
	4	77	60	155	213	101	93	89	124	194
3	1	135	122	270	372	176	162	156	216	338
	2,3	102	92	203	280	132	122	117	163	255
	4	87	78	175	240	113	104	101	140	219
4	1	158	142	317	436	206	190	182	254	396
	2,3	119	107	238	327	155	142	137	190	297
	4	103	93	205	283	133	123	118	164	257
5	1	190	171	380	522	247	228	218	304	475
	2,3	142	128	284	392	185	171	164	228	356
	4	123	111	247	339	160	148	142	199	309
6	1	221	199	443	609	288	266	255	355	554
	2,3	167	150	333	458	217	200	192	266	417
	4	144	130	288	396	187	178	166	230	360
7	1	254	229	508	698	330	306	292	406	635
	2,3	190	171	380	522	247	227	218	304	475
	4	165	149	328	453	214	194	189	264	412

[1] For $100 deductible, add 25% to the table figures.
[2] Age Groups:
1 All autos of current year.
2 All autos of preceding year.
3 All autos of year before last.
4 All others.

(b) What would a $100 deductible policy cost?
(c) He chooses the $100 deductible policy and has two accidents, resulting in repairs costing $350 and $90, respectively. Determine his recovery from the insurance company in each accident.

Solution

(a) Size 5, age group 2, class 10, $200 deductible
Collision premium: $142.00
(b) Add 25% to $142.00: $142.00 + $35.50 = $177.50
(c) First accident: $350 − $100 deductible = $250 recovered
Second accident: $ 90 − $100 deductible = $ 0 recovered

Notes 1. Auto premiums are *not* rounded to the nearest dollar.
2. The deductible clause applies each time.

Comprehensive Coverage

Comprehensive insurance covers losses other than collisions to your own car—theft, vandalism, or fire, for example.

Some rates are shown in Table 9.7. Car size, car age, and the deductible clause affect the premium. Driver class has no effect at all on comprehensive coverage.

Table 9.7

Automobile Comprehensive
Insurance Premiums—
$200 Deductible*

	Age Group		
Size	**1**	**2,3**	**4**
1,2	$ 77	$ 58	$ 50
3	102	77	67
4	128	96	83
5	173	129	113
6	224	168	146
7	282	211	183

* For $100 deductible, add 10% to the table figures.

"Good" Driver and "Bad" Driver Rates

Most auto insurance companies charge reduced rates to "good" drivers and high rates to "bad" drivers. Generally, a "good" rate is available after being insured for three years by the same company.

To determine auto liability insurance premiums for "good" and "bad" drivers, find the *basic* premium from Table 9.4. A multiple is then applied to the basic premium. Assume that the following typical percents correspond to the number of serious "accidents" that a driver has had.

Number of Accidents	Percent of Basic Rate
0	90%
1	110%
2	150%
3	200%
4	250%

The term "accident" has its own meaning to every insurance company. It may be a set money amount such as $100 or over.

SAMPLE PROBLEM 12

Len West's basic liability premium is $110.

(a) Find his actual premium for the first three years during which he had no accidents.
(b) Find his fourth-year premium.
(c) During the fourth year, he has three accidents. Find Mr. West's fifth-year premium.

Solution

	Year	Rate	Premium
(a)	1	Basic	$110
	2	Basic	110
	3	Basic	110
(b)	4	90% × $110	99
(c)	5	200% × $110	220

After three years at the "bad" rate of $220, he will return to the basic $110 in year 8, assuming no additional accidents and that he is still driving.

"Good" and "bad" rates always apply to those policies in which driver classification is a factor. Therefore, these rates apply to (1) liability coverage and (2) collision coverage. The rates might apply to (3) medical payments coverage as well. It is so assumed here. The rates do *not* apply to comprehensive coverage.

Sample Problem 13 summarizes what we know of auto insurance premiums. Notice that the comprehensive coverage is added after multiplying by the "bad" rate.

SAMPLE PROBLEM 13

The Valley Flower Shop, owned by Abdulla Bennett, uses a delivery truck. The truck is 2 years old, size 6, and none of its drivers are under 25. In 1987, the firm had one accident after several good years. Find the 1988 car insurance premium for the following coverages: 100/300/100; $25,000 medical payments; $100 deductible collision; and $100 deductible comprehensive.

Solution

Liability (Table 9.4) class 30, 100/300	$139.00
100,000	86.00
Medical payments (Table 9.5) 25,000	17.00
Collision (Table 9.6) class 30,	
size 6, age 3	192.00
+ 25% for $100 deductible	48.00
	$482.00
Rate with one accident	× 1.10
	$530.20
Comprehensive (Table 9.7)	
size 6, age 3	168.00
+ 10% for $100 deductible	16.80
Total premium	$715.00

No-Fault Insurance

One of the newer trends in car insurance is to make payment regardless of who is at fault in an accident. This is called, naturally enough, *no-fault insurance*. In an accident under no-fault, injured persons are paid by the company insuring the car in which they happen to be riding. Had Sample Problem 9 occurred in a no-fault state, Ms. Ellis's insurance company would have paid for the injuries to Mr. and Ms. Ellis.

A number of states have enacted no-fault laws. Several points should be mentioned.

(1) Each state sets a minimum amount of no-fault coverage. A state may, for example, specify $50,000 as the minimum.

(2) Should damage to those in your car exceed your no-fault coverage, the other driver can still be sued. Thus, liability insurance of 10/20 or higher may still be either advisable or required.

(3) No-fault covers *only* personal injury to those in your car and to pedestrians. It does not cover property damage to your or someone else's car,

nor does it take the place of comprehensive insurance. It does, however, save the need for medical payments coverage.

(4) Premium calculation is virtually the same as that shown in this chapter. Insurance companies claim an average reduction in premiums of 16% with no-fault. The truck in Sample Problem 13, if no-fault were used, would have its $139.00 figure and its $17.00 figure replaced by 84% of the total of the two—84% × $156.00 = $131.04, a savings of about $25.00. Of course, a required 10/20 would add slightly to the cost.

TEST YOUR ABILITY

1. Jane Arden is insured for 20/40/10 liability insurance. She has a tough day, and to top it all she runs into Harris's car, causing injury to Mr. Harris and Ms. Harris, as well as damaging their car. Costs of the injuries to Mr. and Ms. Harris are $13,000 and $23,000, respectively. Damage to the Harris's car costs $4,500. Find the amount to be paid by Ms. Arden's insurance company for the accident.
2. Using Table 9.4, determine the *basic* liability premium for auto insurance in each case:

	Age	Sex	Marital Status	Owner	Driver Training	Coverage	Use
(a)	19	M	S	yes	no	50/100/25	pleasure
(b)	35	F	M	yes	no	20/40/25	15 miles to work
(c)	70	M	M	yes	yes	100/300/100	pleasure
(d)	23	M	M	yes	no	10/20/5	pleasure
(e)	21	F	S	yes	yes	20/40/25	10 miles to work
(f)	19	M	S	no	yes	50/100/100	pleasure

3. Use Table 9.6 to find the *basic* collision premium in each instance:

	Class	Size	Purchased	Deductible
(a)	15	3	this year	$200
(b)	26	6	last year	$100
(c)	31	1	two years ago	$200
(d)	42	7	five years ago	$100
(e)	20	5	yesterday	$200
(f)	12	4	three years ago	$100

4. Amy Lane owns a new size 3 auto and wants to buy a $100 deductible collision policy. She is single, 29 years old, and drives 30 miles to work.
   (a) Calculate her basic collision premium.
   (b) An accident costing $720 in damages occurs. How much will her insurance company pay?
   (c) What is the cost for $200 deductible comprehensive for her car?
5. Basic liability and collision premiums for four drivers are tabulated below. Using the "good" and "bad" rates from this chapter and assuming that each has been insured for at least 3 years, find each driver's premium.

	Basic Liability Cost	Basic Collision Cost	Number of Accidents
(a)	$155.00	$156.00	1
(b)	$118.00	$ 60.00	0
(c)	$270.00	$475.00	3
(d)	$444.00	$380.00	2

6. The delivery truck of Miles Piano Corporation is a brand new, size 5 vehicle. The firm has had no accidents for four years and this year wants the following coverage for its truck: 100/300/100; $20,000 medical; $100 deductible collision; $200 deductible comprehensive. Find the firm's total insurance premium.

Life Insurance

The primary purpose of life insurance is to provide money for the *beneficiaries* (designated survivors) upon an insured person's death. There are different types of life insurance policies, and all of them have various benefits. Four of the types and two of the benefits are discussed in this chapter. For a third benefit (cash surrender value), you can refer to Chapter 15.

Types of Policies

The four main types of life insurance are as follows:

Term insurance provides coverage for a specific time period, usually ten years. Term insurance is the least expensive, and premium payments are made throughout its duration. If the insured dies during the term of the policy, the beneficiary receives the face of the policy. Often, a firm will insure its executives under term policies. Such a procedure is called "keyperson" coverage.

Ordinary whole life insurance covers an individual for life and premiums are paid for life. The insured's death entitles the beneficiary to the face of the policy. This type of policy is the second least expensive.

Limited-payment life insurance, the next highest in cost, provides protection for life but limits payment to a specific time period, usually 20 or 30 years. The individual has life-long protection but pays premiums only during wage-earning years.

Endowment insurance is the most expensive type. Premiums are paid over a specified time period. If the individual dies during that time, the beneficiary receives the face of the policy. However, if the insured is still alive at the end of what is usually a 20-year period, that fortunate person gets the face of the policy.

Calculating Premiums

Life insurance premiums vary according to four factors: (1) age; (2) type of policy; (3) face of the policy; (4) frequency of payment.

For a person looking for insurance, it pays to remember that as an individual becomes older, life expectancy becomes less, and the premiums become higher. We have already noted that some types of life insurance cost more than others. The higher the face, the higher the premium, and the more frequent the terms of payment, the higher the annual cost.

Table 9.8 lists typical life insurance premium rates. As you look over the table, it is important to keep four points in mind:

(1) Because of a longer life expectancy, *a female has to subtract three years from her age.* Thus, the premium for a female, age 47, is the same as for a male, age 44.

Table 9.8

Annual Life Insurance Premiums for Males, per $1,000 Face

Age	10-Year Term	Ordinary Whole Life	20-Payment Life	20-Year Endowment
18	$ 3.98	$ 9.98	$18.18	$42.01
19	3.99	10.32	18.60	42.01
20	4.00	10.67	19.03	42.01
21	4.02	11.01	19.46	42.07
22	4.06	11.34	19.89	42.15
23	4.10	11.69	20.32	42.24
24	4.15	12.05	20.78	42.34
25	4.20	12.44	21.27	42.45
26	4.29	12.86	21.79	42.56
27	4.41	13.30	22.34	42.67
28	4.54	13.77	22.91	42.79
29	4.69	14.26	23.51	42.92
30	4.87	14.78	24.12	43.05
31	5.07	15.33	24.75	43.19
32	5.29	15.90	25.40	43.33
33	5.54	16.50	26.06	43.48
34	5.81	17.13	26.76	43.64
35	6.11	17.81	27.48	43.81
36	6.43	18.52	28.23	44.00
37	6.76	19.26	28.99	44.19
38	7.12	20.04	29.79	44.40
39	7.53	20.87	30.62	44.63
40	8.00	21.75	31.50	44.87
41	8.52	22.69	32.43	45.12
42	9.08	23.69	33.41	45.39
43	9.70	24.73	34.43	45.67
44	10.38	25.82	35.46	45.98
45	11.16	26.94	36.51	46.31
46	12.02	28.09	37.55	46.65
47	12.94	29.27	38.60	46.99
48	13.95	30.50	39.68	47.37
49	15.05	31.78	40.80	47.81
50	16.25	33.14	41.98	48.34
51	17.51	34.55	43.21	48.94
52	18.83	36.00	44.48	49.60
53	20.25	37.53	45.81	50.34
54	21.85	39.17	47.21	51.20
55	23.68	40.95	48.72	52.21

Notes (1) For semiannual premiums, use 51% of the annual rate.
(2) For quarterly premiums, use 26% of the annual rate.
(3) For monthly premiums, use 9% of the annual rate.
(4) For females, use the age of a male three years younger.

(2) Rates are expressed *per $1,000* of face in contrast to fire insurance rates which are quoted per $100.

SAMPLE PROBLEM 14

Determine the annual premium for Sally Washington who is 53 and wants a $20,000 ordinary whole life policy.

Solution

Table 9.8: Age 50 (53 − 3), $33.14 per $1,000
Number of thousands: 20
Annual premium: $33.14 × 20 = $662.80

(3) Age is determined by the nearest birthday. Thomas Berwald, who was born on April 7, 1963, is considered to be 24 up to and including October 6, 1987. As of October 7, he becomes 24½ and is considered to be 25 for life insurance purposes.

SAMPLE PROBLEM 15

Horace Marshall was born on February 22, 1958. If he takes out a $15,000, 20-payment life insurance policy on August 26, 1987, what is his annual premium?

Solution

Age as of August 26, 1987 = 30 to the nearest birthday
Table 9.8: Age 30, $24.12 per $1,000
Number of thousands: 15
Annual premium: $24.12 × 15 = $361.80

(4) Premiums may be paid annually at the stated rates, or semiannually, quarterly, or monthly at stated percents of the annual rate. These percents—51, 26, and 9, respectively—appear at the bottom of Table 9.8. The annual premium is rounded to the nearest cent before application of the percents.

SAMPLE PROBLEM 16

Arthur Lattimore, age 37, takes out a $35,000, ten-year term policy.

(a) Find the monthly premium.
(b) How much more a year is it costing by paying monthly premiums?

Solution

(a) Rate: $6.76
Annual premium: $6.76 × 35 = $236.60
Monthly premium: $236.60 × .09 = $21.29

(b) Total premiums for a year: $21.29 × 12 months =	$255.48
Annual premium	−236.60
Additional cost	$ 18.88

Extended Term and Paid-Up Insurance

As we have seen, the primary purpose of life insurance is protection for survivors after the death of the insured person. But what if the insured person discontinues premium payments? Three options are available. The policyholder can:

(1) Convert the policy to cash. After two or three years, all policies other than term insurance have a cash value. Chapter 15 has a lot to say about this option.

(2) Buy *extended term insurance.* Under this option, the policyholder is insured, at no additional cost, for the face of the policy over a limited time period.

(3) Buy *paid-up insurance.* This option provides coverage for life, in an amount far less than the face of the original policy, at no added cost.

Table 9.9 summarizes both options (2) and (3). It applies only to ordinary whole life policies. Similar tables are available for limited-payment life and endowment insurance. The ages given are for males.

Table 9.9

Paid-up and Extended Term Insurance:
Ordinary Whole Life Policies ($1,000 Face)

	5 Years			10 Years		
		Extended Term			Extended Term	
Age at Issue	Paid-up	Years	Days	Paid-up	Years	Days
18	$ 80	8	328	$227	21	265
19	81	9	79	231	21	160
20	86	9	342	235	21	37
25	102	11	98	260	19	109
30	120	11	46	283	16	363
35	136	9	312	306	14	225
40	149	8	116	325	12	89
45	162	6	312	344	10	26
50	175	5	198	360	8	10

	15 Years			20 Years		
		Extended Term			Extended Term	
Age at Issue	Paid-up	Years	Days	Paid-up	Years	Days
18	$358	25	53	$476	25	224
19	364	24	225	482	25	2
20	369	24	22	487	24	138
25	396	21	111	515	21	113
30	423	18	192	542	18	129
35	447	15	281	568	15	181
40	470	13	58	592	12	308
45	494	10	288	620	10	204
50	512	8	213	635	8	155

SAMPLE PROBLEM 17

At age 25, Jerry Matlich takes out a $10,000 ordinary whole life policy on which he pays premiums for 15 years and then decides to stop.

(a) For what additional time period can he be covered for the face of the policy?
(b) For what amount can be insured for the rest of his life?

Solution

(a) Table 9.9: Age at issue, 25; after 15 years.
Extended term: 21 years, 111 days

(b) Paid-up value: $396 per $1,000 × 10 = $3,960

Many other options are open to the insured. Be sure to consult an insurance agent if you want life insurance.

TEST YOUR ABILITY

1. Calculate the annual life insurance premiums in each of the following cases:

	Age	Sex	Type of Policy	Face
(a)	31	M	10-year term	$ 30,000
(b)	19	M	20-year endowment	$ 10,000
(c)	55	F	ordinary whole life	$ 5,000
(d)	39	F	20-year endowment	$100,000
(e)	49	M	ordinary whole life	$ 7,500
(f)	23	F	20-payment life	$ 18,000
(g)	21	M	10-year term	$ 13,500

2. Find the semiannual cost of premiums in Problems 1(a) and 1(b).
3. Find the quarterly cost of premiums in Problems 1(c) and 1(d).
4. Find the monthly cost of premiums in Problems 1(e) and 1(f).
5. In Problem 1(g), determine the difference in annual cost when premiums are paid quarterly instead of annually.
6. Find the extended term and paid-up values for each of the following ordinary whole life insurance policies issued to men:

	Age at Issue	Age at Cancellation	Face
(a)	45	50	$10,000
(b)	30	45	$ 5,000
(c)	20	30	$40,000
(d)	18	38	$25,000

Other Business Insurance

Lloyd's of London became famous by insuring all types of ventures; these days it is easy to insure almost any business risk. For example, *use and occupancy insurance* protects a firm against losses of profits and from operating expenses during periods when the firm is closed because of such causes as fire or a strike. A firm will estimate its daily expenses and profit and then purchase so many days of insurance.

Business liability insurance protects the firm against injuries to people who are either on the firm's property or who use the firm's products. Workers, on the other hand, are covered by *workmen's compensation insurance.*

Flood insurance protects for losses due to hurricanes, river overflows, or other flooding conditions. This type of insurance is mandated by the federal government in certain states with a high risk of flood damage. Because of the very high dollar loss resulting from a major flood, one insurance company is assigned in each state to write the insurance, but all insurers in the state share the premiums and the losses.

Inland marine insurance protects a firm's products during shipment by land, sea, or air.

Group life insurance protects the families of workers and is generally a low-cost term protection paid either by the employer or by wages withheld from the employee. Rates are usually 20 to 80 cents per $1,000.

Group health insurance is also common, taking the form of either a fringe benefit or a payroll deduction. Benefits usually cover doctor fees, hospital expenses, and, in some plans, dental fees.

Burglary insurance provides protection against thefts. Shoplifting, forgery, and robbing a safe fall into the areas covered by this insurance. The premium is greatly affected by the neighborhood in which a firm is located and the type of business—a jewelry store owner will pay a much higher premium than a dry cleaner.

SUMMARY A wise business owner considers two factors in planning to buy insurance: its cost versus its value. The major slant in this chapter has been on calculating premiums, since this is the primary arithmetic area in insurance. However, Sample Problem 18, a chapter summary problem, emphasizes the value of insurance. What happens is unlikely but possible.

SAMPLE PROBLEM 18

The Warner Watch Company carries the following insurance:

Fire: $100,000 on building
  80% coinsurance policy
Burglary: $500
Auto: 10/20/5, $5,000 medical
Use and occupancy: 10 days × $500 a day = $5,000
Liability: $10,000
Inland marine: $5,000

During February, 1989, the following events occur:

1. Watches in transit worth $3,000 are stolen.
2. Robert Jennings, an elderly customer, slips on ice in the parking lot. The fall inflicts $4,000 in injuries on him.
3. The safe is robbed of $750 in cash.
4. Operations are suspended for 15 working days while everyone is on strike.
5. A fire destroys part of the building, which is valued at $150,000 at the time of destruction. A loss of $60,000 is determined.
6. The company truck hits another car, damaging the car, the other driver, and three other passengers. Costs are $10,000 for the driver and for each passenger, totaling $40,000; $1,000 for the car; $2,000 to repair the company truck; $2,000 for major surgery on the company driver. Warner's state does not have no-fault insurance.

Determine the following:

(a) What part of each loss will be recovered from the insurance company?

(b) What part of each loss will be borne by the Warner Watch Company?
(c) What are the totals for (a) and (b)?

Solution

	(a) Recovered	(b) Expense to Warner	Notes
1. Watches in transit	$ 3,000	$ -0-	Adequately covered on inland marine
2. Liability	4,000	-0-	Sufficient liability coverage
3. Burglary	500	250	Insufficient burglary coverage
4. Use and occupancy	5,000	2,500	Loss of 5 days × $500
5. Fire	50,000	10,000	$\frac{\$100{,}000}{80\% \times \$150{,}000} = \frac{5}{6} \times \text{loss}$
6. Personal injury	20,000	20,000	10/20; $20,000 maximum
Property damage	1,000	-0-	Adequate coverage
Collision	-0-	2,000	None carried
Medical payments	2,000	-0-	Sufficient coverage
(c)	$85,500	$34,750	

Do you think the Warner Watch Company should have carried more insurance, or was it running a reasonable risk? (Or do you think they should start a new business?)

CHECK YOUR READING

1. List and define the eleven types of business insurance discussed in this chapter.
2. What premium is paid for a fire insurance policy if a computation shows (a) $9.34; (b) $15.74?
3. What two factors are considered in using a fire insurance premium table?
4. Why are bargain rates given for two-year and three-year fire insurance policies?
5. Explain the two uses of the short-rate table.
6. If the fire insurance policy of the Marble Greeting Card Shop has a face of $20,000, determine recovery if it has a loss of (a) $10,000; (b) $20,000; (c) $30,000.
7. Under an 80% coinsurance clause: (a) What insurance should be carried on property worth $75,000? (b) If $40,000 is actually carried, what fraction of each loss will the insurance company pay?
8. Determine the driver classification for each person below:
   (a) male, 23, married, pleasure use
   (b) male, not owner, single, 24, driver training
   (c) female, 23, no driver training, single
   (d) male, 70, pleasure, married
   (e) female, 32, 5 miles to work
9. What is meant by "10/20/5"?

10. Compare fire and auto premium rates in terms of (a) policies for more than a year and (b) rounding to the nearest dollar.
11. What four factors affect a collision insurance premium?
12. Which of the factors listed in Question 11 is eliminated when considering comprehensive premiums?
13. Distinguish between a *basic* rate and an *actual* rate for automobile insurance.
14. To what type of coverage do "good" and "bad" driver rates not apply?
15. Distinguish among the four types of life insurance policies discussed in terms of (a) the number of annual premium payments; (b) years of coverage; (c) return of the face to the insured.
16. On an ordinary whole life policy for $1,000, how much less a year is the premium for a 22-year old woman than for a 22-year old man?
17. Find the age to the nearest birthday for Margarita Gonzalez, who was born on May 11, 1959 as of (a) May 11, 1988; (b) October 17, 1988; (c) November 12, 1987.
18. Why is the annual cost for life insurance higher when premiums are paid more frequently?

EXERCISES *Group A*

1. Determine Zelma Grimes's annual premium for a fire insurance policy on her frame building insured for $37,500 and its contents for $10,000 in Territory C.
2. Determine the *basic* liability premiums on auto insurance for each of the following:

	Age	Sex	Marital Status	Owner	Driver Training	Coverage	Use
(a)	19	M	S	yes	yes	20/40/25	pleasure
(b)	35	F	S	yes	yes	50/100/100	15 miles to work
(c)	55	M	M	no	no	100/300/25	corporation
(d)	24	F	S	yes	yes	50/100/100	pleasure
(e)	67	F	M	yes	no	10/20/5	pleasure
(f)	27	M	S	yes	no	100/300/100	5 miles to work

3. Milton Katz owns a four-year-old auto, size 3, and is a class 12 driver. He wants to purchase a $100 deductible collision policy and a $100 deductible comprehensive policy.
   (a) Calculate his basic annual collision cost.
   (b) Calculate his comprehensive cost.
   (c) What will be recovered from his insurance company if $740 of damage is caused to his car in an accident?
4. Find the premiums for each of the following short-rate fire insurance policies:

	Face of Policy	Time	Annual Premium Rate
(a)	$ 32,000	4 months	$0.72
(b)	$ 70,000	20 days	$0.91
(c)	$110,000	65 days	$0.425

5. Determine the annual life insurance premium in each of the following cases:

	Age	Sex	Type of Policy	Face
(a)	57	F	10-year term	$ 8,000
(b)	45	M	ordinary whole life	$ 15,000
(c)	36	M	10-year term	$100,000
(d)	31	F	20-year endowment	$ 10,000
(e)	42	M	20-payment life	$ 7,500
(f)	35	M	ordinary whole life	$ 2,500
(g)	26	F	20-payment life	$ 15,500

6. Find the semiannual premium cost in 5(a) and 5(b).
7. Find the quarterly premium cost in 5(c) and 5(d).
8. Find the monthly premium cost in 5(e) and 5(f).
9. In Exercise 5(g), find the annual savings in premium cost by paying annually instead of semiannually.
10. Ted Mecking decides to cancel his policy for fire insurance on May 10, 1988. A premium of $60 was paid when he took out the policy on March 26. Determine the amounts (a) kept by the insurance company and (b) refunded to Mr. Mecking.
11. Using the figures from Exercise 10, but assuming cancellation by the insurance company, determine the refund.
12. Solve each of the following problems in coinsurance to determine the amount recovered by the insured:

	Face of Policy	Coinsurance	Value of Property	Amount of Loss
(a)	$100,000	80%	$150,000	$60,000
(b)	$ 48,000	80%	$ 90,000	$33,000
(c)	$ 54,000	60%	$100,000	$60,000
(d)	$ 30,000	90%	$ 30,000	$15,000
(e)	$ 28,000	70%	$ 40,000	$40,000
(f)	$ 42,000	70%	$ 60,000	$35,000
(g)	$200,000	60%	$300,000	$10,000
(h)	$ 75,000	80%	$100,000	$80,000

13. Find the *basic* collision premium in each of the following cases:

	Class	Size	Purchased	Deductible
(a)	42	1	last year	$200
(b)	26	5	two years ago	$100
(c)	22	7	today	$200
(d)	10	2	last week	$200
(e)	15	4	four years ago	$100
(f)	31	6	this year	$100

14. Calculate the annual fire insurance premium for each building:

	Face of Policy	Rate
(a)	$50,000	$0.20
(b)	$71,250	$0.305
(c)	$14,500	$0.525

15. Find the extended term and paid-up value in each of the following men's ordinary whole life policies:

	Age at Issue	Age at Cancellation	Face
(a)	35	50	$ 20,000
(b)	20	25	$ 50,000
(c)	18	28	$ 5,000
(d)	50	70	$100,000

16. Determine the total premium for fire insurance on each of the following properties:

	Face of Policy	Construction	Territory	Time
(a)	$ 5,000	Frame	D	2 years
(b)	$ 17,000	Brick	A	3 years
(c)	$225,500	Frame	F	3 years

17. The Anderson Corporation's business delivery truck, size 5, was purchased last year. The firm's coverage is 50/100/25; $10,000 medical; $200 deductible collision; and $100 deductible comprehensive. Determine this year's automobile insurance premium, assuming two accidents last year.
18. Belknap Company's $200,000 building is insured as follows: Company X, $90,000; Company Y, $50,000; and Company Z, $60,000. How much will each company pay toward a loss of $40,000?
19. Determine the annual premium for fire insurance on each of the following properties:

	Face of Policy	Construction	Territory
(a)	$ 14,000	Brick	C
(b)	$ 36,000	Frame	A
(c)	$210,000	Brick	E

20. Selma Juaz holds a 50/100/25 liability policy on her car. She runs into Rick Carter's car causing injury to Carter and his wife for $55,000 and $25,000, respectively. Carter's car is also damaged to the extent of $4,000. How much will Juaz's insurance company pay in this state without no-fault insurance?
21. Phillips Tools insures its brick building in Territory D from October 14 to December 3. Calculate the premium for this $70,000 building.
22. Assuming that each individual has been insured for at least three years, calculate the cost of liability and collision insurance in each case:

	Basic Liability Cost	Basic Collision Cost	Accidents
(a)	$181	$190	1
(b)	$139	$104	4
(c)	$304	$508	0
(d)	$161	$221	2

Group B

1. The Weinberg Glass Manufacturing Company is housed in a modern brick building in Territory B. The firm takes out a $150,000, one-year fire insurance policy on March 16, 1989. On June 4, 1989, the firm reinsures with another company and cancels the first policy. Calculate the amount of the refund from the first company.
2. Determine the recoveries from each insurer in the following coinsurance problems:

	Face of Policy				Value of	
	Co. A	Co. B	Co. C	Coinsurance	Property	Loss
(a)	$40,000	$60,000	$20,000	80%	$150,000	$150,000
(b)	$21,000	$14,000	$ 7,000	70%	$ 70,000	$ 17,500
(c)	$60,000	$30,000	$ 6,000	80%	$110,000	$ 66,000
(d)	$10,000	$20,000	$10,000	60%	$100,000	$ 39,000

3. How much more does 100/300/25 basic liability insurance cost Sylvia Delibes, a class 26 driver, than does 50/100/5?
4. For Jeff Warburg, a class 42 driver, compare the basic costs of collision insurance on a new size 4 auto versus a two-year-old size 5 car.
5. Harriet Sanders, a class 12 driver, carries 20/40/25 insurance on her car.
   (a) Determine the premium for the first three years, during which no accidents occur.
   (b) What is the fourth-year premium?
   (c) There is one accident in the fourth year. What is the fifth-year premium?
   (d) Assuming no additional accidents, find the sixth- and seventh-year premiums.
   (e) What is the eighth-year premium?
6. Determine the annual premiums in each of the following policies:

	Date of Birth	Date of Policy	Sex	Type of Policy	Face of Policy
(a)	May 17, 1967	July 12, 1989	M	10-year term	$ 18,000
(b)	Aug. 9, 1954	Nov. 18, 1989	F	ordinary whole	$ 46,000
(c)	June 12, 1952	April 11, 1989	M	20-payment life	$150,000
(d)	Nov. 19, 1943	May 20, 1988	F	20-year endowment	$ 20,000

7. Ignazio Szanto, age 35, takes out a ten-year term policy. If the face of his policy is $15,000, calculate the savings over ten years by paying annual premiums instead of monthly premiums.

CHAPTER **10**

Depreciation, Other Expenses, and Business Profits

Payroll and insurance are the major but not the only areas of business expenses that we should study. Depreciation is also important because it involves a specific application of business mathematics. This chapter will also make other expenses familiar to you.

Expenses have their ultimate effect on a company's net profit or net loss—in other words, they determine whether a company flourishes or fails. In this chapter, therefore, we will take a closer look at the topic of business profits introduced in Chapter 6.

Your Objectives

After studying Chapter 10, you will be able to

1. Compute depreciation expense by several methods.
2. Prepare a depreciation schedule.
3. Determine the gain or loss on the sale of a tangible asset.
4. Calculate the amounts of expenses for amortization, overhead, bad debts, and property tax.
5. Calculate the costs of administrative functions.
6. Compute net profit or loss for two different types of business firms.

Depreciation

Properties of value owned by a business are called *assets.* Some assets are classified as *tangible,* that is, capable of being touched. Buildings, equipment, furniture, fixtures, and cars are examples of tangible assets.

Tangible assets are purchased for use in the business. As time passes, however, the usefulness of an asset declines, eventually leading to its replacement. The asset may appear to be unchanged, but because of "wear and tear", or technological advances, the asset may be less valuable. The expense that results from the loss of usefulness of an asset is called *depreciation* or *depreciation expense*.

You should be familiar with certain terms before studying the mathematics of depreciation.

Cost of an asset is the total expenditure necessary to purchase the asset and get it into operation. For example, the cost of a car would not only include the "list" price of the car but also other charges for such things as sales taxes, freight, and special options. Likewise, the cost of a large computer would not only include the price of the computer, but also other charges necessary to get it ready for use—special floors or foundations, special wiring, and fees paid to a factory representative to train the company's employees to use the computer.

Estimated useful life (EUL) is the projected length of time an asset will be held for use by a business. EUL can be expressed in years, hours, miles, or capacity for output.

Salvage value, or *scrap value,* is an estimate of what an asset will be worth at the end of its life. What could you get for your car in four years? $1,000, $100, or perhaps, nothing. *Residual value* is another synonym.

Accumulated depreciation is the total depreciation on an asset from the time of its purchase to the time in question. An asset with an annual depreciation of $1,000 will have, after four years, an accumulated depreciation of $4,000.

Book value is the difference between the cost of an asset and its accumulated depreciation. If the asset mentioned in the preceding paragraph had cost $10,000, its book value after four years would be $6,000 ($10,000 − $4,000).

Several methods of calculating depreciation are used in practice. The five most common will be considered here. The cost and life of an asset are needed for all of these methods; the salvage value is needed for most.

Equal Expense Methods

Two of the most common methods of depreciation work on the assumption that the expense for age or use should be spread equally over the life of the tangible asset. These methods are the *straight-line method* and the *units-of-production method.*

The Straight-Line Method

The most popular method for computing depreciation has traditionally been the *straight-line method.* Use of this method assumes an equal amount of depreciation for each full year an asset is used. The popularity of the straight-line method can be attributed to its simplicity. The following formula is easy to use and remember:

$$\text{Annual depreciation} = \frac{C - S}{L}$$

where C = cost; S = salvage value; L = life in years.

SAMPLE PROBLEM 1

Bay Oil bought a truck costing $9,700, which is to be depreciated by the straight-line method. It is estimated that the truck will last for six years, and will have a salvage value of $700. Find (a) the annual depreciation; (b) the book value after the first year; (c) the book value after two years.

Solution

(a) $\frac{C - S}{L} = \frac{\$9{,}700 - \$700}{6 \text{ years}} = \frac{\$9{,}000}{6 \text{ years}} = \$1{,}500$ a year

(b) Book value = Cost − Accumulated depreciation
1st year = $9,700 − $1,500 = $8,200

(c) 2nd year = $9,700 − ($1,500 + $1,500) = $6,700
Or $8,200 − $1,500 = $6,700

Note Accumulated depreciation is subtracted from *cost* ($9,700) to find book value. A very common *error* is to subtract from $9,000.

In order to quickly see the book value of an asset at any point in time, a *depreciation schedule* can be prepared. The form shows the annual depreciation, accumulated depreciation, and book value for each year of an asset's life. The schedule for Bay Oil's truck follows.

Depreciation Schedule

Straight-Line Method
C = $9,700; S = $700; L = 6 years

Year	Annual Depreciation	Accumulated Depreciation	Book Value
			$9,700
1	$1,500	$1,500	8,200
2	1,500	3,000	6,700
3	1,500	4,500	5,200
4	1,500	6,000	3,700
5	1,500	7,500	2,200
6	1,500	9,000	700

Notes (1) The final book value equals the salvage value *in this method.*
(2) On each line, accumulated depreciation + book value equals the cost ($9,700).

Full year depreciation (as calculated above for Bay Oil) should only be used when an asset has been in use for the entire year. When an asset has been in use for only part of a year, it is necessary to calculate depreciation for the period of actual use only. For example, an asset purchased on May 1 would be depreciated for an eight-month period—May 1 to December 31—in the year of purchase. In other words, since the asset was only used for eight months, we can only take 8/12 of the annual depreciation.

Calculations are rounded to the nearest whole month. Thus, we use the following rules:

1. If an asset was purchased on or before the 15th, count the whole month.
2. If an asset was purchased after the 15th, disregard the month completely and start counting from the first day of the following month. For example:

Date of Acquisition	Months of Depreciation in First Calendar Year
January 13	12
January 15	12
January 16	11
July 24	5
December 16	0

SAMPLE PROBLEM 2

A conveyor belt is purchased by Ralston Electronics on September 7, 1988. Its cost is $7,400; life, eight years; salvage value, $0. Using the straight-line method, find (a) the depreciation for 1988; (b) the book value at the end of 1988; (c) the depreciation for 1989; (d) the book value at the end of 1989.

Solution

(a) $\dfrac{C - S}{L} = \dfrac{\$7{,}400 - \$0}{8 \text{ years}} = \dfrac{\$7{,}400}{8 \text{ years}} = \$925$ a year

September 7 to December 31 = 4 months

1988 depreciation = 4/12 or 1/3 × $925 = $308.33

(b) Book value, end of 1988 = $7,400 − $308.33 = $7,091.67

(c) Depreciation for 1989 = $925 for a full year

(d) Book value, end of 1989 = $7,400.00 − ($308.33 + $925.00)
= $6,166.67.

The Units-of-Production Method

The straight-line method is a method of computing depreciation based on time, that is, how long an asset is expected to remain useful. Another way to view the life of an asset is in terms of its output. We can say that an asset will last eight years; we can also say that an asset will:

produce 10,000 units, or
run for 30,000 hours, or
run for 50,000 miles

Each of these expressions is a variation of methods that assume *equal depreciation per unit, hour, or mile.*

We find depreciation *per unit* by calculating (C − S)/L, using the number of units (hours, miles) as L. Annual depreciation is then the product of the units produced multiplied by the rate per unit.

SAMPLE PROBLEM 3

A machine is estimated to produce 40,000 units. If its cost is $8,600 and its salvage value is estimated at $600, find (a) depreciation per unit, and (b) depreciation for the first year if 8,200 units are produced, assuming the units-of-production method.

Solution

(a) $\dfrac{C - S}{L} = \dfrac{\$8{,}600 - \$600}{40{,}000} = \dfrac{\$8{,}000}{40{,}000} = \$0.20$ per unit

(b) Annual depreciation = 8,200 units × $0.20 = $1,640

An interesting question is the following: What if the estimate of 40,000 units is incorrect? What if we actually produce 42,000 units, as follows:

First year	8,200 units
Second year	8,900 units
Third year	8,400 units
Fourth year	8,600 units
Fifth year	7,900 units
Total	42,000 units

The answer is to *adjust the final year's depreciation* so that the final book value is the salvage value. Notice this in the depreciation schedule below.

Depreciation Schedule

Units-of-Production Method
C = $8,600; S = $600; L = 40,000 units ($0.20 per unit)

Year	Units Produced	Annual Depreciation	Accumulated Depreciation	Book Value
				$8,600
1	8,200	$1,640	$1,640	6,960
2	8,900	1,780	3,420	5,180
3	8,400	1,680	5,100	3,500
4	8,600	1,720	6,820	1,780
5	7,900	1,180*	8,000	600

* According to the formula, actual depreciation for the fifth year should have been $0.20 × 7,900 = $1,580, but to use the amount would reduce the final book value to $200, which is not the planned figure.

TEST YOUR ABILITY

1. Determine the annual straight-line depreciation on each of the following assets:

	C	S	L
(a)	$ 7,000	$800	20 years
(b)	$11,000	$-0-	10 years
(c)	$47,200	$600	8 years

2. The F. M. Bakery buys a van, costing $3,650, with an estimated salvage value of $650. It is to be depreciated by the straight-line method over an estimated useful life of six years. Find (a) the annual depreciation; (b) the book value after one year; (c) the book value after two years.
3. Prepare a depreciation schedule for Problem 2.
4. Determine the depreciation for 1988 on each of the following assets by using the straight-line method.

	C	S	L	Date of Purchase
(a)	$14,100	$ 600	10 years	September 3, 1988
(b)	$12,000	$ -0-	20 years	March 17, 1988
(c)	$50,000	$5,000	15 years	July 24, 1988

5. A rock group purchases amplifiers on August 4, 1987. C = \$9,000; S = \$900; L = 9 years. Using the straight-line method, find (a) the depreciation for 1987; (b) the book value at the end of 1987; (c) the depreciation for 1988; (d) the book value at the end of 1988.
6. The Morton Company depreciates its tangible assets by the units-of-production method. A machine costing \$35,000 with an estimated salvage value of \$3,000 has an expected life of 80,000 units. Find the depreciation for each year if production figures are as follows: 1986, 24,000 units; 1987, 30,000 units; 1988, 11,000 units; 1989; 15,000 units.
7. A machine, costing \$40,000 with no salvage value, has an estimated life of 8,000 hours. Actual hours used for 1986–1988 are 2,700, 2,620, and 2,410, respectively. Prepare a depreciation schedule, using the units-of-production method and assuming that the machine is not used after 1988.

Decreasing Expense Methods

In the discussion so far, we have looked at two depreciation methods that assume a constant rate of depreciation. In the discussion that follows, we will look at two methods of depreciation that do not. The *sum-of-the-years' digits method* and the *declining-balance method* are both based on a realistic thought: many of our tangible assets depreciate more rapidly in the early years of their life than they do in later years. In other words, the bulk of depreciation will occur in the early years of an asset's life, and the amount of depreciation will decrease as the asset ages. Calculating the *decreasing expense methods* is more complex than calculating the equal expense methods.

The Sum-of-the-Years' Digits Method

The sum-of-the-years' digits method is an arithmetic device designed to provide greater depreciation estimates in the early years of an asset's life and smaller estimates in later years.

A series of fractions is developed and applied to C − S. The *numerators* of these fractions are based on the number of years the asset is expected to last. For example, if the estimated useful life of an asset is five years, there will be *a first year, a second year, a third year, a fourth year, and a fifth year* in the life of the asset. We arrange these digits in reverse order to obtain our *numerators:* 5, 4, 3, 2, 1.

To obtain our denominator we add the digits $(5 + 4 + 3 + 2 + 1 = 15)$. Thus, the first year's fraction will be $\frac{5}{15}$ and the last year's, $\frac{1}{15}$.

SAMPLE PROBLEM 4

Given C = \$10,000; S = \$2,000; L = 4, find each year's depreciation by the sum-of-the-years' digits method.

Solution

Sum of $4 + 3 + 2 + 1 = 10$; C − S = \$8,000

First year	$\frac{4}{10} \times \$8,000$	= \$3,200
Second year	$\frac{3}{10} \times \$8,000$	= 2,400
Third year	$\frac{2}{10} \times \$8,000$	= 1,600
Fourth year	$\frac{1}{10} \times \$8,000$	= 800
Totals	$\frac{10}{10}$	\$8,000

Notice that each year's depreciation is \$800 less than the previous year's—a constant decrease. For all calculations under this method, the constant decrease will always hold true, assuming that depreciation is being calculated for full years.

What if the life of the asset were 20 years? How could we sum the digits? We could add 20 + 19 + 18... + 1 (a lengthy and dull procedure) or apply the speedy formula:

$$\text{Sum} = \frac{N \times (N + 1)}{2}$$

N is the asset's life. For 20 years:

$$\text{Sum} = \frac{20 \times 21}{2} = 210$$

The Declining-Balance Method

The *declining-balance method* also assumes that there will be greater depreciation in early years and less in later years. This method applies a *constant rate* to each year's book value. If an item costs \$10,000 and a rate of 20% is chosen, the first year's depreciation is:

$$20\% \times \$10{,}000 = \$2{,}000$$

The book value of the asset at the end of the first year is \$8,000 (\$10,000 – \$2,000). The second year's depreciation is:

$$20\% \times \$8{,}000 = \$1{,}600$$

The third year's depreciation will be 20% of (\$8,000 – \$1,600) = 20% × \$6,400.

The rate chosen is generally *twice the straight-line rate.* To find a *straight-line rate,* divide 100% by the years of life. For a five-year life, 100% ÷ 5 = 20% a year straight-line rate. The declining-balance rate is then determined by doubling the straight line rate: 20% × 2 = 40%. Presented below are some common straight-line and declining-balance rates.

Life in Years	Straight-Line Rate	Declining-Balance Rate
4	25%	50%
5	20%	40%
6	$16\frac{2}{3}$%	$33\frac{1}{3}$%
10	10%	20%
20	5%	10%
40	$2\frac{1}{2}$%	5%
50	2%	4%

In the declining-balance method, the rate is applied to cost in the first year, *not* cost minus the salvage value. Salvage value is ignored until the final year when it is usually necessary to adjust the last year's depreciation so that the final book value equals the salvage value.

SAMPLE PROBLEM 5

Given the following data, prepare a depreciation schedule using the declining-balance method: C = \$12,000; S = \$1,100; L = 6 years.

Solution

Straight-line rate = 100% ÷ 6 = $16\frac{2}{3}$%
Declining-balance rate = $16\frac{2}{3}$% × 2 = $33\frac{1}{3}$%
First year = $12,000 × $33\frac{1}{3}$% (use $\frac{1}{3}$) = $4,000

Depreciation Schedule

Declining-Balance Method
C = $12,000; S = $1,100; L = 6 years ($33\frac{1}{3}$% rate)

Year	Annual Depreciation	Accumulated Depreciation	Book Value
			$12,000.00
1	$4,000.00	$ 4,000.00	8,000.00
2	2,666.67	6,666.67	5,333.33
3	1,777.78	8,444.45	3,555.55
4	1,185.18	9,629.63	2,370.37
5	790.12	10,419.75	1,580.25
6	480.25*	10,900.00	1,100.00

* Found by subtracting the salvage value from fifth year's book value.

As stated above, salvage value is not considered when calculating depreciation under the declining-balance method. However, the asset should not be depreciated below a reasonable salvage value. In Sample Problem 5, the actual calculation for the sixth year is $1,580.25 × $33\frac{1}{3}$% = $526.75. Using this amount, however, would have left us with a final book value of $1,053.50, which was not the planned salvage value.

If the asset in Sample Problem 5 had been acquired on October 1 of the first year, the calculation would have to reflect the use of the asset for only part of a year. Thus, the calculation would have been $12,000 × $33\frac{1}{3}$% = $4,000 × $\frac{3}{12}$ = $1,000. The second year would have been $33\frac{1}{3}$% × $11,000.

Accelerated Cost Recovery System (ACRS)

In 1981, Congress passed the 1981 Economic Tax Recovery Act. This Act provided substantial changes in the way depreciation is calculated. The *ACRS method* replaced the old depreciation system for most property placed in service after 1980. However, the old methods still apply to assets placed in service before 1980. Because ACRS is *only* for federal tax purposes, businesses are free to use the old systems when preparing financial reports. Furthermore, since ACRS results in an accelerated rate of depreciation, not all states have adopted it. Georgia, for example, does not permit the use of ACRS. Thus, under current law, businesses located in Georgia have to make two sets of calculations for depreciation, one set for federal tax purposes, and another for their state tax returns.

ACRS is the simplest of the depreciation methods. It ignores salvage value and does not require that depreciation be prorated for assets placed in service during the year. Assets that qualify for ACRS treatment are called "*recovery property.*" Assets that are not classified as recovery property must be depreciated by the old methods.

Under ACRS, tangible assets are grouped into four classes. The classes are based on the *recovery period* of the assets, i.e., the life of the asset. Table 10.1 shows the four recovery periods and their corresponding asset classes.

Table 10.1

ACRS Depreciation Guidelines

Recovery Period	Type of Property
3 years	*3-year property* includes non-real property such as cars, light-duty trucks, and machinery and equipment used in research and experimentation.
5 years	*5-year property* includes most kinds of production machinery, office furniture and equipment, heavy-duty trucks, and other machinery.
10 years	*10-year property* includes some real property, railroad tank cars, residential mobile and prefab homes, and certain coal burning equipment.
15 years	*15-year property* includes public utility property and most forms of real estate.

Calculating Depreciation under ACRS

Depreciation is calculated under ACRS by simply multiplying the cost of an asset by a statutory percentage. *No depreciation is allowed in the year in which an asset is disposed of* if it is discarded before its cost is fully recovered. Recovery percentages are shown below in Table 10.2.

Table 10.2

ACRS Depreciation Rates
(*Applicable Percentages for Class Properties*)

If the Recovery Year is	1	2	3	4	5	6	7	8	9	10	11	12	13	14	15
3-year	25	38	37												
5-year	15	22	21	21	21										
10-year	8	14	12	10	10	10	9	9	9	9					
15-year public utility	5	10	9	8	7	7	6	6	6	6	6	6	6	6	6

To see how calculations are made using ACRS, let's look at an automobile—a three-year property. Referring to the above table we see that an automobile is depreciated 25% in the first year, 38% in the second year, and 37% in the third year. Sample Problem 6 illustrates the ACRS method.

SAMPLE PROBLEM 6

An automobile is purchased in 1987 for $9,000. Calculate the annual depreciation using the ACRS method.

Solution

Year	Calculation		Depreciation
1	$9,000 × .25	=	$2,250.00
2	$9,000 × .38	=	3,420.00
3	$9,000 × .37	=	3,330.00
			$9,000.00

To allow for assets being placed into service during the year, the rate for the first year (25%) is lower than the rates for the second and third years. At the end of the third year, the amount of accumulated depreciation equals the cost of the automobile. Thus, the cost of the asset is said to be "recovered."

Sample Problem 7 illustrates the preparation of a 5-year depreciation schedule under the ACRS method. Notice that salvage value is ignored.

SAMPLE PROBLEM 7

Randle Company acquires office equipment for $17,500 in the current year and estimates a salvage value of $1,240. Prepare a depreciation schedule using schedule under the ACRS method. Notice that salvage value is ignored.

Solution

Year	Rate	Annual Depreciation	Accumulated Depreciation	Book Value
				$17,500
1	15%	$2,625	$ 2,625	14,875
2	22%	3,850	6,475	11,025
3	21%	3,675	10,150	7,350
4	21%	3,675	13,825	3,625
5	21%	3,675	17,500	-0-

The rates used for computing depreciation under ACRS are subject to change by Congress. At this writing, legislation is under consideration that would lower the rates. Lowering the rates would still permit writing off the cost of assets; however, it would take a longer period of time to do so.

Gains and Losses on the Sale of Tangible Assets

Tangible assets are often sold when they are no longer needed by a business. The importance of keeping accurate schedules of depreciation becomes clear when a firm is ready to sell one of its tangible assets.

Gain or loss on the sale is the difference between the cash received for the asset and its book value at the time of sale. Thus, a $500 gain results when an asset with a $7,000 book value is sold for $7,500. Conversely, a $300 loss results when an asset with a $5,800 book value is sold for $5,500. Sample Problem 8 shows this procedure in detail. Notice the reverse procedure for counting months in the case of a sale—July is not counted.

SAMPLE PROBLEM 8

An asset is acquired on April 4, 1986, by the Randolph Shirt Company, at a cost of $6,300. Its life is estimated at eight years, its salvage value at $700. It is sold on July 5, 1988 for $3,700. Assuming the straight-line method, find (a) depreciation for 1986; (b) depreciation for 1987; (c) depreciation for 1988; (d) book value on July 5, 1988; (e) gain or loss on the sale.

Solution

(a) $\frac{C - S}{L} = \frac{\$6{,}300 - \$700}{8 \text{ years}} = \$700$ a year

April 4 to December 31 = 9 months

$\frac{9}{12} \times \$700 = \$525$ for 1986

(b) In 1987 the asset was used for a full year. Depreciation = $700.

(c) 1988: January 1 to July 5 = 6 months (July is omitted since the asset was used for less than half of the month) $\frac{6}{12} \times \$700 = \$350$

(d) Book value = $6,300 − ($525 + $700 + $350) = $4,725

(e)

Cash received	$3,700
Book value	−4,725
Loss on sale	($1,025)

TEST YOUR ABILITY

1. Equipment with a life of nine years, salvage value of $500, and cost of $9,500 is to be depreciated by the sum-of-the-years' digits method. Find (a) the denominator of the fraction; (b) the numerator of the fraction for the first year; (c) the first year's depreciation; (d) the book value after one year.
2. Prepare a depreciation schedule for Problem 1.
3. For each of the following lives, find the declining-balance rate of depreciation at twice the straight-line rate.
   (a) 12 years (c) 25 years
   (b) 30 years (d) 7 years
4. An asset costing $25,000, L = 4 years, S = $2,000, is to be depreciated by the declining-balance method. Prepare a depreciation schedule.
5. Furniture is purchased for $6,200 in 1986. Using ACRS, calculate depreciation for (a) 1986 and (b) 1987.
6. Using the ACRS method prepare a depreciation schedule for a 3-year asset acquired for $30,000 with a salvage value of $2,500.
7. In each of the following problems, find, first, the book value and, next, the gain or loss on the sale.

	Cost	Accumulated Depreciation	Sales Price
(a)	$ 7,500	$ 3,400	$4,200
(b)	$21,600	$19,400	$2,800
(c)	$14,350	$ 5,910	$6,020

8. Equipment with a life estimate of eight years and a salvage estimate of $600, is purchased by the Lousville Developmental Firm for $7,000. It is sold after $3\frac{1}{2}$ years of use for $4,100. Find the gain or loss on the sale.

Other Business Expenses

There are many other types of business expenses besides payroll, insurance, and depreciation. Let's look over five common ones.

Amortization of Intangible Assets

As we have learned, physical assets depreciate. On the other hand, long-lived, nonphysical (intangible) assets *amortize*. *Intangible assets* include such items as patents, copyrights, and trademarks, all of which fit the description of being long-lived, but not physical.

Patents protect ownership rights to processes; copyrights, to written material; and trademarks, to names or symbols of products. Patents are granted for 17 years, copyrights for the author's lifetime plus 50 years, and trademarks for varying periods of time. The cost of each is spread over its maximum lifetime, or useful period if it is less, by the process of amortization. Most amortization is *straight-line*.

SAMPLE PROBLEM 9

A newly issued patent for a milking machine is purchased for $34,000. Find the annual straight-line amortization over its maximum lifetime.

Solution

Maximum life = 17 years
Annual amortization = $34,000 ÷ 17 years = $2,000

General Overhead

The everyday items of running a business—heat, light, power, water, sanitation, rent, repairs, maintenance, and others—can be grouped as *general overhead* expenses.

The most common math problems concerning overhead are the assignment, or *allocation*, of total overhead to departments within a firm and to units of production.

Consider a firm with five departments—sales, production, accounting, personnel, and general office. If the total general overhead is $200,000, how much should be each department's responsibility? One solution is to divide by 5 and assign $40,000 to each department. But, departments vary in size and use different amounts of heat, light, and so forth. A more common procedure is to have the expenses based on area—square footage. Thus if the sales department occupies $\frac{1}{4}$ of the firm's total area, it will be assigned $\frac{1}{4}$ of the general overhead expenses.

SAMPLE PROBLEM 10

The Richmond Tool Company's total overhead expenses are $180,000. Divide this amount over its four departments, based on the square feet given:

Dept. 1	40,000 sq. feet
Dept. 2	100,000 sq. feet
Dept. 3	40,000 sq. feet
Dept. 4	60,000 sq. feet

Solution

Total area: 40,000 + 100,000 + 40,000 + 60,000 = 240,000 sq. feet

$$\text{Dept. 1} \quad \frac{40{,}000}{240{,}000} = \frac{1}{6} \times \$180{,}000 = \$\ 30{,}000$$

$$\text{Dept. 2} \quad \frac{100{,}000}{240{,}000} = \frac{5}{12} \times \$180{,}000 = \quad 75{,}000$$

$$\text{Dept. 3} \quad \frac{40{,}000}{240{,}000} = \frac{1}{6} \times \$180{,}000 = \quad 30{,}000$$

$$\text{Dept. 4} \quad \frac{60{,}000}{240{,}000} = \frac{1}{4} \times \$180{,}000 = \quad 45{,}000$$

Total $180,000

Note The total serves as a check.

A manufacturer also needs to know the overhead cost per unit produced. If Department 1 of the Richmond Tool Company in Sample Problem 10 produced 60,000 pliers, the cost per unit is found by the following calculation:

$$\text{Unit overhead cost} = \frac{\text{Department overhead}}{\text{Units produced}}$$

$$\text{Unit overhead cost} = \frac{\$30{,}000}{60{,}000 \text{ units}} = \$0.50 \text{ a unit}$$

SAMPLE PROBLEM 11

Department 7 of the Fenton Company occupies 17,000 square feet of the firm's 200,000 square foot factory . It produces 10,000 units. Find the unit overhead cost in Department 7 if total overhead is $300,000.

Solution

$$\text{Dept. 7's overhead share} = \frac{17{,}000}{200{,}000} \times \$300{,}000 = \$25{,}500$$

$$\text{Unit overhead cost} = \frac{\$25{,}500}{10{,}000 \text{ units}} = \$2.55 \text{ per unit}$$

One overhead item of increasing concern to business today is the cost of electricity. Electric rates are quoted at so much per 1,000 hours of meter operation, expressed as dollars per *Kilowatt-hour* (Kwh). Usually, the charge is higher when demand is heavy (*peak* rate) and lower when demand is light (*off-peak* rate). Sample Problem 12 illustrates the calculation of a bill for electricity.

SAMPLE PROBLEM 12

Cote Custom Design Company has two electric meters, one for peak and one for off-peak hours. April readings and rates are tabulated below:

Meter	Kwh Start	Kwh End	Rate per Kwh
Peak	77010	99710	$0.08
Off-peak	8878	9612	0.05

Calculate the firm's electric bill for April.

Solution

Peak: 99710 − 77010 = 22,700 Kwh used
Off-peak: 9612 − 8878 = 734 Kwh used

22,700 × $.08 =	$1,816.00
734 × $.05 =	36.70
Total bill	$1,852.70

The firm might consider shifting some of its power needs to the off-peak hours in order to save some electric costs.

Bad Debts

Another common business expense is caused by customers who do not pay their bills. These uncollectible accounts are called *bad debts.*

At the close of each business year, a firm will usually estimate what percent of its charge accounts or sales will not be paid.

SAMPLE PROBLEM 13

Marland, Inc., reports for 1988; $300,000 in sales and $70,000 in outstanding accounts. Determine the bad debts expense if it is estimated at (a) $\frac{1}{2}$% of sales; (b) $2\frac{1}{2}$% of outstanding accounts.

Solution

(a) $\frac{1}{2}\% = .005 \times \$300{,}000 = \$1{,}500$
(b) $2\frac{1}{2}\% = .025 \times \$70{,}000 = \$1{,}750$

Property Tax

A business and a homeowner share many expenses, one of which is tax on real property. Property taxes are paid to state and local governments. Rates and bases of payment vary widely, but the principles of computation are the same.

A firm's property is *assessed,* or valued, by the government taxing authority. Although some states have different plans, an assessment will normally run from 50% to 60% of the property's true value. Property worth $100,000 would be assessed at between $50,000 and $60,000. The taxing authority establishes the property tax *rate* by considering how much money it needs (the *percentage*) and how much the total property is worth (the *base*).

SAMPLE PROBLEM 14

Total taxes to be raised in Starkville are $2,640,000 on property assessed at $80,000,000. Determine the property tax rate, expressed to the nearest tenth of a percent.

Solution

$$\text{Rate} = \frac{\text{Total taxes}}{\text{Assessed valuation}}$$

$$\text{Rate} = \frac{\$2{,}640{,}000}{\$80{,}000{,}000} = \frac{264}{8{,}000} = 3.3\%$$

The rate of 3.3% is then paid by all taxpayers.

SAMPLE PROBLEM 15

Peter Deland lives in Starkville. His property is assessed at $40,000. What property tax will he pay?

Solution

$P = B \times R$
$P = \$40{,}000 \times .033 = \$1{,}320$

A property tax rate can be expressed in several ways:

3.3%
.033 per $1
$3.30 per $100
$33.00 per $1,000
33 mills (thousandths)

Administrative Costs

The office—that part of a business operation that coordinates its administrative functions—is a growing cost factor in modern firms. Studies reveal that 30 to 40 percent of all business costs can be attributed to office operations, including such items as filing, mailroom, bookkeeping, secretarial, and data processing expenses. Thus, any effort to reduce these costs becomes vital to a firm's management.

Cost calculations are made by management in order to enable the comparison of alternatives in reaching a certain goal. The formal name given to a cost analysis of different options is the *feasibility study*. For example, a firm might prepare a feasibility study of the costs of hiring temporary versus full-time help in the office, as illustrated in Sample Problem 16.

SAMPLE PROBLEM 16

Calder Associates is about to hire a new typist and is comparing the costs of hiring a full-time versus a temporary worker. If a new full-time worker were hired, Calder's office manager estimates the expenses would be $4.30 per hour, plus 30% for fringe benefits. A temporary typist would cost $5.00 per hour, but the employment agency, not Calder, would pay fringe benefits. Calculate and compare the earnings for both by the (a) hour; (b) week (assuming a 40-hour week); and (c) year by the more feasible method.

Solution

(a) Full-time per hour: $4.30 + (.30 × $4.30) = $5.59
Temporary per hour: −5.00
Savings per hour: $0.59

(b) Savings per week: $0.59 per hour × 40 hours = $23.60
(c) Savings per year: $23.60 per week × 52 weeks = $1,227.20

As the solution to Sample Problem 16 shows, it is more feasible to hire the temporary worker in this case. There is never one fixed answer to a feasibility study; in any given situation, one alternative can be the better or the best one to choose.

Sample Problem 17 shows one more use of a feasibility study, this time to determine whether or not a firm should invest in a new piece of office equipment.

SAMPLE PROBLEM 17

West Corporation's correspondence costs are as follows for 1988: dictation, $60,000; stenographers, $90,000; typists, $50,000; filing, $50,000; mailroom,

$30,000. The firm produced a total of 25,000 letters during the year, raising the question of whether it would be more feasible to purchase automated word processing equipment to eliminate some of the costs. A consultant estimates a per letter cost of $8.30 with the new equipment. Calculate (a) the cost per letter under the current system; (b) the savings per letter using the new system.

Solution

(a) Current costs: $60,000 + $90,000 + $50,000 + $50,000 + $30,000 = $280,000 ÷ 25,000 letters = $11.20 per letter
(b) $11.20 − $8.30 = $2.90 savings per letter

TEST YOUR ABILITY

1. A newly developed patent for an artificial sweetener cost a firm $20,400. Determine the annual straight-line amortization of the patent over its maximum life.
2. The Wellington Uniform Company has total general overhead expenses of $270,000. Determine the amount to be allocated to each of its four departments, given the following square footage:

Dept. A	320,000 sq. ft.
Dept. B	480,000 sq. ft.
Dept. C	400,000 sq. ft.
Dept. D	240,000 sq. ft.

3. The finishing department of the Wilson Manufacturing Company occupies 12,500 square feet of the firm's 100,000 square feet. If the total overhead is $48,000 and the output of the finishing department is 20,000 units, find the overhead cost per unit.
4. Bourbeau's Boutique shows the following data on its electric bill for July of this year:

Meter	Kwh Start	Kwh End	Rate per Kwh
Peak	46235	51437	$0.075
Off-peak	11210	12994	0.052

Calculate what the firm owes for electricity for the month of July.

5. Pedro International, Inc., had last-year sales of $1,800,000 and unpaid charge accounts of $57,000. Determine the amount of bad debts expense if it is estimated at (a) $\frac{3}{4}$% of sales; (b) 3% of charge accounts.
6. The County of Hanboro wishes to raise taxes of $3,960,000 on property assessed at $264,000,000. Find the tax rate it must impose, to the nearest tenth of a percent.
7. Nora Whitman has real property valued at $90,000, which is assessed at 60% of its value and taxed at a rate of 31 mills. (a) What is the assessed valuation of the property? (b) Calculate Ms. Whitman's property tax.
8. Rourke Industries is considering whether to staff its home office with a full-time receptionist or to hire a worker from Office Temporaries instead. The cost of the full-time worker is estimated to be $5.25 per hour, plus approximately 30% in fringe benefits, while the temporary worker would cost a flat $6.50 per hour. Calculate the savings per (a) hour; (b) week (assuming a 35-hour week); and (c) year, by the more feasible method.

9. Ralph Daiches, sales representative, is considering the alternative ways to acquire a new business car for his sales territory and has narrowed his choices to two: rent or buy. The cost of rental is a flat $0.08 per mile, plus $0.01 per mile insurance. Gas would be bought by Ralph, as it would be if he purchased the car. Purchase costs of an auto are estimated at $9,000; in addition, a three-year life expectancy and annual insurance costs of $500 are anticipated. Determine the following: (a) annual cost of renting; (b) annual cost of buying; (c) annual savings by the more feasible method, assuming that 50,000 miles are driven each year and that depreciation is by the straight-line method.

Business Profits

In Chapter 6, we introduced the calculation of gross and net profits. In this section, we will go one step further—particularly as applied to the profit of service businesses and trading businesses.

Net Profit in a Service Business

A *service business* is one that is not involved in merchandising activities. It usually takes the form of a professional partnership, such as a legal or medical firm, or an individual operation, such as a barber shop.

The basic equation for determining the net profit of a service business is:

Net profit = Fees received − Operating expenses

SAMPLE PROBLEM 18

Jeanne Thomas, M.D., received fees totaling $81,700 for the past calendar year. Her expenses were as follows: salaries of nurse and receptionist, $23,700; other payroll costs, $4,200; office rent, $3,600; depreciation of equipment, $900; supplies, $1,350; utilities, $1,800; miscellaneous expenses, $450. Determine her net profit or loss for the year.

Solution

Fees received		$81,700
Operating expenses:		
Salaries	$23,700	
Payroll costs	4,200	
Rent	3,600	
Depreciation	900	
Supplies	1,350	
Utilities	1,800	
Miscellaneous	450	
Total operating expenses		−36,000
Net profit		$45,700

Net Profit in a Trading Business

The computation of net profit for a *trading business* is a slightly more complex procedure, since merchandising firms require a two-step calculation.

Gross profit = Net sales − Cost of goods sold
Net profit = Gross profit − Operating expenses

Net Sales

Every trading business has a percentage of sales that are returned; consequently, the purchase price has to be refunded. In other instances, an item may be found to be defective, and an allowance is granted to the customer in the form of a refund or a credit slip. *Gross sales*, those before considering returns and allowances, must be reduced to yield *net sales.*

$$\text{Net sales} = \text{Gross sales} - \text{Returns and allowances}$$

Cost of Goods Sold

To find the gross profit, the cost of an item must be subtracted from its sales price, as was shown in Chapter 6. However, at the time of sale, the item's cost is generally not recorded. An item sold to you at $5.99 may have cost $3.75, but it is only the $5.99 that is rung up on the register.

The cost of goods sold is found indirectly. If a firm knows what it originally had, what it bought, and what is left at the end of the year, it can determine what it sold. If you begin the day with $10, receive $30, and wind up with $5, what did you spend? In equation form, the answer of $35 is obtained as follows:

$$\text{Cost of goods sold} = \text{Beginning inventory} + \text{Net purchases} - \text{Ending inventory}$$

Net purchases is found in the same manner as net sales: gross purchases minus any returns and allowances.

SAMPLE PROBLEM 19

From the following data, find the cost of goods sold:

Beginning inventory	$13,500
Purchases	46,300
Purchases returns and allowances	3,900
Ending inventory	12,900

Solution

Beginning inventory	$13,500
Net purchases ($46,300 − $3,900)	42,400
	$55,900
Ending inventory	−12,900
Cost of goods sold	$43,000

Gross and Net Profits

If the firm in Sample Problem 19 had net sales of $72,000 its *gross profit* would be $72,000 − $43,000 = $29,000.

If its operating expenses were $24,000 its *net profit* would be $29,000 − $24,000 = $5,000.

Sample Problem 20 summarizes the calculation of net profit for a trading business.

SAMPLE PROBLEM 20

Andy's Variety Store shows the following data. Calculate the net profit or loss.

Sales	$37,000
Sales returns	1,200
Cost of goods sold	29,400
Operating expenses	8,900

Solution

Net sales ($37,000 − $1,200)	$35,800
Cost of goods sold	−29,400
Gross profit	$ 6,400
Operating expenses	−8,900
Net loss	($ 2,500)

TEST YOUR ABILITY

1. Wendy Huston, CPA, reports the following data for 1989:

Fees received	$36,500
Expenses:	
Office rent	4,200
Salaries	21,600
Payroll taxes	3,800
Depreciation of equipment	500
Supplies	440
Miscellaneous	1,710

Determine her net profit or loss for the year.

2. From the following data, calculate the cost of goods sold:

Inventory, January 1, 1989	$ 19,400
Purchases	136,800
Purchases returns	2,500
Inventory, December 31, 1989	12,450

3. Arnold's Deli reports sales of $27,500, sales returns of $350, purchases of $22,700, purchases returns of $290, a beginning inventory of $2,500, and an ending inventory of $2,300. If operating expenses are $2,700, find the net profit or loss.

CHECK YOUR READING

1. To what type of asset does the term *depreciation* apply?
2. How is the scrap value of an asset determined?
3. Describe two methods of estimating the life of a tangible asset.
4. On a water pump with annual depreciation of $600, how much has accumulated after $4\frac{1}{2}$ years?
5. If the water pump in Question 4 cost $5,000, what is its book value after $4\frac{1}{2}$ years?
6. If C = $8,000, S = $1,000, and accumulated depreciation after three years is $3,000, what is the book value?

7. Determine the number of months of depreciation in the year of purchase for assets acquired on each of the following dates:
   (a) October 12 (b) January 9
   (c) December 20 (d) August 16
8. How is the concept of *equal expense* applied in (a) the straight-line method and (b) the units-of-production method?
9. How is a straight-line rate of depreciation calculated?
10. Find the sum of the digits for (a) 1 to 15; (b) 1 to 30; (c) 1 to 40.
11. Must all tangible assets acquired since 1980 be depreciated by the ACRS method? Explain.
12. Using ACRS, what is the recovery period for (a) mobile homes; (b) most machinery; (c) real estate?
13. What two amounts are compared to find the gain or loss on the sale of tangible assets?
14. Distinguish between depreciation and amortization.
15. What two bases can be used to estimate bad debts expense?
16. If the beginning inventory of the Garcia Lens Corporation is $30,000, net purchases are $70,000, and ending inventory is $20,000, find the cost of goods sold.
17. How is the net profit of a service business determined?

EXERCISES

Group A

1. A building is constructed in 1988 at a cost of $15,000,000. Use the ACRS method to calculate depreciation for 1988 and 1989.
2. The Harvey Tea Company's real property is assessed at 60% of its value of $50,000 and is taxed at a rate of 42 mills.
   (a) What is the assessed valuation of the property?
   (b) What is the tax on the property?
3. A crane costing $18,000 has an estimated life of five years. If its salvage value is estimated at $1,500, prepare a depreciation schedule using the declining-balance method for the asset.
4. Fletcher County needs to raise taxes of $6,480,000. Property in the area is assessed at $120,000,000. Find the tax rate expressed in mills.
5. Determine the annual straight-line depreciation on each of the following assets:

	C	S	L
(a)	$ 4,320	$ -0-	6 years
(b)	$ 7,800	$ 300	15 years
(c)	$94,500	$3,500	50 years

6. The fabricating department of the Roy Corporation occupies 27,000 square feet of the firm's 135,000 square foot area. Total overhead expenses are $17,500. If 20,000 units are produced in this department, find the cost per unit to the nearest penny.
7. Bashoff's Hardware Store shows the following data for 1988: sales, $42,315; sales returns, $1,810; beginning inventory, $8,310; ending inventory, $11,204; purchases, $29,380; purchases returns, $950; operating expenses, $11,800. Calculate Bashoff's net profit or loss for the year.

8. For each of the following lives of new assets, find the declining-balance rate of depreciation:
   (a) 13 years (b) 18 years
   (c) 16 years (d) 22 years
9. A trailer is purchased on August 9, 1988. C = $10,400, S = $1,200, L = 5 years. Using the straight-line method find (a) the depreciation for 1988; (b) the book value at the end of 1988; (c) the depreciation for 1989; (d) the book value at the end of 1989.
10. A car costing $3,800 with an estimated salvage value of $300 and a life of 50,000 miles is used as follows:

1987	21,200 miles
1988	15,300 miles
1989	19,100 miles

Prepare a depreciation schedule, using the machine-hours method.

11. The Star Manufacturing Company allocates its general overhead of $144,000 on the basis of square footage. Assign the proportionate share to each of the following departments:

Accounting Dept.	15,000 sq. ft.
Drafting Dept.	75,000 sq. ft.
Sales Dept.	60,000 sq. ft.
Personnel Dept.	30,000 sq. ft.

12. The Dittman Silverware Company depreciates by the units-of-production method. A machine costing $36,000 with an estimated life of 105,000 units and salvage value of $4,500 has the following production figures:

1986	27,200 units
1987	29,800 units
1988	23,500 units
1989	24,500 units

Find the depreciation expense for each year.

13. Robinson Aircraft's sales are $1,425,000; its charge accounts, $250,000. Calculate bad debts expense at (a) $\frac{1}{2}$% of sales; (b) $3\frac{1}{2}$% of charge accounts.
14. An asset costing $14,500 is to be depreciated by the straight-line method over 11 years, with a salvage value estimated at $1,300. Find (a) the annual depreciation; (b) the book value after one year; (c) the book value after two years.
15. Prepare a depreciation schedule for Exercise 14.
16. Eaton's Office Furnishings electric bill for February shows the following figures:

Meter	Kwh Start	Kwh End	Rate per Kwh
Peak	14386	29124	$0.081
Off-peak	3617	7019	0.047

Calculate the firm's electric bill for February.

17. A barge costing $13,800, with salvage value of $1,200, and a life of six years is to be depreciated by the sum-of-the-years' digits method. Find (a) the

denominator of the fraction; (b) the numerator for the first year; (c) the first year's depreciation; (d) the book value at the end of the first year.

18. Prepare a depreciation schedule for Exercise 17.
19. Malibu Corporation's correspondence costs for 1988 are as follows: dictation, $75,000; stenographic, $95,000; typing, $60,000; filing, $25,000; mailing, $50,000. The firm's letter output is 40,000 letters for the year. Purchase of new equipment for $125,000, with an estimated five-year life and straight-line depreciation, will eliminate the need for stenographic services and halve the cost of dictation. Determine the (a) cost per letter currently; (b) cost per letter if the new equipment were purchased; (c) savings by the more feasible method.
20. Benjamin Anderson, M.D., reports the following data for 1989:

Fees received	$36,220
Expenses:	
Salaries	21,000
Payroll taxes	8% of salaries
Depreciation of equipment	$1,000
Supplies	2,600
Utilities	1,950
Other expenses	425

Find his net profit or loss for the year.

21. A patent for a microwave oven is purchased four years *after* its issue date for $14,300. Calculate the annual straight-line amortization expense over its useful life.
22. In each of the following, find, first, the book value and, second, the gain or loss on sale:

	Cost	Accumulated Depreciation	Sales Price
(a)	$ 4,850	$ 2,150	$2,450
(b)	$19,200	$19,000	$ 360
(c)	$36,800	$34,400	$4,400

23. An industrial engine with a 20-year life is purchased for $37,000, with an estimated salvage value of $1,800. After $5\frac{1}{2}$ years of depreciation by the straight-line method, it is sold to the Oceanic Corporation for $27,500. Calculate the gain or loss on the sale.
24. From the following data, determine the cost of goods sold:

Inventory, July 1, 1988	$ 18,241.50
Purchases	126,145.90
Purchases returns and allowances	4,271.50
Inventory, June 30, 1989	26,307.40

25. Determine the depreciation for 1989 by the straight-line method for each of the following assets:

	C	S	L	Date of Purchase
(a)	$ 4,550	$ 500	9 years	October 6, 1989
(b)	$26,400	$1,800	12 years	March 27, 1989
(c)	$37,500	$ -0-	15 years	December 4, 1989

26. Sturbridge Corporation is considering the purchase of supplies from a variety of vendors and has received the following six bids:

Vendor A	$19,500 per year
Vendor B	$ 9,865 semiannually
Vendor C	$ 425 per week
Vendor D	$ 1,650 per month
Vendor E	$ 860 biweekly
Vendor F	$ 900 semimonthly

Which vendor offers the lowest bid?

27. Prepare a depreciation schedule using the ACRS method for a 3-year asset with a cost of $140,000 and a salvage value of $15,000.

Group B

1. Given the following data: C = $16,500; S = $900; L = 10 years; date acquired, July 20, 1986, prepare a depreciation schedule using the ACRS method.
2. A salt water conversion plant is purchased on February 19, 1986 at a cost of $72,000. Life is estimated at six years; salvage value at $6,600. Prepare a depreciation schedule, using the declining-balance method at twice the straight-line rate.
3. On September 13, 1987, a firm purchases an asset for $6,260, estimating a six-year life and a salvage value of $500. On April 15, 1989, the asset is sold for $4,410. Calculate the gain or loss on the sale. Assume that the straight-line method is used.
4. Total taxes to be raised in Colesville this year are $1,470,000. The property in the area is assessed at $50,000,000. Determine the tax rate expressed as (a) a percent, to the nearest hundredth; (b) a rate per $1 (to four places); (c) a rate per $100; (d) a rate per $1,000; (e) mills (to the nearest tenth).
5. If taxes to be raised in the twin counties of Malveston are $2,460,000 and the rate to be used is 4%, find the total assessed valuation required.
6. Fill in the blanks in each row with the missing figures.

	Net Sales	Beginning Inventory	Net Purchases	Ending Inventory	Cost of Goods Sold	Gross Profit
(a)	$19,000	$6,000	$20,000	$ 4,000	______	______
(b)	______	8,000	______	12,000	$ 19,000	$ 5,000
(c)	25,000	2,000	14,000	______	______	10,000
(d)	71,000	______	83,000	14,000	79,000	______
(e)	______	______	96,000	20,000	100,000	7,000

UNIT 3 REVIEW

Terms You Should Know

A large number of terms has been introduced in Unit 3. Look again through the unit if you cannot define any of the following:

accumulated depreciation
ACRS method
allocation
amortization
assessed valuation
asset
auto liability insurance
bad debts
book value
coinsurance
collision insurance
commission
comprehensive insurance
copyright
cost
cost of goods sold
declining-balance method
deductible
depreciation
differential piecerate
endowment insurance
estimated useful life (EUL)
exemption
extended term insurance
feasibility study
FICA tax
fire insurance
fringe benefits
graduated commission
gross profit
gross wages
incentive plan
indemnity
intangible asset
kilowatt-hour
limited-payment life insurance
machine-hours method
net profit
net sales
net wages
no-fault insurance
ordinary whole life insurance
overhead
overtime
paid-up insurance
patent
payroll deductions
payroll register
premium
property tax
proration
quota
recovery period
residual value
salary
scrap value
service business
short-rate table
standard
straight-line method
straight piecework
sum-of-the-years' digits method
tangible asset
term insurance
trademark
trading business
unemployment tax
units-of-production method
wage
withholding tax

UNIT 3 SELF-TEST

Unit 3 has shown you the many sides of business math. The review presented below covers the highlights. Please supplement your review with the other types of problems not covered. Problems 1–3 deal with Chapter 8.

1. The Briggs Company pays employees as follows: Hourly workers earn time-and-a-half for hours over 40 a week; piecerate workers earn $0.25 a unit up to and including 400 units a week, and get $0.35 a unit if production exceeds 400; sales representatives earn a $12\frac{1}{2}$% commission on all sales over a quota of $800 a week, plus a $240 salary. Find the gross earnings of
   (a) Jones, who works 45 hours, rate $6.00
   (b) O'Hara, who produces 610 pieces
   (c) Adams, who sells $1,400 worth of goods
2. Calculate Ramon Naijorff's net pay if he grosses $210 this week. He is married and claims two exemptions.
3. The Baxter Corporation, with 200 employees, has a weekly gross payroll of $32,000. The firm's state unemployment tax rate is 3.0%, and it contributes 6% of the payroll for pension benefits and $2.20 per worker for life insurance. Find:
   (a) the employer's FICA tax
   (b) state unemployment tax
   (c) federal unemployment tax
   (d) pension cost
   (e) life insurance cost
   (f) total weekly payroll costs

Problems 4 and 5 summarize Chapter 9.

4. Juan Chavez believes in insurance. At age 25, he takes out the following policies:
   Fire, on his $80,000, brick, Territory C home, for three years.
   Car, on his brand new size 4 car, with the following coverages: 10/20/5, $5,000 medical, $200 deductible collision, $100 deductible comprehensive. He just received his license and drives only for pleasure.
   Life, $20,000, 20-year endowment policy.
   Calculate:
   (a) his fire insurance premium
   (b) his auto insurance premium
   (c) his monthly life insurance premium
5. Deborah Weaver, age 32, carries the following insurance policies:
   Fire, $48,000 on property valued at $80,000, under an 80% coinsurance clause.
   Car liability, 10/20/5.
   Life, taken out at age 22, $10,000 ordinary whole policy.
   The following events occur:
   Her property is totally destroyed.

She injures two people in a car accident, causing injuries of $15,000 and $5,000. The other auto is damaged to the extent of $3,000.
She stops paying premiums on her life insurance and elects extended term insurance.

(a) What part of the fire loss must she pay?
(b) What part of the auto damage must she pay?
(c) For what term is her life insurance extended?

Problems 6 and 7 review Chapter 10.

6. G. Lewis Corporation obtains an asset for $20,000, with a salvage value estimated at $2,000 and life at five years or 50,000 units. Find the first year's depreciation by:
   (a) the straight-line method
   (b) the units-of-production method, if 11,000 units are produced
   (c) the declining-balance method
   (d) the sum-of-the-years' digits method
   (e) ACRS
7. A cement-making machine costing $8,000 has a life of eight years. Scrap is estimated to be $500.
   (a) If the asset is purchased on September 5, 1987, calculate normal depreciation to December 31 by the straight-line method.
   (b) What is the 1988 depreciation?
   (c) Determine the book value at the end of 1988.
   (d) If the machine is sold on December 31, 1988, for $6,300, determine the gain or loss on the sale.

KEY TO UNIT 3 SELF-TEST

1. a. $285.00
   b. $213.50
   c. $315.00
2. $179.98
3. a. $2,128
   b. $960
   c. $256
   d. $1,920
   e. $440
   f. $37,672
4. a. $691.00
   b. $409.80
   c. $76.41
5. a. $32,000
   b. $5,000
   c. 21 years, 160 days
6. a. $3,600
   b. $3,960
   c. $8,000
   d. $6,000
   e. $3,000
7. a. $312.50
   b. $937.50
   c. $6,750.00
   d. $450 gain

UNIT 4

The Mathematics of Finance

The everyday money needs of a business are supplied by customers in payment for goods or services. But suppose that company desires a large sum of money for either a plant expansion, a research project, or a major factory overhaul?

The firm must then borrow money to meet these needs. Fortunately, there are many sources available for this. For example, it can take a loan from a bank or other financial institution, issue bonds to the public, or discount a promissory note. In each case, the firm will pay *interest* to the lender.

At the other extreme, rather than borrowing, the firm can try to earn more money. It may sell on the installment plan or lend money. In both cases, the firm will collect *interest.*

Unit 4 explores *interest* problems. In Chapter 11, you will study the facts of simple interest and bank discount. Chapter 12 describes the procedures used in dealing with compound interest over long periods of time. And Chapter 13 introduces you to installment buying and the "Truth-in-Lending" Act, both of which you should find "interest-ing."

CHAPTER 11

Simple Interest and Bank Discount

Just as rent is paid for the use of property and a salary is paid for the use of a person's services, *interest* is paid for the use of someone's money.

The amount of money borrowed is the *principal.* Interest paid only on the principal is *simple interest.* Simple interest is generally charged on overdue customer accounts and on loans to business firms or between individuals.

Your Objectives

After completing Chapter 11, you will be able to

1. Calculate, by various methods, simple interest for various time periods.
2. Select the best method for solving a simple interest problem.
3. Understand the terminology and uses of a promissory note.
4. Calculate the amount of bank discount on a note under various circumstances.

Computing Simple Interest for a Year or More

Simple interest is calculated by the formula:

$$I = P \times R \times T$$

or

$$\textit{I}\text{nterest} = \textit{P}\text{rincipal} \times \textit{R}\text{ate} \times \textit{T}\text{ime}$$

The *rate* is always expressed as a percent or a decimal, and it is always annual.

The *time* is always expressed in years or as a fraction of a year.

SAMPLE PROBLEM 1

Find the simple interest on a principal of $900 for three years at 8%, using the interest formula.

Solution A

$$\begin{aligned} I &= P \times R \times T \\ &= \$900 \times .08 \times 3 \\ &= \$72 \times 3 \\ &= \$216 \end{aligned}$$

Solution B

$$\begin{aligned} I &= P \times R \times T \\ &= \$900 \times 8\% \times 3 \\ &= \frac{\overset{9}{\cancel{900}}}{1} \times \frac{8}{\underset{1}{\cancel{100}}} \times \frac{3}{1} = \$216 \end{aligned}$$

The first method (Solution A) is a two-step process: (1) multiply the principal by the rate to find one year's interest; (2) multiply the answer to (1) by the actual time period.

The second method (Solution B) is the *cancellation method.* Each factor is expressed in fraction form, and then canceled and reduced.

Sample Problem 2 shows both methods again, this time with a fractional rate and a fractional year.

SAMPLE PROBLEM 2

Find the simple interest, by formula, on Nancy Burke's loan of $600 for $5\frac{1}{2}$ years at $7\frac{1}{4}\%$.

Solution A

$$\begin{aligned} I &= P \times R \times T \\ &= \$600 \times .0725 \times 5\tfrac{1}{2} \\ &= \$43.50 \times 5\tfrac{1}{2} \\ &= \$239.25 \end{aligned}$$

Solution B

$$\begin{aligned} I &= P \times R \times T \\ &= \$600 \times 7\tfrac{1}{4}\% \times 5\tfrac{1}{2} \\ &= \frac{\overset{3}{\cancel{600}}}{1} \times \frac{29}{\underset{2}{\cancel{400}}} \times \frac{11}{2} = \frac{957}{4} = \$239.25 \end{aligned}$$

Computing Simple Interest for Less than a Year

To find simple interest for time periods of less than one year, the time is expressed as a fraction of a year. For eight months, for example, the time would be:

$$\frac{8 \text{ months}}{12 \text{ months}} = \frac{2}{3} \text{ of a year}$$

If the annual interest is $30, then eight months' interest is $\frac{2}{3} \times \$30$, or $20.

If, however, we are to find the interest between two specific dates, it is necessary to know the time in days.

Finding Time in Days

To find the time between two dates, two methods can be used. Sample Problem 3 shows the method used for counting the days.

SAMPLE PROBLEM 3

Find the number of days from May 13, 1988 to October 3, 1988.

Solution

Days remaining in May (31 − 13)	18
Days in June	30
Days in July	31
Days in August	31
Days in September	30
Days in October until October 3	3
Total	143 days

The first day, May 13, is not counted; however, the last day—October 3—is counted.

The second method of finding time in days is very fast. Simply use a time table such as Table 11.1. The table lists the number of each day of the year in consecutive order. To solve Sample Problem 3 using Table 11.1, look up May 13, and October 3, May 13 is day 133; October 3 is day 276. Therefore:

$$276 - 133 = 143 \text{ days}$$

Table 11.1—and similar tables—should be used with caution. You can see how it is easy to go wrong from this computation of the time from January 17, 1988 to April 3, 1988:

April 3 is day	93
January 17 is day	−17
Time	76 days

But, 1988 is divisible by 4; therefore, it is a *leap year*. Since the month of February has 29 days and it is included in the time period, add one day. The answer, therefore, is 77 days. The moral: "Look before you leap."

Sample Problem 4 shows how to use Table 11.1 to find the time between dates in two different years.

SAMPLE PROBLEM 4

Find the time from August 3, 1988 to March 10, 1989, using Table 11.1.

Solution

August 3 is day 215 of 1988
March 10 is day 69 of 1989

Days left in 1988 (365 − 215)	150
Days in 1989 to March 10	69
Total	219 days

Table 11.1

Numbers of the Days of the Year

Day of Month	Jan.	Feb.	Mar.	Apr.	May	June	July	Aug.	Sept.	Oct.	Nov.	Dec.
1	1	32	60	91	121	152	182	213	244	274	305	335
2	2	33	61	92	122	153	183	214	245	275	306	336
3	3	34	62	93	123	154	184	215	246	276	307	337
4	4	35	63	94	124	155	185	216	247	277	308	338
5	5	36	64	95	125	156	186	217	248	278	309	339
6	6	37	65	96	126	157	187	218	249	279	310	340
7	7	38	66	97	127	158	188	219	250	280	311	341
8	8	39	67	98	128	159	189	220	251	281	312	342
9	9	40	68	99	129	160	190	221	252	282	313	343
10	10	41	69	100	130	161	191	222	253	283	314	344
11	11	42	70	101	131	162	192	223	254	284	315	345
12	12	43	71	102	132	163	193	224	255	285	316	346
13	13	44	72	103	133	164	194	225	256	286	317	347
14	14	45	73	104	134	165	195	226	257	287	318	348
15	15	46	74	105	135	166	196	227	258	288	319	349
16	16	47	75	106	136	167	197	228	259	289	320	350
17	17	48	76	107	137	168	198	229	260	290	321	351
18	18	49	77	108	138	169	199	230	261	291	322	352
19	19	50	78	109	139	170	200	231	262	292	323	353
20	20	51	79	110	140	171	201	232	263	293	324	354
21	21	52	80	111	141	172	202	233	264	294	325	355
22	22	53	81	112	142	173	203	234	265	295	326	356
23	23	54	82	113	143	174	204	235	266	296	327	357
24	24	55	83	114	144	175	205	236	267	297	328	358
25	25	56	84	115	145	176	206	237	268	298	329	359
26	26	57	85	116	146	177	207	238	269	299	330	360
27	27	58	86	117	147	178	208	239	270	300	331	361
28	28	59	87	118	148	179	209	240	271	301	332	362
29	29	—*	88	119	149	180	210	241	272	302	333	363
30	30	—	89	120	150	181	211	242	273	303	334	364
31	31	—	90	—	151	—	212	243	—	304	—	365

* Add one day after February 28 for *leap years* (years divisible by 4, such as 1984 and 1988).

Try one more on your own: October 4, 1987 to March 2, 1988. Is the answer 150 days? (Did you add one for the leap year 1988?)

Exact Interest

Interest based on a 365-day year is *exact interest,* which is always used by the federal government. To express time in the interest formula, the number of days is placed over a denominator of 365.

SAMPLE PROBLEM 5

Find the exact interest on a loan to Cynthia Barth of $600 for 90 days at 12%, using the cancellation method.

Solution

$$I = P \times R \times T$$
$$= \$600 \times 12\% \times 90 \text{ days}$$
$$= \frac{\overset{6}{\cancel{600}}}{1} \times \frac{12}{\underset{1}{\cancel{100}}} \times \frac{\overset{18}{\cancel{90}}}{\underset{73}{\cancel{365}}} = \$17.75$$

Bankers' Interest

Bankers' interest is based on a 360-day year and is used by banks and business firms. Calculating bankers' interest is generally simpler than calculating exact interest because 360 has many more factors than 365 (which is used for exact interest).

For the same principal, rate, and time, bankers' interest will always be *higher* than exact interest. This is because a numerator over 360—as opposed to over 365—will result in a higher percent. Let's do Sample Problem 5 again, but this time, Ms. Barth will borrow from a bank.

$$I = \frac{\overset{6}{\cancel{600}}}{1} \times \frac{\overset{3}{\cancel{12}}}{\underset{1}{\cancel{100}}} \times \frac{\overset{1}{\cancel{90}}}{\underset{\underset{1}{\cancel{4}}}{\cancel{360}}} = \$18.00$$

She has paid \$0.25 more. It may not seem like much, but as the loan becomes larger, so does the interest.

For bankers' interest, a month is always considered to be 30 days or $\frac{1}{12}$ of a year. Thus, eight months' interest can be calculated by expressing time as $\frac{8}{12} = \frac{2}{3}$ of a year—or $8 \times 30 = 240$ days, giving us $\frac{240}{360}$, which reduces to $\frac{2}{3}$ of a year.

The Meaning of Amount

In any discussion about interest, the term *amount* is the sum that must be repaid by the user of the principal at the end of the time period. Amount is the sum of the principal and the interest, and can be expressed in symbols as:

$$A = P + I$$

In Sample Problem 5, the amount is \$600.00 + \$17.75 = \$617.75.

A term that is synonymous with amount is *maturity value*. The maturity value of a loan is the sum due when the loan must be repaid.

In Sample Problem 6, we review (1) how to find time and (2) how to calculate bankers' interest and amount.

SAMPLE PROBLEM 6

Find (a) the bankers' interest and (b) the amount due on a loan of \$975 to the Proust Dramatic Company from April 17 to June 16 of the same year, if the interest rate is 6%.

Solution

(a) From Table 11.1:

June 16 is day	167
April 17 is day	−107
Time	60 days

$$I = P \times R \times T$$
$$= \$975 \times 6\% \times 60 \text{ days}$$

$$= \frac{975}{1} \times \frac{\overset{1}{\cancel{6}}}{100} \times \frac{\overset{1}{\cancel{60}}}{\underset{1}{\cancel{360}}} = \frac{975}{100} = \$9.75$$

(b) $A = P + I$
$= \$975.00 + \$9.75 = \$984.75$

TEST YOUR ABILITY

1. Find the simple interest in each of the following problems:

	P	R	T
(a)	$ 900	8%	4 years
(b)	$1,000	10%	$2\frac{1}{2}$ years
(c)	$ 700	$7\frac{1}{2}$%	5 years
(d)	$ 500	11%	$3\frac{1}{4}$ years
(e)	$3,600	$7\frac{1}{4}$%	$2\frac{3}{4}$ years

2. Anthony Zang borrows $750 from Oscar Brown for three years at $7\frac{1}{4}$%. (a) How much simple interest will be due? (b) What amount must Zang repay after three years?
3. Without using a time table, find the number of days from:
   (a) June 25, 1988 to September 16, 1988
   (b) January 18, 1988 to May 17, 1988
   (c) July 7, 1989 to November 1, 1989
   (d) December 8, 1988 to January 2, 1989
   (e) October 17, 1988 to August 3, 1989
4. Go back and do each part of Problem 3, using Table 11.1 this time.
5. Use the cancellation method to find the exact interest on each of the following loans:

	P	R	T
(a)	$ 500	8%	73 days
(b)	$ 600	9%	219 days
(c)	$1,460	$8\frac{1}{2}$%	60 days
(d)	$ 850	7%	292 days
(e)	$ 800	$12\frac{1}{2}$%	60 days

6. Use the data from Problem 5 and the cancellation method to find the bankers' interest in each case.
7. Jerome Smith lends Josette Lyons $800 on March 17, 1988. The 11% loan is repaid on July 15. Find (a) the bankers' interest and (b) the amount on this loan.

The 60-Day, 6% Method

Let's stop briefly in order to make two observations:

(1) If you tried the preceding Test Your Ability problems, you did a lot of calculating. The cancellation method you used is accurate but requires too much work and time.

(2) Look back at Sample Problem 6. Compare in particular the figures of the principal and the interest. The principal is $975; the interest, $9.75. Both sets of

digits are the same, but the decimal point differs. Is this coincidence? If so, we have a lot of work still ahead of us. However, consider the rate and time factors.

When the rate is 6% and the time is 60 days, bankers' interest on any principal would be calculated as:

$$I = P \times \frac{\overset{1}{\cancel{6}}}{100} \times \frac{\overset{1}{\cancel{60}}}{\underset{\underset{1}{\cancel{60}}}{\cancel{360}}} = P \times \frac{1}{100}$$

The interest is $\frac{1}{100}$, or 1% of the principal; fortunately the solution to Sample Problem 6 is no coincidence. We could have found the bankers' interest by moving the decimal point of the principal two places to the left ($9.75).

A rule for *bankers' interest* by the 60-day, 6% method is: Whenever the time is 60 days and the rate is 6%, the interest is $\frac{1}{100}$, or 1% of the principal. This rule permits us to mentally calculate bankers' interest. Six percent is a common rate and 60 days is a common time for a business loan. But, as you will see below, the rule is also adaptable to other rates and other time periods.

To begin your success with the 60-day, 6% method of finding bankers' interest, look at the following list of problems:

P	I for 60 days at 6%
$ 500.00	$ 5.00
623.45	6.23
623.50	6.24
1.90	0.02
1,736,842.95	17,368.43

Next, let's consider a principal of $2,400, lent at 6% for 60 days. The interest is $24. What if the time were 30 days instead of 60? Since 30 is $\frac{1}{2}$ of 60, the interest for 30 days would be $\frac{1}{2}$ of $24, or $12. In other words, we can use *aliquot parts* of 60 days to find the interest for other time periods.

Here is a list of some common aliquot parts of 60 days:

Number of Days	Aliquot Part of 60 Days	Number of Days	Aliquot Part of 60 Days
30	$\frac{1}{2}$	40	$\frac{2}{3}$
20	$\frac{1}{3}$	45	$\frac{3}{4}$
15	$\frac{1}{4}$	24	$\frac{2}{5}$
12	$\frac{1}{5}$	36	$\frac{3}{5}$
10	$\frac{1}{6}$	48	$\frac{4}{5}$
6	$\frac{1}{10}$	50	$\frac{5}{6}$
5	$\frac{1}{12}$	3	$\frac{1}{20}$
4	$\frac{1}{15}$	2	$\frac{1}{30}$

The list is not all-inclusive, but for your own convenience, you can expand it: place the number of days over 60 and reduce to lowest terms. Sample Problem 7 uses aliquot parts.

SAMPLE PROBLEM 7

Using the 60-day, 6% method, find the bankers' interest at 6% on $1,200 for (a) 60 days; (b) 30 days; (c) 6 days; (d) 24 days; (e) 45 days; (f) 120 days.

Solution

(a) Interest for 60 days = 1% × \$1,200 = \$12.00

(b) Interest for 30 days = $\frac{1}{2}$ × \$12.00 = \$6.00

(c) Interest for 6 days = $\frac{1}{10}$ × \$12.00 = \$1.20

(d) Interest for 24 days = $\frac{2}{5}$ × \$12.00 = \$4.80

(e) Interest for 45 days = $\frac{3}{4}$ × \$12.00 = \$9.00

(f) Interest for 120 days = 2 × \$12.00 = \$24.00

In (f), notice that the 60-day, 6% method can be applied to multiples of 60 days as well as to parts of it. Since 120 is twice 60, the interest for 120 days is twice the interest for 60 days.

Aliquot parts of 60 days can be *added* or *subtracted* to find the interest for other time periods. For 96 days, for example, add the interest figures for 60, 30, and 6 days. For 57 days, subtract 3 days' interest from 60 days' interest. The variations are almost endless.

SAMPLE PROBLEM 8

Using the 60-day, 6% method, find the bankers' interest for the Weston Jewelry Company on \$1,800.00 at 6% for (a) 88 days; (b) 134 days; (c) 54 days.

Solution

(a) Interest for 60 days = \$18.00
20 days = 6.00 ($\frac{1}{3}$ × \$18.00)
6 days = 1.80 ($\frac{1}{10}$ × \$18.00)
2 days = 0.60 ($\frac{1}{30}$ × \$18.00)
Interest for 88 days = \$26.40

(b) Interest for 60 days = \$18.00
Interest for 120 days = \$36.00 (2 × \$18.00)
10 days = 3.00 ($\frac{1}{6}$ × \$18.00)
4 days = 1.20 ($\frac{1}{15}$ × \$18.00)
Interest for 134 days = \$40.20

(c) Interest for 60 days = \$18.00
−6 days = −1.80 ($\frac{1}{10}$ × \$18.00)
Interest for 54 days = \$16.20

Notice that in part (b) we took $\frac{1}{6}$ of the *60-day* interest figure to find ten days' interest and $\frac{1}{15}$ of the 60-day figure to find four days' interest. A very frequent mistake is to base these calculations on the 120-day figure.

Finally, we can apply the 60-day, 6% method to other rates. If P = \$2,400, then I for 60 days at 6% is \$24. And if the rate were 3%? Our rate is $\frac{1}{2}$ of 6%, so $\frac{1}{2}$ × \$24 = \$12. At 9%? We add 6% (\$24) + 3% (\$12) to get 9%, or \$36. Common aliquot parts of 6% are listed below.

Rate	Aliquot Part of 6%	Rate	Aliquot Part of 6%
1%	$\frac{1}{6}$	4%	$\frac{2}{3}$
$1\frac{1}{2}$%	$\frac{1}{4}$	$4\frac{1}{2}$%	$\frac{3}{4}$
2%	$\frac{1}{3}$	5%	$\frac{5}{6}$
3%	$\frac{1}{2}$		

SAMPLE PROBLEM 9

Using the 60-day, 6% method, find the bankers' interest for Starship Hamburgers on $840 for 60 days at (a) 6%; (b) 2%; (c) $10\frac{1}{2}$%, (d) 12%; (e) 8%.

Solution

(a) Interest at 6% = 1% × $840 = $8.40
(b) Interest at 2% = $\frac{1}{3}$ × $8.40 = $2.80
(c) Interest at $10\frac{1}{2}$% = $\frac{3}{4}$ × $8.40 = $6.30 + $8.40 = $14.70
(d) Interest at 12% = 2 × $8.40 = $16.80
(e) Interest at 8% = $8.40 + $2.80 = $11.20

Sample Problem 10 shows how to apply the rule to situations in which the time is *not* 60 days *and* the rate is *not* 6%. Two different solutions are shown.

SAMPLE PROBLEM 10

Find the bankers' interest on a loan of $700 to the Halingwood Choir Association for 85 days at 8%.

Solution A

Interest for 60 days, 6%	=	$ 7.00
2%	=	2.33 ($\frac{1}{3}$ × $7.00)
Interest for 60 days, 8%	=	$ 9.33
20 days	=	3.11 ($\frac{1}{3}$ × $9.33)
5 days	=	0.78 ($\frac{1}{12}$ × $9.33)
Interest for 85 days, 8%	=	$13.22

Solution B

Interest for 60 days, 6%	=	$ 7.00
20 days	=	2.33 ($\frac{1}{3}$ × $7.00)
5 days	=	0.58 ($\frac{1}{12}$ × $7.00)
Interest for 85 days, 6%	=	$ 9.91
2%	=	3.30 ($\frac{1}{3}$ × $9.91)
Interest for 85 days, 8%	=	$13.21

In the two situations we had to adjust *both* the rate and the time. In Solution A, the rate was adjusted first; then the days were adjusted. The aliquot parts were applied to the $9.33 figure—not the original $7.00 figure.

In Solution B, the days were adjusted first, and then the rate was adjusted, based on the interest for 85 days at 6%.

As in all cases whenever you have a choice of methods, choose the easiest one, that is, the method that involves the least calculating. However, avoid rounding, if you can, during the first series of adjustments. In Sample Problem 10, it was not possible to do so. Notice also that there were two different answers. If you differ from the Test Your Ability answers by a penny, it will be merely the consequence of taking alternate routes to solving the problems.

The Six-Day, 6% Method

A variation of the 60-day, 6% method for finding *bankers' interest* is the *six-day, 6% method.* Many accountants and business executives find this method easier to apply. Move the decimal point of the principal *three* places to the left to find six days' interest at 6%. For $1,200, I at 6% for six days = $1.20.

Then apply aliquot parts of six days to find the interest for other time periods. What would the interest be on $1,200 at 6% for three days? The answer is $0.60 ($\frac{1}{2}$ × $1.20). For nine days? Yes, $1.20 + $0.60 = $1.80. Sample Problem 11 shows a further application of this method.

SAMPLE PROBLEM 11

Find the interest for the Finmore Cheese Company on $2,400 at 6% for (a) 10 days and (b) 17 days.

Solution

(a) Interest for 6 days, 6% = $2.40
4 days = 1.60 ($\frac{2}{3}$ × $2.40)
Interest for 10 days, 6% = $4.00

(b) Interest for 6 days, 6% = $2.40

Interest for 12 days, 6% = $4.80 (2 × $2.40)
5 days = 2.00 ($\frac{5}{6}$ × $2.40)
Interest for 17 days, 6% = $6.80

TEST YOUR ABILITY

1. Find the bankers' interest for 60 days at 6% on (a) $825.00; (b) $345.86; (c) $2,017.92; (d) $2,047.31; (e) $8.50; (f) $35.00.
2. Find the bankers' interest on $2,160 at 6%, using the 60-day, 6% method, for (a) 12 days; (b) 50 days; (c) 5 days; (d) 90 days; (e) 180 days; (f) 55 days.
3. Using the 60-day, 6% method, find the bankers' interest on $720 for 60 days at (a) 5%; (b) 3%; (c) 8%; (d) 4$\frac{1}{2}$%; (e) 13%; (f) 10%.
4. Find the bankers' interest and amount in each of the following problems, using the 60-day, 6% method.

	P	R	T
(a)	$ 645	6%	2 months
(b)	$ 410	6%	180 days
(c)	$ 650	6%	36 days
(d)	$1,200	9%	60 days
(e)	$ 840	10%	60 days
(f)	$ 750	8%	60 days
(g)	$ 900	7%	80 days
(h)	$ 600	7$\frac{1}{2}$%	110 days
(i)	$ 800	4$\frac{1}{2}$%	75 days
(j)	$1,500	8%	21 days

5. Repayment of a loan for $820 made to the Phelps Company was due on October 1, 1988. It is paid, with bankers' interest at 8%, on December 30, 1988. Find the amount paid, using the 60-day, 6% method.
6. Find the bankers' interest on $960 at 6%, using the six-day, 6% method, for (a) 5 days; (b) 8 days; (c) 11 days; (d) 21 days.

Table 11.2

Bankers' Interest
($100 for a 360-day Year)

Time	4%	$4\frac{1}{2}\%$	5%	$5\frac{1}{2}\%$	6%	7%	$7\frac{1}{2}\%$	8%
1 day	.0111	.0125	.0139	.0153	.0167	.0194	.0208	.0222
2 days	.0222	.0250	.0278	.0306	.0333	.0389	.0417	.0444
3 days	.0333	.0375	.0417	.0458	.0500	.0583	.0625	.0667
4 days	.0444	.0500	.0556	.0611	.0667	.0778	.0833	.0889
5 days	.0556	.0625	.0694	.0764	.0833	.0972	.1042	.1111
6 days	.0667	.0750	.0833	.0917	.1000	.1167	.1250	.1333
7 days	.0778	.0875	.0972	.1069	.1167	.1361	.1458	.1556
8 days	.0889	.1000	.1111	.1222	.1333	.1556	.1667	.1778
9 days	.1000	.1125	.1250	.1375	.1500	.1750	.1875	.2000
10 days	.1111	.1250	.1389	.1528	.1667	.1944	.2083	.2222
11 days	.1222	.1375	.1528	.1681	.1833	.2139	.2292	.2444
12 days	.1333	.1500	.1667	.1833	.2000	.2333	.2500	.2667
13 days	.1444	.1625	.1806	.1986	.2167	.2528	.2708	.2889
14 days	.1556	.1750	.1944	.2139	.2333	.2722	.2917	.3111
15 days	.1667	.1875	.2083	.2292	.2500	.2917	.3125	.3333
16 days	.1778	.2000	.2222	.2444	.2667	.3111	.3333	.3556
17 days	.1889	.2125	.2361	.2597	.2833	.3306	.3542	.3778
18 days	.2000	.2250	.2500	.2750	.3000	.3500	.3750	.4000
19 days	.2111	.2375	.2639	.2903	.3167	.3694	.3958	.4222
20 days	.2222	.2500	.2778	.3056	.3333	.3889	.4167	.4444
21 days	.2333	.2625	.2917	.3208	.3500	.4083	.4375	.4667
22 days	.2444	.2750	.3056	.3361	.3667	.4278	.4583	.4889
23 days	.2556	.2875	.3194	.3514	.3833	.4472	.4792	.5111
24 days	.2667	.3000	.3333	.3667	.4000	.4667	.5000	.5333
25 days	.2778	.3125	.3472	.3819	.4167	.4861	.5208	.5556
26 days	.2889	.3250	.3611	.3972	.4333	.5056	.5417	.5778
27 days	.3000	.3375	.3750	.4125	.4500	.5250	.5625	.6000
28 days	.3111	.3500	.3889	.4278	.4667	.5444	.5833	.6222
29 days	.3222	.3625	.4028	.4431	.4833	.5639	.6042	.6444
30 days	.3333	.3750	.4167	.4583	.5000	.5833	.6250	.6667
60 days	.6667	.7500	.8333	.9167	1.0000	1.1667	1.2500	1.3333
90 days	1.0000	1.1250	1.2500	1.3750	1.5000	1.7500	1.8750	2.0000
120 days	1.3333	1.5000	1.6667	1.8333	2.0000	2.3333	2.5000	2.6667
150 days	1.6667	1.8750	2.0833	2.2917	2.5000	2.9167	3.1250	3.3333
180 days	2.0000	2.2500	2.5000	2.7500	3.0000	3.5000	3.7500	4.0000

Interest by Table

The 60-day, 6% method is a convenient way of solving many business problems, as is the cancellation method of finding bankers' interest. For more complex problems, simple interest tables, such as Tables 11.2 and 11.3, are available.

Table 11.2 is based on a 360-day year and is therefore used for *bankers' interest.* Table 11.3 is based on a 365-day year and is used to find *exact interest.* Both tables give the interest on $100 for specific rates and for specific time periods. For example, on $100 for 120 days at 7%, the *bankers' interest*

Table 11.3

Exact Interest
($100 for a 365-day Year)

Time	4%	$4\frac{1}{2}$%	5%	$5\frac{1}{2}$%	6%	7%	$7\frac{1}{2}$%	8%
1 day	.0110	.0123	.0137	.0151	.0164	.0192	.0205	.0219
2 days	.0219	.0247	.0274	.0301	.0329	.0384	.0411	.0438
3 days	.0329	.0370	.0411	.0452	.0493	.0575	.0616	.0658
4 days	.0438	.0493	.0548	.0603	.0658	.0767	.0822	.0877
5 days	.0548	.0616	.0685	.0753	.0822	.0959	.1027	.1096
6 days	.0658	.0740	.0822	.0904	.0986	.1151	.1233	.1315
7 days	.0767	.0863	.0959	.1055	.1151	.1342	.1438	.1534
8 days	.0877	.0986	.1096	.1205	.1315	.1534	.1644	.1753
9 days	.0986	.1110	.1233	.1356	.1479	.1726	.1849	.1973
10 days	.1096	.1233	.1370	.1507	.1644	.1918	.2055	.2192
11 days	.1205	.1356	.1507	.1658	.1808	.2110	.2260	.2411
12 days	.1315	.1479	.1644	.1808	.1973	.2301	.2466	.2630
13 days	.1425	.1603	.1781	.1959	.2137	.2493	.2671	.2849
14 days	.1534	.1726	.1918	.2110	.2301	.2685	.2877	.3068
15 days	.1644	.1849	.2055	.2260	.2466	.2877	.3082	.3288
16 days	.1753	.1973	.2192	.2411	.2630	.3068	.3288	.3507
17 days	.1863	.2096	.2329	.2562	.2795	.3260	.3493	.3726
18 days	.1973	.2219	.2466	.2712	.2959	.3452	.3699	.3945
19 days	.2082	.2342	.2603	.2863	.3123	.3644	.3904	.4164
20 days	.2192	.2466	.2740	.3014	.3288	.3836	.4110	.4384
21 days	.2301	.2589	.2877	.3164	.3452	.4027	.4315	.4603
22 days	.2411	.2712	.3014	.3315	.3616	.4219	.4521	.4822
23 days	.2521	.2836	.3151	.3466	.3781	.4411	.4726	.5041
24 days	.2630	.2959	.3288	.3616	.3945	.4603	.4931	.5260
25 days	.2740	.3082	.3425	.3767	.4110	.4795	.5137	.5479
26 days	.2849	.3205	.3562	.3918	.4274	.4986	.5342	.5699
27 days	.2959	.3329	.3699	.4068	.4438	.5178	.5548	.5918
28 days	.3068	.3452	.3836	.4219	.4603	.5370	.5753	.6137
29 days	.3178	.3575	.3973	.4370	.4767	.5562	.5959	.6356
30 days	.3288	.3699	.4110	.4521	.4931	.5753	.6164	.6575
60 days	.6575	.7397	.8219	.9041	.9863	1.1507	1.2329	1.3151
90 days	.9863	1.1096	1.2329	1.3562	1.4794	1.7260	1.8493	1.9726
120 days	1.3151	1.4794	1.6438	1.8082	1.9726	2.3014	2.4657	2.6301
150 days	1.6438	1.8493	2.0548	2.2603	2.4657	2.8767	3.0822	3.2877
180 days	1.9726	2.2192	2.4657	2.7123	2.9589	3.4520	3.6986	3.9452

(Table 11.2) is \$2.3333, or \$2.33 cents; the *exact interest* (Table 11.3) is \$2.3014, or \$2.30 cents.

In simple interest tables, rates and/or times can be *combined.* For example, the exact interest on \$100 for 19 days at 10% can be found as follows, using Table 11.3:

Interest for 19 days at 6%	=	\$0.3123
4%	=	0.2082
Interest for 19 days at 10%	=	\$0.5205

To find the bankers' interest on \$100 at $5\frac{1}{2}$% for 63 days, using Table 11.2:

Interest for 60 days at $5\frac{1}{2}$%	=	\$0.9167
3 days	=	0.0458
Interest for 63 days at $5\frac{1}{2}$%	=	\$0.9625

Tables 11.2 and 11.3 give interest figures for \$100. If the principal is \$500, simply multiply the table figure by 5. On \$500 for 60 days at 5%, the exact interest from Table 11.3 is \$0.8219 × 5 = \$4.1095 = \$4.11.

Sample Problems 12 and 13 illustrate the uses of Tables 11.2 and 11.3.

SAMPLE PROBLEM 12

Using Table 11.2, find the bankers' interest on a loan to Gem Lights, Inc., of \$735 for 37 days at $7\frac{1}{2}$%.

Solution

Bankers' interest on \$100 for 30 days at $7\frac{1}{2}$%	=	\$0.6250
7 days	=	0.1458
Bankers' interest on \$100 for 37 days at $7\frac{1}{2}$%	=	\$0.7708
× Number of hundreds (\$735 ÷ \$100)		×7.35
Bankers' interest on \$735 for 37 days at $7\frac{1}{2}$%	=	\$5.67

SAMPLE PROBLEM 13

Using Table 11.3, find the exact interest on \$735 borrowed by Utopia Land Development for 37 days at $7\frac{1}{2}$%.

Solution

Exact interest on \$100 for 30 days at $7\frac{1}{2}$%	=	\$0.6164
7 days	=	0.1438
Exact interest on \$100 for 37 days at $7\frac{1}{2}$%	=	\$0.7602
× Number of hundreds (\$735 ÷ \$100)		×7.35
Exact interest on \$735 for 37 days at $7\frac{1}{2}$%	=	\$5.59

Notice that in each problem, the number of hundreds was found by moving the decimal point of the principal two places to the left. Rounding was done after multiplication.

Observe once again, by comparing Sample Problems 12 and 13, that bankers' interest will always exceed exact interest for the same P, R, and T.

TEST YOUR ABILITY

1. Using Table 11.2, find the bankers' interest and amount in each of the following problems:

	P	R	T
(a)	\$ 500	8%	14 days
(b)	\$ 800	$5\frac{1}{2}$%	4 days
(c)	\$1,200	6%	19 days
(d)	\$ 750	$7\frac{1}{2}$%	27 days
(e)	\$ 200	8%	157 days
(f)	\$3,000	$10\frac{1}{2}$%	134 days
(g)	\$8,200	10%	26 days
(h)	\$2,100	$11\frac{1}{2}$%	10 days
(i)	\$ 450	9%	23 days
(j)	\$ 560	14%	229 days

2. Using Table 11.3, find the exact interest and amount for each of the cases presented in Problem 1.

Summary of Simple Interest Methods

For a year or more, use the cancellation method.

For less than a year, after finding the time in days:

(1) For exact interest, use a 365-day year and the formula $I = P \times R \times T$, *or* use Table 11.3.

(2) For bankers' interest, use a 360-day year and the formula $I = P \times R \times T$, *or* use the 60-day, 6% method (or the six-day variation), *or* use Table 11.2.

Sample Problem 14 summarizes simple interest calculation by three methods to find bankers' interest.

SAMPLE PROBLEM 14

Find the bankers' interest on \$840 borrowed by the Middlemarch Factory for 36 days at 8% using (a) the interest formula; (b) the 60-day, 6% method; (c) Table 11.2.

Solution

(a) $I = P \times R \times T$

$= \$840 \times 8\% \times 36 \text{ days}$

$$= \frac{\overset{7}{\cancel{840}}}{1} \times \frac{8}{100} \times \frac{\overset{12}{\cancel{36}}}{\underset{1}{\underset{\cancel{3}}{\cancel{360}}}} = \frac{672}{100} = \$6.72$$

(b)
Interest for 60 days, 6%	= \$ 8.40
2%	= 2.80 ($\frac{1}{3} \times \$8.40$)
Interest for 60 days, 8%	= \$11.20
Interest for 36 days, 8%	= \$ 6.72 ($\frac{3}{5} \times \$11.20$)

(c) Table 11.2

Interest on \$100 for 30 days at 8%	= \$0.6667
6 days	= 0.1333
Interest on \$100 for 36 days at 8%	= \$0.8000
× Number of hundreds (\$840 ÷ \$100)	× 8.4
Interest on \$840 for 36 days at 8%	= \$6.72

Once again, the best method is the easiest method for you. So, unless otherwise specified, choose whichever method you prefer.

The Promissory Note

When borrowing money from a friend, you might sign an IOU ("I Owe You"), which is a written promise to pay a specific sum of money by a specific date. When a business or an individual borrows from a bank, the borrower has to sign a *promissory note,* which is a formal IOU. A typical promissory note is shown below, along with a list of special terms.

[1] September 12, 1988

[2] Sixty days AFTER DATE I PROMISE TO PAY TO THE ORDER OF

[5] Stanley Burns, [4] $650.00 (Six hundred fifty and no/100--DOLLARS)

PAYABLE AT THE Second National Bank

(SIGNED) [6] William Kelly

[7] VALUE RECEIVED

[8] WITH INTEREST AT 0% [3] DUE November 11, 1988

1. The *date of the note* is September 12, 1988.
2. The *time,* or term, *of the note* is 60 days.
3. The *due date,* or *maturity date,* is November 11, 1988.
4. The *face* is $650. The face is written in both figures and words to discourage anyone from changing the amount.
5. The *payee* is Stanley Burns. After 60 days, Burns will be paid at or through the Second National Bank.
6. The *maker* is William Kelly, who owes the debt to Burns.
7. *Value received* means that the note was given in exchange for something, such as a loan or a purchase of merchandise.
8. The note is *noninterest-bearing.* Kelly will pay Burns $650 on November 11. It is more common for a note to be *interest-bearing.* If this were the case, Kelly would pay $650 plus interest on the due date.

An important feature of a promissory note is that it is *negotiable*—it can be transferred to another payee (a person, firm, or bank) by the signature (*endorsement*) of the present payee. The words *pay to the order of* permit the transfer. If Burns needs the $650 prior to November 11, he can endorse the note to a bank and receive cash for it. Kelly will then pay the bank—not Burns—on November 11. Cashing a note at a bank is called *discounting a note.*

Finding the Due Date of a Note

The time of a note is expressed in either months or days. If the time is in months, the note is due on the same date so many months later. For example, a note dated April 9 with a time of four months is due on August 9. A note dated March 31 with a time of three months would be due on June 31; since June has only 30 days, however, June 30 becomes the due date.

When time is given in days, the exact number of days must be counted.

SAMPLE PROBLEM 15

A note is dated July 30, 1988. Find the due date if the time of the note is (a) 1 month; (b) 30 days; (c) 2 months; (d) 60 days.

Solution

(a) July 30, 1988 + 1 month = August 30, 1988.
(b) Thirty days from July 30, 1988 is August 29, 1988 (one day left in July + 29 more in August make 30).
(c) July 30, 1988 + 2 months = September 30, 1988.
(d) Sixty days from July 30, 1988 is September 28, 1988 (one day left in July + 31 in August make 32; 28 more are needed in September to make 60).

The due date can also be found by using Table 11.1. In (d) of Sample Problem 15, for example, July 30 is day 211. Sixty days later is day 271 (211 + 60), which is September 28.

Bank Discount

As we have already seen, discounting a note means cashing it at a bank. The charge made by the bank for this service is *bank discount.* Although in ordinary business transactions, it is customary to collect interest at the time a debt is due (the maturity date), banks collect interest in advance. *Bank discount* is simple bankers' interest on the *maturity value* of a note (the sum due when the note must be repaid) and is *deducted in advance.* The cash received for the discounted note is called the *proceeds.*

Bank discount is used by some for personal bank loans, as well as for loans given in exchange for notes endorsed by a business firm to a bank.

To see the difference between a loan at simple interest and one at bank discount, let's look at a loan of $6,000 for 60 days at 8%.

Simple interest:

Interest on $6,000 for 60 days at 8%	= $80
Cash received on the loan	= $6,000
Cash to be repaid (maturity value)	= $6,000 + $80 = $6,080

Bank discount:

Interest as above	= $80
Cash received on the loan (proceeds)	= $6,000 − $80 = $5,920
Cash to be repaid (maturity value)	= $6,000

In both cases, the amount of simple interest is the same. But for a note owed at simple interest, the borrower receives the full principal; at bank discount, the borrower receives less than the full principal. Therefore, at bank discount, a higher rate of interest is actually charged, based on the amount borrowed.

Term of Discount

The period of time for which bank discount is charged is the *term of discount.* It runs from the date of discount until the date of maturity (due date)—the time during which a bank's money is on loan.

For a loan made directly to the maker of a note (a personal bank loan, for example), the term of discount is the same as the time of the note.

In the case of a customer's note that the payee endorses over to a bank *after* the original date, the term of discount and the time of the note are *not* the same.

A note with a time of 60 days that is transferred to a bank ten days after being made has a term of discount of 50 days—the time remaining to maturity. The following diagram shows this idea:

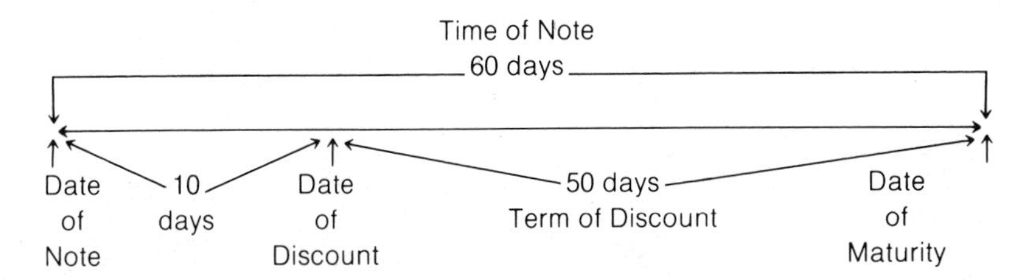

To find the term of discount, two steps are needed:

(1) Determine the due date.
(2) Count the days from the date of discount to the due date.

SAMPLE PROBLEM 16

Find the term of discount for the McGuire Corporation's note, dated August 22 and discounted on September 13, if the time of the note is (a) 60 days; (b) 2 months.

Solution

(a) Due date = August 22 + 60 days = October 21
Term of discount runs from September 13 to October 21 = 38 days
(b) Due date = August 22 + 2 months = October 22
Term of discount runs from September 13 to October 22 = 39 days

TEST YOUR ABILITY

1. Find the due date of the Bartlett Company's note dated June 16, 1988, if the time of the note is (a) 1 month; (b) 2 months; (c) 3 months; (d) 30 days; (e) 60 days; (f) 90 days.
2. Salvatore Toscano's own 45-day note is discounted at 8%. If the face of the note is $1,440, find the proceeds.
3. Find the maturity date and the term of discount in each of the following problems:

	Date of Note	Time of Note	Date of Discount
(a)	May 17	30 days	May 24
(b)	January 20	1 month	February 4
(c)	November 3	75 days	December 16
(d)	December 21	2 months	January 6
(e)	June 16	80 days	July 3

Discounting a Customer's Note

Bank discount is based on the *maturity value* of a promissory note, the *rate of discount*, and the *term of discount*. These three factors correspond to principal, rate, and time in simple interest problems.

If a note is noninterest-bearing, the maturity value is the same as the face of the note.

SAMPLE PROBLEM 17

The Jones Company's one-month, noninterest-bearing note, dated March 17, is discounted at 8% by the Smith Company on March 30. If the face of the note is $720, find the proceeds received by the Smith Company.

Solution

Maturity value of note: $720
Rate of discount: 8%
Term of discount:
  Due date = March 17 + one month = April 17
  Term runs from March 30 to April 17 = 18 days
Bank discount:
  $720 at 8% for 18 days
  Table 11.2: $0.4000 × 7.2 = $2.88
Proceeds = Maturity value − Bank discount
Proceeds: $720.00 − $2.88 = $717.12

The Smith Company leaves the bank with $717.12. Had it waited until April 17, it would have received $720.00. But, when you need the cash, it is often worth paying a small fee.

If a note is interest-bearing, the maturity value must first be calculated. Sample Problems 18 and 19 show two outcomes from discounting customers' interest-bearing notes.

SAMPLE PROBLEM 18

Find the proceeds of the Vineland Company's $900, 8%, two-month note, discounted at 8%, if the term of discount is 30 days.

Solution

Maturity value of note: $900 + interest for two months (60 days) at 8% = $912
Rate of discount: 8%
Term of discount: 30 days
Bank discount:
  $912 for 30 days at 8%
  $912 for 60 days at 6% = $9.12
  $912 for 30 days at 6% = $\frac{1}{2}$ × $9.12 = $4.56
  $912 for 30 days at 2% = $\frac{1}{3}$ × $4.56 = 1.52
  $912 for 30 days at 8% = $6.08
Proceeds: $912.00 − $6.08 = $905.92

SAMPLE PROBLEM 19

The Ellis Corporation receives a 90-day, 6%, $800 note from the Barnes Company on March 25, 1988, and discounts the note at 9% on April 4, 1988. Find the proceeds of the note.

Solution

Maturity value of note: $800 + interest for 90 days at 6% = $812.00
Rate of discount: 9%

Term of discount:

Due date is March 25 + 90 days = June 23

Term is from April 4 to June 23 = 80 days

Bank discount:

$812 at 9% for 80 days

$$\frac{812}{1} \times \frac{\overset{1}{\cancel{9}}}{100} \times \frac{\overset{2}{\cancel{80}}}{\underset{\underset{1}{\cancel{40}}}{\cancel{360}}} = \frac{1624}{100} = \$16.24$$

Proceeds: $812.00 – $16.24 = $795.76

Notice that in Sample Problem 18, the proceeds exceeded the face, while in Sample Problem 19, the face exceeded the proceeds. Why? Because in the first case, the interest and the discount rates were the same; in the second case, the discount rate exceeded the interest rate, *and* the note was discounted for a much longer period of time.

TEST YOUR ABILITY

1. Alexander, Inc., receives a 90-day, noninterest-bearing note from the Mays Company on July 14, 1988. The note is discounted at 9% on July 24. If the face is $360, find the proceeds.
2. The Cook Company issues a three-month, noninterest-bearing note to the West Company on January 17, 1988. West holds the note until February 13, at which time it discounts the $660 note at 10%. Find the proceeds.
3. A 6%, 90-day note for $4,200 is discounted by the Plains Loan Association for 60 days at 10%. Find the proceeds.
4. Gonzalez Industries issues a four-month, $900, 7% note to the Bellmore Company on May 24, 1988. The note is discounted on June 26 at 8%. Find the proceeds.

CHECK YOUR READING

1. What two business situations call for the payment of simple interest?
2. Describe the cancellation method of finding simple interest.
3. What is the purpose of a time table, such as Table 11.1?
4. Distinguish between bankers' interest and exact interest by means of (a) their uses and (b) their computation.
5. Why does bankers' interest always exceed exact interest for the same principal, rate, and time?
6. State the rule of the 60-day, 6% method of finding bankers' interest.
7. What combinations of aliquot parts of 60 days could be used to find the interest for (a) 37 days; (b) 138 days; (c) 208 days?
8. How would you find the interest on $700 for 65 days at 2%, using Table 11.2 or 11.3?
9. List three methods for finding bankers' interest.
10. What are two methods for finding exact interest?
11. What is a *promissory note*?
12. How does the feature of negotiability make a promissory note a valuable possession?

13. Examine the promissory note that follows and answer the questions.
    (a) What is the date of the note?
    (b) What is the time of the note?
    (c) What is its due date?
    (d) What is its face?
    (e) Is the note interest-bearing?
    (f) Who is the payee?
    (g) Who is the maker?
    (h) What is the maturity value of the note?

July 18, 1988

Ninety days AFTER DATE I PROMISE TO PAY TO THE ORDER OF

Lewis Adams , $425.00 (Four hundred twenty-five and no/100 DOLLARS)

PAYABLE AT THE State Bank

(SIGNED) Cindy Robinson

VALUE RECEIVED

WITH INTEREST AT 7% DUE ____

14. Are notes due in 60 days and in two months due on the same date? Explain.
15. How does bank discount differ from simple interest?
16. Complete this statement: The term of discount runs from the date of ________ to the date of ________.
17. What additional step is required in working a bank discount problem involving a customer's interest-bearing note, compared to one with a noninterest-bearing note?
18. Compare the face of a note to its proceeds when discounted if the note is (a) a firm's own note; (b) a customer's noninterest-bearing note; (c) a customer's interest-bearing note.

EXERCISES *Group A*

1. Find the bankers' interest on $1,440 at 6%, using the 60-day, 6% method, for (a) 15 days; (b) 40 days; (c) 4 days; (d) 80 days; (e) 240 days; (f) 84 days.
2. Bellows Company receives a $540, 120-day note from Jane Lewis dated June 7, 1988. The note is discounted at 9% on August 22, 1988. Find the proceeds.
3. Find the simple interest in each of the following problems:

	P	R	T
(a)	$ 600	8%	3 years
(b)	$ 800	9%	$2\frac{1}{2}$ years
(c)	$2,000	$7\frac{1}{2}$%	4 years

(d) $7,600	14%	$2\frac{1}{4}$ years
(e) $1,000	$9\frac{1}{2}$%	$3\frac{3}{4}$ years

4. Using Table 11.2, find the bankers' interest and amount in each problem:

	P	R	T
(a)	$ 500	5%	21 days
(b)	$ 400	8%	16 days
(c)	$ 2,000	$4\frac{1}{2}$%	180 days
(c)	$ 5,000	7%	15 days
(e)	$ 800	6%	92 days
(f)	$ 1,200	7%	37 days
(g)	$ 550	13%	33 days
(h)	$ 320	$7\frac{1}{2}$%	124 days
(i)	$ 4,500	9%	7 days
(j)	$10,000	$11\frac{1}{2}$%	13 days

5. Using Table 11.3, find the exact interest and amount for each item in Exercise 4.
6. Use the six-day, 6% method to find the bankers' interest on $3,000 at 6% for (a) 3 days; (b) 11 days; (c) 18 days; (d) 37 days.
7. Bellini Suits, Inc., receives a $900, 7%, 90-day note from McMillan dated April 3, 1988. The note is discounted on April 27 at 8%. Find the proceeds.
8. Rice's account with Argonaut, Inc., is paid on July 17, 1988, after being due on May 8, 1988. Interest at 8% is charged for the overdue amount of $650. How much does Rice pay?
9. Find the due date of Lorraine Humbolt's note dated October 16, 1988, if the time of the note is (a) 60 days; (b) 90 days; (c) 360 days; (d) 2 months; (e) 3 months; (f) 1 year.
10. Without using a time table, find the number of days from
    - (a) August 17, 1988 to November 11, 1988
    - (b) January 12, 1988 to April 1, 1988
    - (c) March 8, 1988 to October 3, 1988
    - (d) December 29, 1988 to January 15, 1989
    - (e) November 12, 1988 to April 8, 1989
11. Solve each part of Exercise 10, using Table 11.1.
12. Using the 60-day, 6% method, find the bankers' interest and amount in each problem below:

	P	R	T
(a)	$ 835	6%	2 months
(b)	$ 375	6%	240 days
(c)	$ 525	6%	24 days
(d)	$ 310	9%	60 days
(e)	$ 420	8%	60 days
(f)	$ 846	10%	60 days
(g)	$ 990	8%	70 days
(h)	$3,000	$7\frac{1}{2}$%	130 days
(i)	$5,100	$4\frac{1}{2}$%	80 days
(j)	$ 810	9%	12 days

13. Using the cancellation method, find the exact interest on each of the following loans:

	P	R	T
(a)	\$ 900	9%	146 days
(b)	\$ 730	7%	50 days
(c)	\$2,920	12%	60 days
(d)	\$ 425	$7\frac{1}{2}$%	219 days
(e)	\$1,200	11%	60 days

14. Use the data from Exercise 13 and find, by formula, the bankers' interest in each example.
15. Find the bankers' interest that the Ellis Home Bank will receive on the following 60-day, 6% notes: (a) \$428.46; (b) \$85.47; (c) \$1,462.81; (d) \$10,949.81; (e) \$79.90; (f) \$5.25.
15. Alan Bowle's own 75-day note for \$630 is discounted at 7%. Find the proceeds.
17. Use the interest formula to find the bankers' interest and amount on a loan of \$625 at $10\frac{1}{2}$% from April 11, 1988 to June 28, 1988.
18. The Briggs Corporation's three-month note, dated October 30, 1988, is discounted at 6% on November 14, 1988. Find the proceeds of this \$740, 7% note.
19. Using the 60-day, 6% method, find the bankers' interest on \$120 for 60 days at (a) 2%; (b) 4%; (c) 8%; (d) $7\frac{1}{2}$%; (e) 10%; (f) $11\frac{1}{2}$%.
20. Sylvan Brothers, Inc., receives a three-month, noninterest-bearing note with a face of \$1,300 on September 24 of this year and discounts the note on October 5 at 9% at the Travelers Savings Bank. Find the proceeds.
21. Find the maturity date and the term of discount in each example:

	Date of Note	Time of Note	Date of Discount
(a)	April 27	80 days	July 9
(b)	July 30	4 months	July 30
(c)	March 2	300 days	August 15
(d)	May 9	5 months	May 12
(e)	August 11	120 days	September 8

22. Peter Sargent borrows \$550 from Axel Jordan for $3\frac{1}{2}$ years at $7\frac{1}{4}$%.
    (a) How much simple interest does Sargent owe?
    (b) What amount will he repay after $3\frac{1}{2}$ years?

Group B

1. Solve each of the following problems for bankers' interest. Use any method, but try to choose the most efficient one.

	P	R	T
(a)	\$ 720.00	8%	47 days
(b)	\$ 268.00	9%	34 days
(c)	\$ 135.00	10%	97 days
(d)	\$ 471.00	7%	68 days
(e)	\$ 219.25	$7\frac{1}{2}$%	142 days

(f)	$ 837.61	$10\frac{1}{2}\%$	71 days
(g)	$ 401.29	5%	69 days
(h)	$2,417.18	6%	210 days
(i)	$ 364.18	6%	7 days
(j)	$5,020.50	10%	19 days

2. Using Table 11.2, find the bankers' interest and amount on a loan of $750 to Bogus Party Supplies, Inc., for 239 days at 9%.
3. On a loan of $850 for 86 days at 7%, how much greater is the bankers' interest than the exact interest?
4. Sheila Tarkington's note is discounted for a term of 46 days on June 29, 1988. It is dated June 14, 1988. Find the time in days.
5. Find the proceeds of the following note, discounted at 7% on its issue date.

June 27, 1988

Three months AFTER DATE I PROMISE TO PAY TO THE ORDER OF

Fay Cote, $540.00 (Five hundred forty and no/100- DOLLARS)

PAYABLE AT THE Lakeland Trust Company

(SIGNED) Larry Cutshaw

VALUE RECEIVED

WITH INTEREST AT 0% DUE

6. If the note in Exercise 5 bore interest at 7%, find the proceeds.

CHAPTER **12**

Accumulation of Funds

*B*oth individuals and businesses need to put aside money for the future. An individual may want a sufficient retirement income or money for a child's college education. A business may want a new building or need to repay a large bond issue.

In either case, it is desirable that any money put aside for such purposes should be used to earn interest—but not simple interest. The best kind of interest is *compound interest.*

Your Objectives

After completing Chapter 12, you will be able to

1. Understand how money accumulates with compound interest.
2. Interpret several compound interest tables.
3. Calculate the future value of a present sum and the present value of a future sum.
4. Prepare schedules of money accumulation at compound interest.
5. Compute the balance of a savings account.

Compound Interest

If interest earned on the original principal at the end of one time period is added to that principal, then the interest earned during the next time period will be based on a compound or combined figure. While simple interest is earned on the principal only, *compound interest* is interest paid on *both the principal and past interest.*

Let's look at an investment of \$1 for three years at 8% simple interest. (A dollar is a good unit to study, since all other sums are based on it.)

$$I = \$1.00 \times .08 \times 3 = \$0.24$$

The *amount* after three years will be:

$$A = \$1.00 + \$0.24 = \$1.24$$

In other words, \$1 invested for three years at 8% *simple interest* will amount to \$1.24. This is what happens, step-by-step.

Original investment	\$1.00
First-year interest (\$1 × .08)	0.08
Balance, end of first year	\$1.08
Second-year interest (\$1 × .08)	0.08
Balance, end of second year	\$1.16
Third-year interest (\$1 × .08)	0.08
Balance, end of third year	\$1.24

Now, compare the above with the *compound interest* tabulation that follows. Interest is compounded once a year (annually).

Original investment	\$1.00
First-year interest (\$1 × .08)	0.08
Balance, end of first year	\$1.08
Second-year interest (\$1.08 × .08)	0.0864
Balance, end of second year	\$1.1664
Third-year interest (\$1.1664 × .08)	0.093312
Balance, end of third year	\$1.259712

The simple interest earned is \$0.24 (\$1.24 – \$1.00); the compound interest earned is \$0.259712 (\$1.259712 – \$1.00). After the first interest period, compound interest will always be more than simple interest.

Now, consider a principal of \$800 invested for three years at 8% compounded annually. We might work it out step-by-step, as we did for \$1, or reason that since \$800 is 800 × \$1, it will amount to 800 times as much.

$$\$1.259712 \times 800 = \$1{,}007.7696 = \$1{,}007.77$$

SAMPLE PROBLEM 1

Spruce Lumber Company invests \$4,500 for three years at 8% compounded annually. Find (a) the amount to which it will accumulate and (b) the compound interest earned.

Solution

(a) \$1 for three years at 8% compounded annually is \$1.259712; \$4,500 is 4,500 × \$1.259712 = \$5,668.70

(b) Interest = \$5,668.70 – \$4,500.00 = \$1,168.70

In a compound interest problem, we find the amount first. We find the interest by subtracting the sum invested from the amount.

Interest can be compounded for periods other than annually. Quarterly compounding (every three months) is typical in any savings bank. Interest can also be compounded semiannually (every six months) and monthly. Let's examine the difference between annual and semiannual compounding of $1 invested at 8% for one year. Annual compounding yielded $1.08. Semiannual compounding will work out as follows:

Original investment	$1.00
First semiannual interest	
($1 × .08 × $\frac{1}{2}$ year = $1 × .04)	0.04
Balance, end of first period	$1.04
Second semiannual interest	
($1.04 × .08 × $\frac{1}{2}$ year = $1.04 × .04)	0.0416
Balance, end of first year	$1.0816

The *greater* the number of interest periods in a year, the *higher* the annual interest. In the problem above, with an 8% *nominal rate,* the interest is being earned at an *effective* (actual) *rate* of 8.16% (R = P ÷ B). Effective interest rates are covered in Chapter 15 as part of our discussion of investments.

Compound interest calculation can be cumbersome, so a table can be very handy. Table 12.1 shows what $1 will amount to at various interest rates for various time periods. Tables such as Table 12.1 are constructed by using the same procedure that we followed to find what $1 would amount to at 8%, compounded annually for three years. In fact, looking across the "3" line to the 8% column, you will find $1.2597120, the amount calculated, plus a seventh decimal place for greater accuracy.

Compound interest tables differ from simple interest tables in two ways:

(1) They are based on $1, while simple interest tables are based on $100.

(2) Rates and days *cannot* be combined.

When interest is compounded *annually,* the number of interest periods is the same as the number of years.

SAMPLE PROBLEM 2

Alice Gallagher invests $5,000 at 6%, compounded annually, for nine years. (a) What amount will she have accumulated after nine years? (b) What compound interest will she have earned?

Solution

(a) Table 12.1: $1 at 6% for 9 periods = $1.6894790
Amount = $1.6894790 × 5,000 = $8,447.40

(b) Compound interest = $8,447.40 − $5,000.00 = $3,447.40

In nine years, Ms. Gallagher's interest is well over half of the original principal. Her investment will have doubled by the end of 12 years at 6%, when $1 becomes $2.0121965.

When interest is compounded *semiannually,* there are two interest periods a year. For a four-year period, there would be eight interest periods. If the annual rate is 8%, the semiannual rate would be half of that or 4%. In other words:

For 8%, compounded *annually* for four years, look up 8%, four periods and find $1.3604890;

Table 12.1

Amount of $1 at Compound Interest

Number of Interest Periods	2%	3%	4%	6%	8%	12%
1	1.0200000	1.0300000	1.0400000	1.0600000	1.0800000	1.1200000
2	1.0404000	1.0609000	1.0816000	1.1236000	1.1664000	1.2544000
3	1.0612080	1.0927270	1.1248640	1.1910160	1.2597120	1.4049280
4	1.0824322	1.1255088	1.1698586	1.2624770	1.3604890	1.5735194
5	1.1040808	1.1592741	1.2166529	1.3382256	1.4693281	1.7623417
6	1.1261624	1.1940523	1.2653190	1.4185191	1.5868743	1.9738227
7	1.1486857	1.2298739	1.3159318	1.5036303	1.7138243	2.2106814
8	1.1716594	1.2667701	1.3685690	1.5938481	1.8509302	2.4759632
9	1.1950926	1.3047732	1.4233118	1.6894790	1.9990046	2.7730788
10	1.2189944	1.3439164	1.4802443	1.7908477	2.1589250	3.1058482
11	1.2433743	1.3842339	1.5394541	1.8982986	2.3316390	3.4785500
12	1.2682418	1.4257609	1.6010322	2.0121965	2.5181701	3.8959760
13	1.2936066	1.4685337	1.6650735	2.1329283	2.7196237	4.3634931
14	1.3194788	1.5125897	1.7316764	2.2609040	2.9371936	4.8871123
15	1.3458683	1.5579674	1.8009435	2.3965582	3.1721691	5.4735657
16	1.3727857	1.6047064	1.8729812	2.5403517	3.4259426	6.1303936
17	1.4002414	1.6528476	1.9479005	2.6927728	3.7000181	6.8660409
18	1.4282462	1.7024331	2.0258165	2.8543392	3.9960195	7.6899658
19	1.4568112	1.7535061	2.1068492	3.0255995	4.3157011	8.6127617
20	1.4859474	1.8061112	2.1911231	3.2071355	4.6609571	9.6462931
25	1.6406060	2.0937779	2.6658363	4.2918707	6.8484752	17.0000643
30	1.8113616	2.4272625	3.2433975	5.7434912	10.0626569	29.9599220
40	2.2080397	3.2620378	4.8010206	10.2857179	21.7245215	93.0509698
50	2.6915880	4.3839060	7.1066834	18.4201543	46.9016125	289.0021868

For 8%, compounded *semiannually* for four years, look up 4%, eight periods and find $1.3685690, a higher figure.

What would you look up for *quarterly* compounding for four years at 8%? The answer is 2% and 16 periods (one-fourth the rate and four times the number of years). The figure from Table 12.1 is $1.3727857, a still higher figure.

SAMPLE PROBLEM 3

Find the amount and the compound interest earned on Winsor Candle Company's $3,000 investment at 12% for 10 years if the interest is compounded (a) annually; (b) semiannually; (c) quarterly.

Solution

(a) Annual compounding, 12%, 10 periods:
$3.1058482 × 3,000 = $9,317.54
Compound interest = $9,317.54 − $3,000.00 = $6,317.54

(b) Semiannual compounding, 6%, 20 periods:
$3.2071355 × 3,000 = $9,621.41
Compound interest = $9,621.41 − $3,000.00 = $6,621.41

(c) Quarterly compounding, 3%, 40 periods:
$3.2620378 × 3,000 = $9,786.11
Compound interest = $9,786.11 − $3,000.00 = $6,786.11

The Annuity

So far, we have been investing a single sum of money and letting it grow. Consider now the *annuity,* which is a series of equal investments at regular intervals—for example, $100 a month for 20 years or $2,000 a year for five years.

Suppose you put away $1 a year for the next five years at 8%, compounded annually. What would be the amount of your annuity? This problem can be solved using Table 12.1:

First dollar will earn interest for 5 years at 8%	= $1.4693281
Second dollar will earn interest for 4 years at 8%	= 1.3604890
Third dollar will earn interest for 3 years at 8%	= 1.2597120
Fourth dollar will earn interest for 2 years at 8%	= 1.1664000
Fifth dollar will earn interest for 1 year at 8%	= 1.0800000
Total amount accumulated after 5 years	$6.3359291

This procedure is accurate but lengthy. Think about a 100-period problem! Fortunately, for almost every tedious method, there is a convenient table: in this case, Table 12.2. Look for five periods, 8% and find $6.3359290, a difference due to rounding.

SAMPLE PROBLEM 4

Richard Riley saves $500 a year for his daughter's college education. If interest is compounded annually at 6%, what will he have accumulated after 14 years?

Solution

Table 12.2: 6%, 14 periods = $22.2759699
$22.2759699 × 500 = $11,137.98

Riley will have deposited $500 for 14 years, or a total of $7,000. The interest is $11,137.98 − $7,000.00 = $4,137.98.

Use Table 12.2 for quarterly and semiannual deposits in the same way you used Table 12.1. What is the amount of semiannual deposits of $1 for three years at 6%, compounded semiannually? Yes, 3% for six periods is $6.6624622. The dollar has been invested six times.

SAMPLE PROBLEM 5

Millie Crawford conscientiously deposits $150 every three months into a fund paying interest at 12%, compounded quarterly. What amount will she have in her fund after ten years?

Solution

Table 12.2: 3%, 40 periods = $77.6632975
$77.6632975 × 150 = $11,649.49

There are two limitations to Table 12.2:

(1) The table is not adaptable to all compounding since not all rates are included.

Table 12.2

Amount of Annuity of $1 at Compound Interest

Number of Interest Periods	2%	3%	4%	6%	8%	12%
1	1.0200000	1.0300000	1.0400000	1.0600000	1.0800000	1.1200000
2	2.0604000	2.0909000	2.1216000	2.1836000	2.2464000	2.3744000
3	3.1226080	3.1836270	3.2464640	3.3746160	3.5061120	3.7793280
4	4.2040402	4.3091358	4.4163226	4.6370930	4.8666010	5.3528474
5	5.3081210	5.4684099	5.6329755	5.9753186	6.3359290	7.1151891
6	6.4342834	6.6624622	6.8982945	7.3938376	7.9228034	9.0890118
7	7.5829691	7.8923360	8.2142263	8.8974679	9.6366276	11.2996932
8	8.7546284	9.1591061	9.5827953	10.4913160	11.4875578	13.7756564
9	9.9497210	10.4638793	11.0061071	12.1807949	13.4865625	16.5487352
10	11.1687154	11.8077957	12.4863514	13.9716426	15.6454875	19.6545834
11	12.4120897	13.1920296	14.0258055	15.8699412	17.9771265	23.1331334
12	13.6803315	14.6177904	15.6268377	17.8821377	20.4952966	27.0291094
13	14.9739382	16.0863242	17.2919112	20.0150659	23.2149203	31.3926025
14	16.2934169	17.5989139	19.9135876	22.2759699	26.1521139	36.2797148
15	17.6392853	19.1568813	20.8245311	24.6725281	29.3242830	41.7532805
16	19.0120710	20.7615877	22.6975124	27.2128798	32.7502257	47.8836741
17	20.4123124	22.4144354	24.6454129	29.9056526	36.4502437	54.7497150
18	21.8405586	24.1168684	26.6712294	32.7599917	40.4462632	62.4396808
19	23.2973698	25.8703745	28.7780786	35.7855912	44.7619643	71.0524425
20	24.7833172	27.6764857	30.9692017	38.9927267	49.4229214	80.6987356
25	32.6709057	37.5530423	43.3117446	58.1563827	78.9544152	149.3339336
30	41.3794408	49.0026782	58.3283353	83.8016774	122.3458680	270.2926046
40	61.6100228	77.6632975	98.8265363	164.0476836	279.7810402	859.1423886
50	86.2709895	116.1807733	158.7737670	307.7560589	619.6717689	2688.0204340

(2) All problems assume that the frequency of deposits coincides with the frequency of interest compounding—quarterly deposits and quarterly compounding, for example. This is not realistic, since you might deposit once a year, yet have interest compounded quarterly. However, we will not explore that far. Appropriate details and tables will be found in finance texts.

Retirement Plans

The principles of the annuity also cover retirement pensions. Many employees have a retirement plan at work in which equal amounts are withheld from each paycheck and deposited in a fund earning compound interest.

Increasing importance has been given to retirement plans in recent years; it is "an idea whose time has come." Two government approved substitutes to date are the Individual Retirement Account (*IRA*) plan and the *Keogh* plan. Tax reforms recently legislated have placed restrictions on the ability of the planholder to defer the amounts contributed to these plans. For example, an IRA is deductible only up to certain income levels. A Keogh (401K) plan is deductible only if your company has such a plan. Other technical changes make it necessary for you to check with your tax advisor before using either plan.

Both plans are *tax-deferred,* meaning that no income tax is paid on the amounts deposited until retirement, when tax rates are lower. If you earn $19,000 and deposit $1,000 in such a plan, your income for tax purposes is only $18,000. If you simply "save" $1,000, your taxable income is $19,000 *plus* the interest on the savings.

Both plans are based on *earned income* only; you must have worked to earn the income you deposit. For example, dividends received on stock are not earned income.

Watch the possible growth of an IRA annuity in Sample Problem 6's *annuity schedule.*

SAMPLE PROBLEM 6

Myron Salcedo opens an IRA account paying interest at 11%, compounded annually. He works very steadily and makes annual deposits of $1,000. Prepare an annuity schedule for the first six years.

Solution

Year	Beginning Balance	+	Annual Contribution	=	Total	+	Interest at 11%	=	Ending Balance
1			$1,000.00		$1,000.00		$110.00		$1,110.00
2	$1,110.00		1,000.00		2,110.00		232.10		2,342.10
3	2,342.10		1,000.00		3,342.10		367.63		3,709.73
4	3,709.73		1,000.00		4,709.73		518.07		5,227.80
5	5,227.80		1,000.00		6,227.80		685.06		6,912.86
6	6,912.86		1,000.00		7,912.86		870.41		8,783.27

Notes

(1) The contribution in an annuity is made at the *beginning* of the period and earns interest for that period.
(2) Interest was calculated at 11% on the total of beginning balance plus the annual contribution.
(3) The ending balance (the sum of the total plus the interest) becomes the next year's beginning balance.
(4) The actual accumulation in a bank compounding *monthly* or *quarterly* would have been higher.

TEST YOUR ABILITY

1. If $1 is compounded annually for seven years at 8%, it yields $1.7138243. Use this fact to determine the accumulation of each of the following principals, assuming the same conditions: (a) $10; (b) $600; (c) $500,000.
2. Rita Anderson deposits $4,000 in a fund with an interest rate of 12%, for three years. Find the amount and the compound interest earned if the interest is compounded (a) annually; (b) semiannually; (c) quarterly.
3. Use Table 12.1 to find the amount and the compound interest in each problem.

	P	R	T	Compounded
(a)	$ 200	4%	13 years	annually
(b)	$ 500	8%	11 years	annually
(c)	$ 700	8%	8 years	semiannually
(d)	$ 1,000	12%	15 years	semiannually
(e)	$ 5,000	12%	2 years	quarterly
(f)	$10,000	8%	2 years	quarterly

4. An investment of $1 every year for nine years at 8% yields a sum of $13.4865625. Using this fact and the same circumstances, find the sum to which the following annuities would accumulate: (a) $500; (b) $20; (c) $10,000.
5. Use Table 12.2 to find the amount of each of the following annuities:

	Annuity	Invested and Interest Compounded	R	T
(a)	$1,000	annually	8%	7 years
(b)	$ 300	annually	12%	9 years
(c)	$ 200	semiannually	8%	8 years
(d)	$ 50	semiannually	6%	9 years
(e)	$ 100	quarterly	8%	10 years
(f)	$ 450	quarterly	12%	3 years

6. Prepare an annuity schedule for Selma Bennet's contributions of $500 a year at 7%, compounded annually, for four years in an IRA account.

Present Value of a Future Amount

The accumulation of funds can be viewed as two questions. One question is: "How much will I have in so many periods at a given rate of interest if I invest X dollars today or X dollars every period?" After our study of compound interest and the annuity, we can answer this question.

The other question is: "What must I invest today to obtain X dollars in the future, after so many periods at a given rate?" For example, "If I want $1,000 after five years at 8%, compounded annually, then what sum *less than* $1,000 must I invest today?" In other words, what is the *present value* of the future $1,000?

Present value refers to the dollar value *now* of a future sum of money. The spending power of a future dollar might be only 90¢ or 20¢ in today's terms, depending on how far off the future is and what rate of interest we are assuming.

Present value problems can be solved using Table 12.1, but dividing instead of multiplying, since we are going in the reverse direction. If we want $1 in three years at 8%, compounded annually, we divide $1 by $1.2597120 to determine the sum to be invested today.

$$\$1 \div \$1.2597120 = \$0.7938322 = \$0.79$$

Investing $0.79 today for three years, with interest compounded annually at 8%, will yield $1.00. This computation, with necessary rounding, can be checked as follows:

Original deposit	$0.79	
First-year interest ($0.79 × .08)	0.0632	
Balance, end of first year	$0.8532	
Second-year interest ($0.8532 × .08)	0.0682	
Balance, end of second year	$0.9215	
Third-year interest ($0.9215 × .08)	0.0737	
Balance, end of third year	$0.9952	= $1.00

Honestly, this method is about the most tedious we have yet encountered. Division by a seven-digit number increases the likelihood of error and is also time-consuming. For these reasons, present value tables, such as Table 12.3, are available. Notice on the "3" line in the 8% column the quotient of our division, $0.7938322.

SAMPLE PROBLEM 7

Find the amount that Orlando Grasso has to invest today to obtain $10,000 in seven years, if 12% interest, compounded annually, is earned.

Solution

Table 12.3: 7 periods, 12% = $0.4523492
$0.4523492 × 10,000 = $4,523.49

Check What will $4,523.49 invested at 12%, compounded annually, amount to in seven years?

Table 12.1: 7 periods, 12% = $2.2106814
$2.2106814 × 4,523.49 = $9,999.995 = $10,000

Table 12.3

Present Value of $1 at Compound Interest

Number of Periods	2%	3%	4%	6%	8%	12%
1	0.9803922	0.9708738	0.9615385	0.9433962	0.9259259	0.8928571
2	0.9611688	0.9425959	0.9245562	0.8899964	0.8573388	0.7971939
3	0.9423223	0.9151417	0.8889964	0.8396193	0.7938322	0.7117802
4	0.9238454	0.8884870	0.8548042	0.7920937	0.7350299	0.6355181
5	0.9057308	0.8626088	0.8219271	0.7472582	0.6805832	0.5674269
6	0.8879714	0.8374843	0.7903145	0.7049605	0.6301696	0.5066311
7	0.8705602	0.8130915	0.7599178	0.6650571	0.5834904	0.4523492
8	0.8534904	0.7894092	0.7306902	0.6274124	0.5402689	0.4038832
9	0.8356443	0.7664167	0.7025856	0.5918985	0.5002490	0.3606100
10	0.8203483	0.7440939	0.6755642	0.5583948	0.4631935	0.3219732
11	0.8042630	0.7224213	0.6495809	0.5267875	0.4288829	0.2874761
12	0.7884932	0.7013799	0.6245970	0.4969694	0.3971138	0.2566751
13	0.7730325	0.6809513	0.6005741	0.4688390	0.3676979	0.2291742
14	0.7578750	0.6611178	0.5774751	0.4423010	0.3404610	0.2046198
15	0.7430147	0.6418619	0.5552645	0.4172651	0.3152417	0.1826963
16	0.7284458	0.6231669	0.5339082	0.3936463	0.2918905	0.1631217
17	0.7141626	0.6050164	0.5133732	0.3713644	0.2702690	0.1456443
18	0.7001594	0.5873946	0.4936281	0.3503438	0.2502490	0.1300396
19	0.6864308	0.5702860	0.4746424	0.3305130	0.2317121	0.1161068
20	0.6729713	0.5536758	0.4563870	0.3118047	0.2145482	0.1036668
25	0.6095309	0.4776056	0.3751168	0.2329986	0.1460179	0.0588233
30	0.5520709	0.4119868	0.3083187	0.2313774	0.0993773	0.0333779
40	0.4528908	0.3065568	0.2082890	0.0972222	0.0460309	0.0107468
50	0.3715279	0.2281071	0.1407126	0.0542884	0.0213212	0.0034602

Note As you see, the answer to a present value calculation can be checked by finding the amount of the present value, using Table 12.1.

In a present value problem, the amount of interest earned is found by subtracting the present value from the future amount. In Sample Problem 7:

Future amount	$10,000.00
Present value	−4,523.49
Interest	$ 5,476.51

This interest is called *compound discount.* Sample Problem 8 reviews its calculation and illustrates quarterly compounding of the interest.

SAMPLE PROBLEM 8

Arnold Reissman wants to have $20,000 in ten years. Interest can be earned at 8%, compounded quarterly. Find (a) the sum to be invested today and (b) the compound discount.

Solution

(a) Table 12.3: 2%, 40 periods = $0.4528908
$0.4528908 × 20,000 = $9,057.82

(b) Compound discount = Future amount − Present value
= $20,000.00 − $9,057.82
= $10,942.18

Present Value of an Annuity

In the previous section, we found the *single amount* that must be invested today in order to get back a larger *single amount* at some time in the future.

Here is another logical question: "What *single amount* can I put aside today in order to get back a *series* of amounts in the future?" Perhaps you would like to receive $10,000 a year after your 65th birthday. What one sum must you invest now in order to have this annuity? In other words, what is the *present value of an annuity?*

Since an annuity is a series of regular payments, we must calculate the present value of a series of amounts. If $1 is to be earned in one year, we must find the present value of $1 for one period; if a second $1 is to be earned in the second year, we find the present value of $1 for two periods, and so on. Using Table 12.3 to find the sum to be invested today, in order to receive $1 per year for the next four years at 8%, compounded annually, we can set up the following solution:

Present value of $1 due in one year at 8%	= $0.9259259
Present value of $1 due in two years at 8%	= 0.8573388
Present value of $1 due in three years at 8%	= 0.7938322
Present value of $1 due in four years at 8%	= 0.7350299
Amount to be invested today to receive $1 per year for the next four years	$3.3121268

Investing $3.31 today, with interest compounding at 8% annually, permits withdrawals of $1 a year for the next four years.

Table 12.4

Present Value of Annuity of $1
at Compound Interest

Number of Periods	2%	3%	4%	6%	8%	12%
1	0.9803922	0.9708738	0.9615385	0.9433962	0.9259259	0.8928571
2	1.9415609	1.9134697	1.8860947	1.8333927	1.7832648	1.6900510
3	2.8838833	2.8286114	2.7750910	2.6730120	2.5770970	2.4018312
4	3.8077287	3.7170984	3.6298952	3.4651056	3.3121268	3.0373493
5	4.7134595	4.5797072	4.4518223	4.2123638	3.9927100	3.6047762
6	5.6014309	5.4171914	5.2421369	4.9173243	4.6228797	4.1114073
7	6.4719911	6.2302830	6.0020547	5.5823814	5.2063701	4.5637565
8	7.3254814	7.0196922	6.7327449	6.2097938	5.7466389	4.9676397
9	8.1622367	7.7861089	7.4353316	6.8016923	6.2468879	5.3282497
10	8.9825850	8.5302028	8.1108958	7.3600870	6.7100814	5.6502229
11	9.7868480	9.2526241	8.7604767	7.8868746	7.1389643	5.9376990
12	10.5753412	9.9540040	9.3850738	8.3838439	7.5360780	6.1943741
13	11.3483737	10.6349553	9.9856478	8.8526830	7.9037759	6.4235483
14	12.1062488	11.2960731	10.5631229	9.2949839	8.2442370	6.6281681
15	12.8492635	11.9379351	11.1183874	9.7122490	8.5594787	6.8108644
16	13.5777093	12.5611020	11.6522956	10.1058953	8.8513692	6.9739861
17	14.2918719	13.1661185	12.1656688	10.4772597	9.1216381	7.1196304
18	14.9920313	13.7535131	12.6592970	10.8276035	9.3718871	7.2496700
19	15.6784620	14.3237991	13.1339394	11.1581165	9.6035992	7.3657768
20	16.3514333	14.8774749	13.5903263	11.4699212	9.8181474	7.4694436
25	19.5234565	17.4131477	15.6220799	12.7833562	10.6747762	7.8431396
30	22.3964556	19.6004414	17.2920333	13.7648312	11.2577833	8.0551846
40	27.3554792	23.1147720	19.7927739	15.0462969	11.9246133	8.2437776
50	31.4236059	25.7297640	21.4821846	15.7618606	12.2334846	8.3044976

Table 12.4 eliminates adding a series of present values, as shown. On the "4" line, for 8%, you find $3.3121268. For 10 years at 8%, compounded quarterly, 2% for 40 periods = $27.3554792. By depositing $27.36 today, you can withdraw $1 every three months for the next 10 years.

SAMPLE PROBLEM 9

How much must Alicia Godwin invest today to receive an annuity of $500 every six months for the next two years, if interest is earned at 6%, compounded semiannually?

Solution

Table 12.4: 3%, four periods = $3.7170984
$3.7170984 × 500 = $1,858.55

Let's follow Sample Problem 9 a bit further. Investing $1,858.55 today permits semiannual withdrawals of $500.00 for two years. The listing below shows this in action.

Original balance	$1,858.55
First-period balance ($1,858.55 × .03)	55.76
Balance, end of first period	$1,914.31
First withdrawal	−500.00
Balance, start of second period	$1,414.31
Second-period interest ($1,414.31 × .03)	42.43
Balance, end of second period	$1,456.74
Second withdrawal	−500.00
Balance, start of third period	$ 956.74
Third-period interest ($956.74 × .03)	28.70
Balance, end of third period	$ 985.44
Third withdrawal	−500.00
Balance, start of fourth period	$ 485.44
Fourth-period interest ($485.44 × .03)	14.56
Balance, end of fourth period	$ 500.00
Final withdrawal	−500.00
Final balance	$ -0-

TEST YOUR ABILITY

1. In order to have $1 in eight years at 8% interest, compounded annually, you must invest $0.5402689 today. Use this fact to determine the investment needed under the same circumstances to accumulate (a) $1,000; (b) $50; (c) $30,000.
2. Use Table 12.3 to find the present value and compound discount in each problem.

	Amount Desired	R	T	Compounded
(a)	$ 4,000	12%	17 years	annually
(b)	$ 2,000	8%	25 years	annually
(c)	$ 600	12%	9 years	semiannually
(d)	$ 3,000	6%	15 years	semiannually
(e)	$20,000	12%	5 years	quarterly
(f)	$ 850	8%	4 years	quarterly

3. (a) If interest at 8% is compounded semiannually, find the present value of $3,000, 15 years from now. (b) Check your answer to (a) by using Table 12.1.
4. Today's investment of $7.8868746 will permit withdrawals of $1 a year for the next 11 years, with interest compounding annually at 6%. Determine the investment needed to permit withdrawals, under the same terms, of (a) $100; (b) $2,000; (c) $35.
5. Use Table 12.4 to find the present value of each of the following annuities:

	Annuity	R	T	Interest Compounded and Withdrawals Made
(a)	$ 200	8%	5 years	annually
(b)	$ 500	6%	9 years	annually
(c)	$2,000	8%	2 years	semiannually

	Annuity	R	T	Interest Compounded and Withdrawals Made
(d)	$ 150	6%	8 years	semiannually
(e)	$ 25	12%	5 years	quarterly
(f)	$ 600	8%	3 years	quarterly

The Sinking Fund

Your goal is $20,000 in ten years. You want to invest a single sum of money that will accumulate to this goal, using Table 12.3, *but* you cannot afford a large deposit. A series of small equal investments is possible, however, setting up a *sinking fund*—a gradual accumulation of money, with interest, in equal periodic installments. In the same way, a business firm can start a sinking fund to redeem bonds, erect a building, or provide pension payments.

A sinking fund differs from the annuity situations, presented earlier, in two ways:

(1) Deposits are made into the sinking fund at the *end* of each period. No interest is earned for the first period. In the annuity, interest was earned in the first period.

(2) In our sinking fund problems, we will be looking for the amount of the periodic contribution needed to yield a known sum. In the annuity problems, we knew the amount of the periodic contribution or withdrawal and were seeking either the final amount or the present value.

Table 12.5 is a sinking fund table, entitled "Annuity That Amounts to $1 at Compound Interest." It shows what sums must be deposited for each period of time at given rates of interest in order to yield $1 in the future. For example, if you want $1 in three years and interest of 8% is compounded annually, Table 12.5 shows $0.3080335, or 31¢. Depositing 31 cents a year for three years will return $1, as shown below:

First investment	$0.31	
First-year interest	–0–	(deposit at end of year)
Second-year interest ($0.31 × .08)	0.02	
Second investment	0.31	
Balance, end of second year	$0.64	
Third-year interest ($0.64 × .08)	0.05	
Third investment	0.31	
Balance, end of third year	$1.00	

As you can see, the main distinction between a sinking fund and an annuity is in the first year. No interest is credited until the second year.

SAMPLE PROBLEM 10

Find the amount that the Malone Detective Agency invests at the end of every three months for the next four years to accumulate $3,000, if interest is compounded quarterly at 8%.

Solution

Table 12.5: 2%, 16 periods = $0.0536501
$0.0536501 × 3,000 = $160.95

Table 12.5

Annuity That Amounts to $1
at Compound Interest
(Sinking Fund Table)

Number of Periods	2%	3%	4%	6%	8%	12%
1	1.0000000	1.0000000	1.0000000	1.0000000	1.0000000	1.0000000
2	0.4950495	0.4926108	0.4901961	0.4854369	0.4807692	0.4716981
3	0.3267547	0.3235304	0.3203485	0.3141098	0.3080335	0.2963490
4	0.2426238	0.2390271	0.2354901	0.2285915	0.2219208	0.2092344
5	0.1921584	0.1883546	0.1846271	0.1773964	0.1704565	0.1574097
6	0.1585258	0.1545975	0.1507619	0.1433626	0.1363154	0.1232257
7	0.1345120	0.1305064	0.1266096	0.1191350	0.1120724	0.0991177
8	0.1165098	0.1124564	0.1085278	0.1010359	0.0940148	0.0813028
9	0.1025154	0.0984339	0.0944930	0.0870222	0.0800797	0.0676789
10	0.0913265	0.0872305	0.0832909	0.0758680	0.0690295	0.0569842
11	0.0821779	0.0780775	0.0741490	0.0667929	0.0600763	0.0484154
12	0.0745596	0.0704621	0.0675522	0.0592770	0.0526950	0.0414368
13	0.0681184	0.0670295	0.0601437	0.0529601	0.0465218	0.0356772
14	0.0626020	0.0585263	0.0546690	0.0475849	0.0412969	0.0308712
15	0.0578255	0.0537666	0.0499411	0.0429628	0.0368295	0.0268242
16	0.0536501	0.0496109	0.0458200	0.0389521	0.0329769	0.0233900
17	0.0499698	0.0459525	0.0421985	0.0354448	0.0296294	0.0204567
18	0.0467021	0.0427087	0.0389933	0.0323565	0.0267021	0.0179373
19	0.0437818	0.0398139	0.0361386	0.0296209	0.0241276	0.0157630
20	0.0411567	0.0372157	0.0335818	0.0271846	0.0218522	0.0138788
25	0.0312204	0.0274279	0.0240120	0.0182267	0.0136788	0.0075000
30	0.0246499	0.0210193	0.0178301	0.0126489	0.0088274	0.0041437
40	0.0165558	0.0132624	0.0105235	0.0064615	0.0038602	0.0013036
50	0.0118232	0.0088655	0.0065502	0.0034443	0.0017429	0.0004167

With quarterly deposits of $160.95 for four years, and interest at 8%, compounded quarterly, the detective agency will accumulate $3,000.

The Sinking Fund Schedule

Another common feature in business is a *sinking fund schedule,* which shows how a firm's money accumulates. Sample Problem 11 goes over calculation of a periodic deposit and illustrates a sinking fund schedule.

SAMPLE PROBLEM 11

(a) Find the amount that Douglas Severin invests at the end of each of the next seven years, with interest at 6%, compounded annually, to accumulate $10,000.

(b) Prepare a sinking fund schedule.

Solution

(a) Table 12.5: 6%, 7 periods = $0.1191350
$0.1191350 × 10,000 = $1,191.35 each year

(b) Year	Amount at Beginning	Interest at 6%	+	Annual Contribution	=	Total Increase	Amount at End
1				$1,191.35		$1,191.35	$ 1,191.35
2	$1,191.35	$ 71.48		1,191.35		1,262.83	2,454.18
3	2,454.18	147.25		1,191.35		1,338.60	3,792.78
4	3,792.78	227.57		1,191.35		1,418.92	5,211.70
5	5,211.70	312.70		1,191.35		1,504.05	6,715.75
6	6,715.75	402.95		1,191.35		1,594.30	8,310.05
7	8,310.05	498.60		1,191.35[1]		1,689.95	10,000.00

[1] The last payment is sometimes rounded by a few cents in order to make the final balance exact.

Compare this form to the annuity schedule shown earlier and notice again the differences in the first year and in the base for the interest calculations.

Summary of Interest Tables

One of the most important things in this chapter is to know *when* to use a particular table. With practice, choosing the right one becomes natural. Meanwhile, the following summary will be helpful.

Table	Title	Typical Question
12.1	Amount of $1 at Compound Interest	What will $1 today amount to in five years, at 8% annually?
12.2	Amount of Annuity of $1 at Compound Interest	What will deposits of $1 a year for five years amount to, at 8% annually?
12.3	Present Value of $1 at Compound Interest	What must be invested today to have $1 in five years, at 8% annually?
12.4	Present Value of Annuity of $1 at Compound Interest	What must be invested today to withdraw $1 a year for the next five years, at 8% annually?
12.5	Annuity That Amounts to $1 at Compound Interest (Sinking Fund Table)	What must be invested each year to have $1 in five years, at 8% annually?

TEST YOUR ABILITY

1. An investment of $0.0354448 a year for 17 years at 6%, compounded annually, will accumulate to $1. Using this fact, determine the annual deposit needed to accumulate, under the same terms, (a) $100; (b) $5,000; (c) $300,000.
2. Using Table 12.5, find the periodic sinking fund deposit in each situation:

	Amount Desired	R	T	Deposits Made and Interest Compounded
(a)	$ 1,000	8%	13 years	annually
(b)	$ 500	12%	14 years	annually

(c)	$ 6,000	12%	9 years	semiannually
(d)	$ 4,000	8%	20 years	semiannually
(e)	$ 1,200	8%	3 years	quarterly
(f)	$25,000	12%	10 years	quarterly

3. James Flynn wants to have $1,000 in five years. Interest can be earned at 6%, compounded annually. (a) Find the amount of his annual contribution. (b) Prepare a sinking fund schedule.
4. You have the opportunity to invest at 8%, compounded annually, over a time span of five years.
   (a) What will annual deposits of $300 amount to in five years?
   (b) What sum must be invested today to receive $300 a year for the next five years?
   (c) What will $300 invested today amount to in five years?
   (d) What annual deposit must be made to accumulate $300 in five years?
   (e) What sum must be invested today to receive $300 in five years?

The Savings Account

An individual or a business can earn compound interest by opening a savings account in a lending institution—for example, a savings bank, a savings and loan association, a commercial bank, or a credit union.

A savings account has many features which affect the computation of compound interest.

(1) Interest is normally credited quarterly or monthly. If an account pays interest at 5% a year, a rate of $1\frac{1}{4}$% (5% ÷ 4) is applied every three months or $\frac{5}{12}$% each month. Some banks today compound interest daily, a technique that became practical with the development of modern computers. If the annual rate is 5%, the daily rate is $\frac{5}{365}$ (a change from bankers' 360-day year) which equals .0137%, or .000137, a day. Interest is still added quarterly or monthly to the depositor's account.

(2) Savings accounts generally earn interest on whole dollar amounts only, disregarding cents. For example, $100.000 and $100.99 earn the same interest.

(3) Typical savings accounts earn interest on full quarters (three-month periods) only. A deposit made on March 1 will not earn interest for the quarter ending March 31. Similarly, a withdrawal on June 29 would earn no interest for the second quarter, which ends on June 30. Another modern trend, however, is for savings banks to pay interest from the day of deposit to the day of withdrawal; therefore, a deposit earns interest for all the days it is in the account. A lower rate of interest is earned on the "modern" account, usually $\frac{1}{2}$% less.

(4) Some savings accounts have a built-in limitation: they are ineligible for interest if the balance falls under a minimum.

(5) A modern bank innovation that features aspects of the checking and the savings account is the *NOW account*. Checks can be written without incurring a service charge, so long as the balance remains at or above a certain level; in addition, interest is earned on any balance in the account, no matter how small. A typical NOW account agreement might read: "Interest of $5\frac{1}{2}$% is earned on any balance in this account. No service charge will be made for checks written during any month when the average balance is $500 or more. When an average monthly balance falls below $500, a fee of 15 cents will be charged for each check written during that month."

(6) Another innovation offered by savings banks, the savings certificate, is discussed under Bonds and Other Investments (Chapter 15).

Compound interest tables are not used to find the interest on savings accounts, primarily because deposits and/or withdrawals are so frequent.

Sample Problems 12, 13, and 14 show the features of a savings account.

SAMPLE PROBLEM 12

Find the amount in the savings account of the Martinez Company after one year, if the interest at 6% is compounded quarterly. A balance of $645.25 was on hand at the start of the year, and no additions or withdrawals were made during the year.

Solution

Beginning balance	$645.25
First-quarter interest ($1\frac{1}{2}\%$ × $645)	9.68
Balance, end of first quarter	$654.93
Second-quarter interest ($1\frac{1}{2}\%$ × $654)	9.81
Balance, end of second quarter	$664.74
Third-quarter interest ($1\frac{1}{2}\%$ × $664)	9.96
Balance, end of third quarter	$674.70
Fourth-quarter interest ($1\frac{1}{2}\%$ × $674)	10.11
Balance, end of year	$684.81

Note Interest was earned on whole dollar amounts only at one-fourth of the annual rate each quarter.

SAMPLE PROBLEM 13

Find the year-end balance for the following account of Matty Henshaw, earning 8% a year, compounded quarterly, with interest given only for full quarters and only if a minimum balance of $300 is maintained:

Balance, January 1	$ 700
Deposit, March 1	500
Withdrawal, November 1	−1,000

Solution

Balance, January 1	$ 700.00
Deposit, March 1	500.00
Total	$1,200.00
First-quarter interest (2% × $700)	14.00
Balance, end of first quarter	$1,214.00
Second-quarter interest (2% × $1,214)	24.28
Balance, end of second quarter	$1,238.28
Third-quarter interest (2% × $1,238)	24.76
Balance, end of third quarter	$1,263.04
Withdrawal, November 1	1,000.00
Balance	$ 263.04
Fourth-quarter interest	−0−
Balance, end of year	$ 263.04

The March 1 deposit was not in the account for a full quarter, so no interest was earned on it. The fourth-quarter balance went below the minimum; no interest was recorded. Had interest been paid from the day of deposit to the day of withdrawal (as in the next problem), interest on the $1,000 would have been earned from October 1 to November 1—as well as one month's interest on the March 1 deposit.

SAMPLE PROBLEM 14

The Crusoe Company deposits $800 in a bank paying 5% interest, compounded quarterly, from the day of deposit to the day of withdrawal. The deposit is made on September 12 and withdrawn on November 9. Find the interest earned.

Solution

One quarter = 90 days
Rate = $1\frac{1}{4}$%
Interest from September 12–September 30 (end of quarter) =
$1\frac{1}{4}$% for 18 days on $800
$.0125 \times \frac{18}{90} \times \$800 = \$2.00$
Interest from September 30–November 9 = $1\frac{1}{4}$% for 40 days on $802
$.0125 \times \frac{40}{90} \times \$802 = \$4.46$
Total interest earned = $2.00 + $4.46 = $6.46

The second calculation was based on the compounded amount of $802.

TEST YOUR ABILITY

1. Find the year-end balance in the savings account of the Milady Escort Company if the firm began the year with a balance of $521.75 and made no additions or withdrawals. Interest at 6% is compounded quarterly.
2. The Gray Company has a savings account paying interest at 8%, compounded quarterly, for full quarters only, provided a minimum balance of $500.00 is kept. Figures for the year are as follows:

Balance, January 1	$ 600.00
Deposit, February 1	900.00
Withdrawal, October 15	−1,200.00

   Find the year-end balance.
3. The Friburg National Bank pays interest at 6%, compounded quarterly, from day of deposit to day of withdrawal. Find the interest earned on a deposit of $900 made on June 10 and withdrawn on August 29.

CHECK YOUR READING

1. Distinguish between simple and compound interest.
2. At what point in a compound interest problem is the interest calculated?
3. Explain how to use a compound interest table when quarterly compounding is used.
4. Write a definition of *annuity*.
5. Could "Amount of Annuity" problems be solved from Table 12.1, the "Amount of $1" table? Why do you think so?
6. What is the meaning of *present value*?

7. Explain how a present value of $1 table (Table 12.3) and an amount of $1 table (Table 12.1) can be used as a check on each other in problems.
8. Distinguish between the terms *compound interest* and *compound discount*.
9. How would you describe a sinking fund?
10. Why is the first line in Table 12.5 the same for all rates? And on the other hand, why is the first line in Table 12.2 not the same for all rates?
11. How does the compound interest computation on a savings account differ from that on a fund?
12. Is a NOW account a checking or a savings account? Explain.

EXERCISES *Group A*

1. One dollar invested today, compounded annually at 6% for 15 years, amounts to $2.3965582. Use this fact to find the accumulation of each of the following principals: (a) $20; (b) $7,000; (c) $850.
2. Marilyn Leibman decides to begin contributing annual deposits of $1,000 to an IRA plan. Prepare an annuity schedule to show the first five years' accumulation, if interest is compounded annually at 7%.
3. Using Table 12.4, find the present value of each of the following annuities:

	Annuity	R	T	Interest Compounded and Withdrawals Made
(a)	$ 1,000	8%	16 years	annually
(b)	$ 5,000	12%	3 years	quarterly
(c)	$ 300	12%	8 years	semiannually
(d)	$10,000	8%	5 years	quarterly
(e)	$ 800	6%	30 years	annually
(f)	$ 1,500	16%	1 year	semiannually

4. You have an opportunity to invest at 8%, compounded quarterly.
   (a) What must you invest today to have $600 every three months for the next two years?
   (b) What will investments of $600 every quarter for the next two years amount to?
   (c) What must you invest today to have $600 in two years?
   (d) What will an investment of $600 today amount to in two years?
   (e) What equal quarterly deposits must be made for the next two years to accumulate $600?
5. Using Table 12.3, find the present value and compound discount in each situation:

	Amount Desired	R	T	Compounded
(a)	$ 1,000	8%	7 years	semiannually
(b)	$ 700	6%	19 years	annually
(c)	$ 900	8%	3 years	quarterly
(d)	$ 6,000	12%	13 years	annually
(e)	$ 3,400	12%	8 years	semiannually
(f)	$15,000	16%	10 years	quarterly

6. The Sewanhaka National Bank pays interest at 6%, compounded quarterly, from the day of deposit to the day of withdrawal. Find the interest earned on a $2,000 deposit made on July 17 and withdrawn on October 20 of the same year.
7. If you are a very modest investor and wish to have $1 in 15 years at 6% interest, compounded annually, you must invest $0.4172651 today. On a larger scale, what must one invest under the same terms to accumulate (a) $1,000; (b) $800; (c) $50,000?
8. Using Table 12.1, find the amount and the compound interest in each of the following problems.

	P	R	T	Compounded
(a)	$ 900	8%	10 years	quarterly
(b)	$ 3,000	8%	9 years	annually
(c)	$ 500	16%	10 years	quarterly
(d)	$40,000	6%	25 years	semiannually
(e)	$ 2,500	8%	40 years	annually
(f)	$ 150	12%	25 years	semiannually

9. An annual investment of $0.0429628 for 15 years at 6%, compounded annually, will accumulate to $1. What must be invested in each year, under the same circumstances, to accumulate (a) $50; (b) $2,000; (c) $50,000?
10. Find the amount in the savings account of the Jester Corporation after one year, if interest of 6% is compounded quarterly. The firm began the year with a balance of $409.50 and made no additions or withdrawals during the year.
11. Using Table 12.5, find the amount of the periodic deposit in each of the following situations:

	Amount Desired	R	T	Deposits Made and Interested Compounded
(a)	$ 1,000	12%	19 years	annually
(b)	$ 800	6%	9 years	semiannually
(c)	$ 3,000	6%	10 years	annually
(d)	$15,000	12%	5 years	quarterly
(e)	$ 500	8%	15 years	semiannually
(f)	$60,000	16%	5 years	quarterly

12. By investing $1 every year for the next 15 years at 6%, compounded annually, you will have $24.6725281. Use this fact to determine the account accumulated on annual investments of (a) $50; (b) $1,000; (c) $25.
13. Central Bank pays 8% a year, compounded quarterly, if a minimum balance of $1,000 is maintained. Interest is payable for full quarters only. Calculate the year-end balance of the following account:

Balance, January 1	$500
Deposit, March 1	800
Withdrawal, August 1	−200

14. Marsha Kartzen deposits $6,000 in an account paying 8% interest for ten years. Find the amount and the compound interest earned if interest is compounded (a) annually; (b) semiannually; (c) quarterly.

15. Interest is compounded semiannually at a rate of 8% in the Fawcett Loan Association.
    (a) Find the present value of $5,000 nine years from now.
    (b) Check your answer to (a) by using Table 12.1.
16. Bruce Applebaum will need $10,000 in six years. His money can earn interest at 8%, compounded annually. (a) Find the amount of his annual contribution. (b) Prepare a schedule of sinking fund accumulation.
17. Using Table 12.2, find the amount of each of the following annuities:

Annuity	R	T	Invested and Interest Compounded
(a) $ 600	16%	7 years	semiannually
(b) $10,000	12%	5 years	quarterly
(c) $ 300	8%	11 years	annually
(d) $ 50	8%	15 years	semiannually
(e) $ 150	6%	7 years	annually
(f) $ 750	16%	5 years	quarterly

18. Jimmy Horton's investment of $9.7122490 today with interest accumulating at 6%, compounded annually, will permit withdrawals of $1 a year for the next 15 years. What investment is necessary, under the same circumstances, to permit annual withdrawals of (a) $100; (b) $500; (c) $2,500.

Group B

1. The Henry Repair Shop invests $5,000 in a fund at 12%, compounded quarterly. At the end of five years, the shop withdraws the interest. How much will the shop withdraw?
2. William Reese sets up a trust fund for his eight-year-old son, to be used for college expenses when the child reaches 18.
    (a) If Mr. Reese makes quarterly deposits of $100 and the deposits earn interest at 8%, compounded quarterly, what will have accumulated after 10 years?
    (b) Assume that he can only make semiannual deposits of $200 for 10 years, with interest compounded semiannually. Determine the amount of the annuity.
3. Paula White deposits $500 every six months for $3\frac{1}{2}$ years in an account paying interest at 8%, compounded semiannually. Prepare an annuity schedule to show the growth of her investments.
4. The Hudson Water Corporation needs $20,000 in eight years to repay a bond investment. What single deposit must be made today to accumulate this sum if interest is available at 8%, compounded semiannually?
5. Electronics, Inc., wants to deposit a sufficient sum of money today at 8%, compounded quarterly, to be able to withdraw $5,000 every three months for the next four years. What amount should be deposited?
6. Frances Hopper's goal is a computer, so she needs $4,000 in five years, with deposits made semiannually and interest at 8%, compounded semiannually. Prepare a schedule of her sinking fund accumulation.

CHAPTER **13**

Consumer Credit

For a long time, consumers who bought on the installment plan or who used credit cards or charge accounts were often unable to obtain accurate information about the interest charges from their creditors. On July 1, 1969, the Federal Reserve Board adopted *Regulation Z,* more commonly known as the *Truth-in-Lending Act,* ending a period of "let the buyer beware." Subsequent revisions to Regulation Z, most recently the 1980 *Truth-in-Lending Simplification and Reform Act,* have defined the rights of the consumer even further.

Regulation Z requires that when a consumer borrows money from a lending institution or buys on credit, the creditor must supply a statement of the true annual rate of interest being charged and/or specify which method is being used to calculate the interest charges.

As a consumer and/or a future business person, you should be familiar with the operations and laws of credit plans.

Your Objectives

After studying Chapter 13, you will be able to

1. Identify some of the key rules in Regulation Z.
2. Calculate the interest charges on credit cards and charge accounts by various methods.
3. Compute finance charges and monthly payments on installment buying.
4. Prepare a schedule of debt-reduction for a bank loan, mortgage loan, or other type of financial loan.
5. Calculate the true annual rate of interest on various types of credit.
6. Begin to choose among various sources of consumer credit.

Terms of Regulation Z

Regulation Z applies to banks, credit unions, department stores, car dealers, consumer finance companies—in fact, any business or individual extending credit for a sum payable in more than four installments and for which there is a finance charge. A creditor is required by Regulation Z to supply a customer with the details of the *finance charge*—all the costs that the consumer must pay the creditor. Common costs include interest, loan fees, insurance, and discounts. To make our work in this chapter easier, let's assume interest to be the only finance charge.

Regulation Z distinguishes between two kinds of credit: *open-end credit* and *other credit*. In *open-end credit*, the finance charge is calculated monthly, based on unpaid amounts. Charge accounts and credit card accounts, including cash advances, are examples of open-end credit.

Other credit includes installment sales and loans, that is, when a specific sum is to be repaid in a series of payments. Finance charges are part of each payment and are based on the original amount borrowed.

If you are interested in further details about Regulation Z, you can get a copy by writing to your district's Federal Reserve Bank.

Open-End Credit

Open-end credit, as shown by the popularity of bank, gasoline, and department store charge accounts, is very common today. Let's take a brief look at three of the methods of calculating interest on such accounts.

Generally, interest is calculated monthly at $\frac{1}{12}$ of the annual rate. Typical annual rates are:

19.8% on the first $500
12% on amounts over $500

Monthly rates are then:

1.65% on the first $500
1% on amounts over $500

In some plans, a flat rate is charged on all amounts; in others, a charge—sometimes as high as 24%—is made only on larger amounts, such as $1,000 or more. Plans vary from state to state.

Some charge plans use a daily rate, similar to a savings account rate for daily compounding. Dividing the annual rate by 365 yields the daily rate. We will avoid this extra work and assume monthly rates in our problems.

A common method of calculating the interest charge is the *previous balance method*. The monthly rate is applied to the *unpaid balance* at the start of the month.

SAMPLE PROBLEM 1

Alex Bolduc's account at the Central Department Store had a June 1 balance of $750. During June, he paid $150 towards his balance on the 9th, and purchased $100 more on the 22nd. Find the finance charge for June, using an 18% annual rate and the previous balance method.

Solution

Previous balance	\$750
Monthly rate $1\frac{1}{2}\%$	×.015
Finance charge	\$11.25

If this method is used, both *purchases and payments made during the current month are ignored.* It is to the consumer's advantage to ignore purchases, but not to ignore payments. Had the account been paid in full, the finance charge would still have remained at \$11.25. The *adjusted balance method,* shown in Sample Problem 2, helps: Purchases are ignored, but payments are not.

SAMPLE PROBLEM 2

Assume, for the data in Sample Problem 1, the adjusted balance method. Calculate Bolduc's finance charge.

Solution

Previous balance	\$750
Payment	−150
Adjusted balance	\$600
Monthly rate $1\frac{1}{2}\%$	×.015
Finance charge	\$9.00

The third method is the *average daily balance method* in which both purchases and payments are considered, as well as the days involved. An average daily balance of a charge account is calculated by multiplying each balance by the number of days it was maintained. In Bolduc's case, he had a balance of:

\$750 from June 1 through 8 or	8 days,
\$600 from June 9 through 21 or	13 days, and
\$700 from June 22 through 30 or	9 days
	30 days

The balance changes *on the day* of payment or purchase. The average daily balance is calculated as follows:

$$
\begin{aligned}
\$750 \times 8 &= \$\ 6{,}000 \\
\$600 \times 13 &= \quad 7{,}800 \\
\$700 \times 9 &= \quad 6{,}300 \\
& \quad \$20{,}100 \div 30 \text{ days} = \$670
\end{aligned}
$$

The finance charge is then .015 × \$670 = \$10.05. The average daily balance method seems to be the most prevalent method today.

Installment Buying

Installment buying is an important feature of our economy. A person who cannot afford the price of an item can usually "buy now, pay later." In this type of purchase, the buyer gets the use of the item (such as a car) and agrees to pay a specific amount over a specific number of weeks, months, or years. Often, a certain sum is required as a *down payment* on the purchase.

SAMPLE PROBLEM 3

Find the total cost of a portable copying machine purchased for $75 down and 24 monthly payments of $52.50.

Solution

Total cost = $75 + (24 × $52.50) = $75 + $1,260 = $1,335

Is $1,335 expensive? This can be decided by comparing $1,335 with the price of the copying machine if it had been purchased for cash. If the cash price was $1,200, then the finance charge, or *carrying charge*, would be $135.

Computing the Finance Charge

Finance charges were defined earlier as the total costs of credit to the consumer. Finance charge can also be defined as the difference between the cash price and the installment price.

SAMPLE PROBLEM 4

Find the finance charge for a generator with a cash price of $730 that is purchased for $50 down and 18 monthly payments of $40.

Solution

Installment price = $50 + (18 × $40) =	$770
Cash price	−730
Finance charge	$ 40

The finance charge is often expressed as a percentage of the cash price. In such a case, it is treated as simple interest for the time period required.

SAMPLE PROBLEM 5

A car with a cash price of $5,200 is to be paid for over 36 months at an annual charge of 8%. Find the finance charge.

Solution

$$
\begin{aligned}
I &= P \times R \times T \\
&= \$5{,}200 \times 8\% \times 36 \text{ months} \\
&= \$5{,}200 \times .08 \times 3 \text{ years} \\
&= \$1{,}248
\end{aligned}
$$

Computing the Periodic Payment

Another way math is used in installment buying is to compute the amount of each payment. In Sample Problem 6, the finance charge is included in the installment price. In Sample Problem 7, the charge is added on to the cash price.

SAMPLE PROBLEM 6

A stove can be purchased on the installment plan for $825, with $90 down and the remainder in seven equal monthly installments. Find the amount of each installment.

Solution

Balance = $825 − $90 = $735

Monthly payment = $735 ÷ 7 = $105

SAMPLE PROBLEM 7

Sophia Magnani is going to buy a stereo system on the installment plan for \$2,000, plus a 7% "add-on" charge. Calculate her monthly payment if installments are to be paid over (a) 15 months; (b) 24 months.

Solution

(a) Finance charge = \$2,000 × 7% × $1\frac{1}{4}$ years = \$175
Total cost = \$2,000 + \$175 = \$2,175
Monthly payment = \$2,175 ÷ 15 = \$145

(b) Finance charge = \$2,000 × 7% × 2 years = \$280
Total cost = \$2,000 + \$280 = \$2,280
Monthly payment = \$2,280 ÷ 24 = \$95

The choice of time periods is the consumer's responsibility.

TEST YOUR ABILITY

1. Ellen McNiff's account at the Wilkes Department Store carries an annual interest rate of 18%. Her June 1 balance is \$300. During June, she makes a \$200 purchase on the 11th and a \$300 payment on the 21st. Determine her finance charge under (a) the previous balance method; (b) the adjusted balance method; (c) the average daily balance method.
2. Find the total cost of each of the following installment purchases:

	Down Payment	Number of Payments	Amount of Each Payment
(a)	\$ 5.00	10	\$ 7.50
(b)	\$ 8.50	20	\$11.00
(c)	\$ 25.00	36	\$40.00
(d)	\$100.00	19	\$55.00
(e)	\$ 40.00	17	\$40.00

3. Using the answers to Problem 2, find the finance charge in each case, given the following cash prices: (a) \$70.00; (b) \$200.00; (c) \$1,200.00; (d) \$1,000.00; (e) \$649.50.
4. A set of lamps with a cash price of \$2,800 is sold on the installment plan, with annual interest at 8%, to be paid over 24 months. Find the finance charge.
5. Find the monthly payment for each of the following installment purchases:

	Installment Price	Down Payment	Number of Payments
(a)	\$ 850.00	\$ 50.00	16
(b)	\$1,045.00	\$185.00	20
(c)	\$ 289.95	\$ 5.00	40
(d)	\$1,440.00	\$144.00	72
(e)	\$ 410.00	\$ 80.00	36

6. Judith Greissman buys a car for \$4,500, to be paid over a 36-month period, with an 8% add-on charge. Find (a) the finance charge; (b) the monthly payment.

Actuarial Loans and Debt-Reduction Schedules

A consumer can also "buy now, pay later" by obtaining a loan from a bank or other lending organization and then repaying it in installments. Loans may vary from a five-month "bill-paying" loan to a thirty-year home mortgage loan.

Lending institutions differ from retail installment-credit stores in a very basic way—the method of computing periodic payments. While a retail store might use either of the methods shown in Sample Problems 6 and 7, a lending institution—a bank, a credit union, a consumer finance company, a savings and loan association—will use the *actuarial method* to calculate the loan's periodic payments. The actuarial method, or "U.S. Rule," is based on compound interest and annuity tables.

The lending institution first decides on the monthly rate of interest. The rate varies with the type of institution. For example, a large credit union might charge $\frac{3}{4}$% a month, while a small loan company might charge $2\frac{1}{2}$% a month for the same loan. The rate may also vary with the type of loan. A loan made with your savings pledged as security would be charged a lower rate than an unsecured loan. However it is chosen, the rate is applied to the *unpaid balance* at the start of each month.

The lender then refers to a table, such as Table 13.1, which shows the amount of each monthly payment at various *monthly* rates of interest to repay a loan of $1. For instance, in order to repay a loan of $1 in 15 months at a rate of $\frac{7}{10}$% monthly, you would pay $0.070461 each month, or roughly 7 cents a month. This payment covers principal and interest.

As with other compound interest tables based on $1, multiplying the appropriate table figure by the amount of the loan will give the monthly payment.

SAMPLE PROBLEM 8

Using Table 13.1, calculate the monthly payment needed to repay John Stuart's loan of $1,000 in 26 months at a monthly rate of $\frac{2}{3}$%.

Solution

26 months at $\frac{2}{3}$% = $0.042019
$0.042019 × 1,000 = $42.019 = $42.02 a month

The Finance Charge on an Actuarial Loan

In retail installment sales, the finance charge can be found *before* the monthly payment is computed. In actuarial loans, the finance charge is found *after* computing the monthly payment.

SAMPLE PROBLEM 9

Using the information given in Sample Problem 8, find Mr. Stuart's finance charge.

Solution

Total payments = $42.02 × 26 =	$1,092.52
Amount of loan	−1,000.00
Finance charge	$ 92.52

Debt-Reduction Schedules

What is the monthly payment for a loan of $400 at 1%, to be repaid in four months? Returning to Table 13.1, we find $0.256281, which multiplied by 400 = $102.5124.

Table 13.1

Equal Monthly Payment per $1 Borrowed
to Reduce a Debt (Rates per Month)

Months	$\frac{1}{2}$%	$\frac{6}{10}$%	$\frac{2}{3}$%	$\frac{7}{10}$%	$\frac{3}{4}$%	$\frac{8}{10}$%	1%
1	1.005000	1.006000	1.006667	1.007000	1.007500	1.008000	1.010000
2	.503753	.504504	.505005	.505256	.505632	.506008	.507512
3	.336672	.337341	.337788	.338011	.338346	.338681	.340022
4	.253133	.253761	.254181	.254390	.254705	.255020	.256281
5	.203010	.203614	.204018	.204220	.204522	.204825	.206040
6	.169595	.170184	.170577	.170774	.171069	.171364	.172548
7	.145729	.146306	.146692	.146885	.147175	.147465	.148628
8	.127829	.128399	.128779	.128970	.129256	.129542	.130690
9	.113907	.114471	.114848	.115036	.115319	.115603	.116740
10	.102771	.103330	.103703	.103890	.104171	.104453	.105582
11	.093659	.094214	.094586	.094772	.095051	.095331	.096454
12	.086066	.086619	.086988	.087173	.087451	.087730	.088849
13	.079642	.080192	.080561	.080745	.081022	.081299	.082415
14	.074136	.074685	.075051	.075235	.075511	.075788	.076901
15	.069364	.069911	.070277	.070461	.070736	.071013	.072124
16	.065189	.065735	.066100	.066284	.066559	.066835	.067945
17	.061506	.062051	.062415	.062598	.062873	.063149	.064258
18	.058232	.058776	.059140	.059323	.059598	.059873	.060982
19	.055303	.055846	.056210	.056393	.056667	.056943	.058052
20	.052666	.053210	.053574	.053756	.054031	.054306	.055415
21	.050282	.050825	.051188	.051371	.051645	.051921	.053031
22	.048114	.048657	.049020	.049203	.049477	.049753	.050864
23	.046135	.046677	.047041	.047224	.047498	.047774	.048886
24	.044321	.044863	.045227	.045410	.045685	.045961	.047073
25	.042652	.043195	.043559	.043742	.044017	.044293	.045407
26	.041112	.041655	.042019	.042202	.042477	.042753	.043869
27	.039686	.040229	.040593	.040776	.041052	.041328	.042446
28	.038362	.038905	.039270	.039453	.039729	.040006	.041124
29	.037129	.037673	.038038	.038221	.038497	.038774	.039895
30	.035979	.036523	.036888	.037072	.037348	.037626	.038748
31	.034903	.035447	.035813	.035997	.036274	.036551	.037676
32	.033895	.034439	.034805	.034989	.035266	.035545	.036671
33	.032947	.033492	.033859	.034043	.034320	.034599	.035727
34	.032056	.032602	.032968	.033153	.033431	.033710	.034840
35	.031216	.031762	.032129	.032314	.032592	.032871	.034004
36	.030422	.030969	.031336	.031521	.031800	.032080	.033214
37	.029671	.030219	.030587	.030772	.031051	.031331	.032468
38	.028960	.029508	.029877	.030062	.030342	.030622	.031762
39	.028286	.028834	.029204	.029389	.029669	.029950	.031092
40	.027646	.028194	.028564	.028750	.029030	.029312	.030456
41	.027036	.027586	.027956	.028142	.028423	.028705	.029851
42	.026456	.027006	.027377	.027563	.027845	.028127	.029276
43	.025903	.026454	.026825	.027012	.027293	.027577	.028727
44	.025375	.025927	.026298	.026485	.026768	.027051	.028204
45	.024871	.025423	.025795	.025983	.026265	.026550	.027705
46	.024389	.024942	.025314	.025502	.025785	.026070	.027228
47	.023927	.024481	.024854	.025042	.025325	.025611	.026771
48	.023485	.024039	.024413	.024601	.024885	.025173	.026334

Normal rounding would make it $102.51 per month; but in reducing actuarial loans, we round up in favor of the lender—in this case, $102.52. This procedure does not give them an unfair advantage; it does avoid being short on the last payment, which is adjusted downward if necessary. One exception is when the third digit after the decimal point is a zero, such as $32.0709. This amount is kept at $32.07.

To sum up, the loan of $400.00 plus 1% interest is to be repaid in four installments of $102.52. Let's follow this through.

The first payment is made when $400.00 is owed. In the actuarial method, payments are first applied to interest on the unpaid balance and then to the reduction of the principal. The interest at 1% on $400.00 is $4.00; the remainder of the payment, $102.52 − $4.00 = $98.52, is used to reduce the principal from $400.00 to $400.00 − $98.52 = $301.48.

The second month's interest is 1% on $301.48 = $3.01, so the principal reduction is $102.52 − $3.01 = $99.51. The new principal is $301.48 − $99.51 = $201.97.

In the third month, after allowing interest of $2.02, the principal is reduced by $100.50 to $101.47.

In the last month, the payment must cover interest (on $101.47) of $1.01 and reduce the principal by $101.47 to zero. Thus, a payment of $1.01 + $101.47 = $102.48 is the final payment. The data can be summarized in a table called a *debt-reduction schedule.*

Schedule of Debt Reduction

Loan of $400 at 1%
Repaid in Four Monthly Installments

Payment	Amount	Interest at 1%	Reduction of Principal	Principal Balance
				$400.00
1	$102.52	$4.00	$ 98.52	301.48
2	102.52	3.01	99.51	201.97
3	102.52	2.02	100.50	101.47
4	102.48*	1.01	101.47	-0-

* To adjust final balance to zero.

Sample Problem 10 shows another debt-reduction schedule, which is also known as an *amortization* schedule.

SAMPLE PROBLEM 10

Prepare a debt-reduction schedule for the Nivana Cosmetics Company's loan of $2,000 at 1%, repaid in 13 monthly installments.

Solution

Table 13.1: 1%, 13 months = $0.082415
$0.082415 × 2,000 = $164.83 per month

Schedule of Debt Reduction

Loan of $2,000 at 1%
Repaid in 13 Monthly Installments

Payment	Amount	Interest at 1%	Reduction of Principal	Principal Balance
				$2,000.00
1	$164.83	$20.00	$144.83	1,855.17
2	164.83	18.55	146.28	1,708.89
3	164.83	17.09	147.74	1,561.15
4	164.83	15.61	149.22	1,411.93
5	164.83	14.12	150.71	1,261.22
6	164.83	12.61	152.22	1,109.00
7	164.83	11.09	153.74	955.26
8	164.83	9.55	155.28	799.98
9	164.83	8.00	156.83	643.15
10	164.83	6.43	158.40	484.75
11	164.83	4.85	159.98	324.77
12	164.83	3.25	161.58	163.19
13	164.82*	1.63	163.19	-0-

* The final payment of $164.82 is the sum of $1.63 interest plus $163.19 to reduce the principal to zero.

The Amortized Mortgage

Another very common debt that is reduced by the actuarial method is a mortgage on a home or a business. A *mortgage* is a long-term promissory note given to a bank or other lender by the borrower of money. In return for a loan, the borrower agrees to make a series of monthly PI (principal and interest) payments. Since repayment is by the actuarial method, all payments are first applied to interest on the unpaid balance and then to reduction of the principal. A mortgage amortization schedule is simply a longer version of the debt-reduction schedules we have looked at previously.

Table 13.2 is an example of a typical table of monthly payments on a mortgage. Notice that the amount of each payment figure depends on two factors: the annual rate of interest and the term of the mortgage. As you study Table 13.2, observe that it is based on $1,000 of face value. Sample Problem 11 illustrates how a monthly payment is calculated.

SAMPLE PROBLEM 11

Roger Goodson secures a five-year, 10%, $60,000 amortized mortgage loan from the Harrington National Bank. (a) Calculate the monthly payment. (b) Calculate the annual cost. (c) Calculate the total cost over five years.

Solution

(a) Table 13.2, $21.25 per $1,000 per month
$60,000 = 60 thousands
$21.25 × 60 = $1,275 per month
(b) $1,275 × 12 months = $15,300 per year
(c) $15,300 × 5 years = $76,500 total cost

Table 13.2

Amortized Mortgage
Monthly Payment per $1,000 Loan

Annual Interest Rate	5 Years	10 Years	15 Years	20 Years	25 Years	30 Years
8%	$20.28	$12.14	$ 9.56	$ 8.37	$ 7.72	$ 7.34
9	20.76	12.67	10.15	9.00	8.40	8.05
$9\frac{1}{2}$	21.01	12.94	10.45	9.33	8.74	8.41
10	21.25	13.22	10.75	9.66	9.09	8.78
$10\frac{1}{2}$	21.50	13.50	11.06	9.99	9.45	9.15
11	21.75	13.78	11.37	10.33	9.81	9.53
$11\frac{1}{2}$	22.00	14.06	11.69	10.67	10.17	9.91
12	22.25	14.35	12.01	11.02	10.54	10.29
$12\frac{1}{4}$	22.38	14.50	12.17	11.19	10.72	10.48
$12\frac{1}{2}$	22.50	14.64	12.33	11.37	10.91	10.68
$12\frac{3}{4}$	22.63	14.79	12.49	11.54	11.10	10.87
13	22.76	14.94	12.66	11.72	11.28	11.07
14	23.27	15.53	13.32	12.44	12.04	11.85
15	23.79	16.14	14.00	13.17	12.81	12.65

The $60,000 loan costs Roger $76,500, an amount $16,500 higher, all of which is interest paid for the use of the mortgage loan.

A mortgage payment table can be used for several purposes other than calculating monthly payments. Sample Problem 12 illustrates how you can use Table 13.2 to determine how large a mortgage you can afford.

SAMPLE PROBLEM 12

Lillian McCoy can afford a monthly PI payment of $525.00. If she can obtain a 30-year, $12\frac{1}{4}$% mortgage, how large a mortgage will she be able to take out?

Solution

30-year, $12\frac{1}{4}$% factor from Table 13.2 is $10.48
Divide $525.00 by $10.48 to find mortgage amount
$525.00 ÷ $10.48 = 50.095 thousands
50.095 × $1,000 = $50,000 rounded

Rounding to the nearest thousand is the most logical way to solve a problem of this type. Lillian will now be able to see how large a down payment she can make, adding it to the $50,000 mortgage amount to determine just how much she can afford to pay for a home. Sample Problem 13 shows another way to use Table 13.2: determining the rate of interest that an individual can afford.

SAMPLE PROBLEM 13

Ernest Williams wants to pay a maximum of $900.00 a month for an $80,000 mortgage. If he is able to secure a 20-year term, what is the maximum rate he can afford to pay?

Solution

Divide $900.00 by 80 thousands = $11.25
Locate figure in Table 13.2 under 20 years nearest to $11.25 = $11.19. The answer is a $12\frac{1}{4}$% maximum rate.

Sample Problem 14 concludes this section on amortized mortgages by illustrating how to use Table 13.2 to find the mortgage term.

SAMPLE PROBLEM 14

Donna Amato wants to pay a maximum of $700.00 a month for a $70,000 mortgage. If the prevailing market rate of interest is $11\frac{1}{2}$%, what mortgage term should she seek?

Solution

Divide $700.00 by 70 thousands = $10.00
Locate figure in Table 13.2 on $11\frac{1}{2}$% line nearest to $10.00 = $9.91. The answer is a 30-year term.

TEST YOUR ABILITY

1. Using Table 13.1, find the monthly payment for each of the following loans:

	Amount of Loan	Monthly Interest Rate	Number of Payments
(a)	$ 600.00	$\frac{2}{3}$%	14
(b)	$2,500.00	$\frac{1}{2}$%	11
(c)	$ 475.00	$\frac{7}{10}$%	48
(d)	$ 350.00	$\frac{3}{4}$%	12
(e)	$1,400.00	$\frac{8}{10}$%	30

2. For each part of Problem 1, find the finance charge.
3. Harvey Sedlacek's loan of $720 is to be repaid in 20 months at $\frac{7}{10}$%. Find (a) the monthly payment; (b) the finance charge.
4. Prepare a schedule of debt reduction for your $650 loan at 1% to be repaid in seven monthly installments.
5. Mary Brisson secures a 30-year, $10\frac{1}{2}$%, $80,000 amortized mortgage loan from the First Savings Bank. (a) Calculate the monthly payment. (b) Calculate the annual cost. (c) What is the total cost over the life of the loan?
6. Sylvester Brown can afford a monthly principal and interest mortgage payment of $420.00. If he can secure a 25-year, 9% amortized mortgage loan, what amount of mortgage can he afford?
7. Lori Hoidahl wants to pay no more than $700.00 a month for a $70,000 amortized mortgage. If she is able to secure a 20-year term, what is the maximum rate she can pay?
8. Fred Wallace wants to pay a maximum of $540.00 a month for a $60,000 mortgage. If he is able to borrow at a time when the interest rate is 9%, what mortgage term should he seek to obtain?

Finding the True Annual Rate of Interest

You, as a consumer, consider many factors when choosing a source of credit. You might think about either the ease of obtaining the credit or the length of the repayment period.

A third factor, one often overlooked, is the rate of interest—not the stated rate, but the *true annual rate of interest* that is actually being charged.

This rate cannot be calculated by a simple formula for rate because the base (the principal) keeps changing. There are, however, three methods for finding the true annual rate.

One method is by using tables prepared by the Federal Reserve Board. Table 13.3 is taken from the full table found in Regulation Z. To use Table 13.3, first divide I by P and multiply by 100; then go to the table and locate on the line showing the number of monthly payments the number closest to your answer. The column heading is the true annual rate of interest.

For example, if $I = \$50.00$ and $P = \$125.00$, then $I \div P = .40$. And $.40 \times 100 = 40.00$. If the number of payments is 42, follow the 42 line to the number closest to 40.00 – 39.85, which is closer than 40.40. The answer is a rate of 20.00%.

What if $I = \$49.40$, $P = \$100.00$, and the time is 84 months? The answer to our calculation is 49.40, and the true annual rate is 12.25%.

In conclusion, we must know *principal, interest* (finance charge), and *time* in order to find the true annual rate of interest.

The Rate on Installment Buying

On an installment purchase, the finance charge is the interest. The principal is the *cash price less any down payment.* If a shed can be purchased for $600 cash and $50 is paid down, $550 is being "borrowed" and is treated as the principal. Sample Problem 15 shows how to calculate the true annual rate of interest on an installment purchase.

SAMPLE PROBLEM 15

A cassette recorder is purchased on the installment plan for $20 down and ten monthly payments of $23.88 each. The cash price would have been $220. Find the true annual rate of interest, using Table 13.3.

Solution

$$\begin{aligned} \text{Finance charge} &= \text{Installment price} - \text{Cash price} \\ &= \$20.00 + (10 \times \$23.88) - \$220.00 \\ &= \$20.00 + \$238.80 - \$220.00 \\ &= \$38.80 \\ \text{Principal} &= \text{Cash price} - \text{Down payment} \\ &= \$220.00 - \$20.00 \\ &= \$200.00 \end{aligned}$$

$I \div P = \$38.80 \div \$200.00 = .1940 \times 100 = 19.40$

Table 13.3: 19.40, 10 payments = 40.25%

In an add-on situation, the interest is calculated on the principal amount. If there is no down payment, the principal is equal to the cash price. If a down payment is made, the principal equals the cash price minus the down payment.

Table 13.3

Annual Percentage Rate
Charges per $100 Borrowed

Number of Payments	12.00%	12.25%	12.50%	12.75%	20.00%
1	1.00	1.02	1.04	1.06	1.67
2	1.50	1.53	1.57	1.60	2.51
3	2.01	2.05	2.09	2.13	3.35
4	2.51	2.57	2.62	2.67	4.20
5	3.02	3.08	3.15	3.21	5.06
6	3.53	3.60	3.68	3.75	5.91
7	4.04	4.12	4.21	4.29	6.78
8	4.55	4.65	4.74	4.84	7.64
9	5.07	5.17	5.28	5.39	8.52
10	5.58	5.70	5.82	5.94	9.39
11	6.10	6.23	6.36	6.49	10.28
12	6.62	6.76	6.90	7.04	11.16
13	7.14	7.29	7.44	7.59	12.05
14	7.66	7.82	7.99	8.15	12.95
15	8.19	8.36	8.53	8.71	13.85
16	8.71	8.90	9.08	9.27	14.75
17	9.24	9.44	9.63	9.83	15.66
18	9.77	9.98	10.19	10.40	16.57
19	10.30	10.52	10.74	10.96	17.49
20	10.83	11.06	11.30	11.53	18.41
21	11.36	11.61	11.85	12.10	19.34
22	11.90	12.16	12.41	12.67	20.27
23	12.44	12.71	12.97	13.24	21.21
24	12.98	13.26	13.54	13.82	22.15
30	16.24	16.60	16.95	17.31	27.89
36	19.57	20.00	20.43	20.87	33.79
42	22.96	23.47	23.98	24.49	39.85
48	26.40	26.99	27.58	28.18	46.07
54	29.91	30.58	31.25	31.93	52.44
60	33.47	34.23	34.99	35.75	58.96
72	40.76	41.70	42.64	43.59	72.46
84	48.28	49.41	50.54	51.67	86.53
96	56.03	57.35	58.68	60.01	101.15
108	63.99	65.52	67.05	68.59	116.29
120	72.17	73.90	75.65	77.41	131.91

(continued)

Table 13.3 (continued)

Annual Percentage Rate
Charges per $100 Borrowed

Number of Payments	20.25%	20.50%	40.00%	40.25%	40.50%
1	1.69	1.71	3.33	3.35	3.37
2	2.54	2.57	5.03	5.06	5.09
3	3.39	3.44	6.74	6.78	6.82
4	4.25	4.31	8.47	8.52	8.58
5	5.12	5.18	10.22	10.28	10.35
6	5.99	6.06	11.99	12.06	12.14
7	6.86	6.95	13.77	13.86	13.95
8	7.74	7.84	15.57	15.67	15.77
9	8.63	8.73	17.39	17.51	17.62
10	9.51	9.63	19.23	19.36	19.48
11	10.41	10.54	21.09	21.23	21.37
12	11.31	11.45	22.97	23.12	23.27
13	12.21	12.36	24.86	25.02	25.19
14	13.11	13.28	26.77	26.95	27.13
15	14.03	14.21	28.70	28.89	29.08
16	14.94	15.13	30.65	30.85	31.06
17	15.86	16.07	32.61	32.83	33.05
18	16.79	17.01	34.59	34.83	35.06
19	17.72	17.95	36.59	36.84	37.09
20	18.66	18.90	38.61	38.87	39.14
21	19.60	19.85	40.64	40.92	41.20
22	20.54	20.81	42.69	42.99	43.28
23	21.49	21.77	44.76	45.07	45.38
24	22.44	22.74	46.85	47.17	47.50
30	28.26	28.64	59.73	60.15	60.57
36	34.25	34.71	73.20	73.72	74.25
42	40.40	40.95	87.24	87.87	88.51
48	46.71	47.35	101.82	102.58	103.33
54	53.17	53.91	116.93	117.80	118.67
60	59.80	60.64	132.51	133.51	134.51
72	73.51	74.55	165.00	166.26	167.51
84	87.80	89.07	199.03	200.56	202.09
96	102.65	104.15	234.36	236.16	237.96
108	118.03	119.78	270.74	272.82	274.90
120	133.90	135.90	307.98	310.33	312.69

Source: U.S. Federal Reserve Board (Vol. 1, *Regulation Z*, 1969).

SAMPLE PROBLEM 16

A panel van with a cash price of $5,500 is purchased on a deferred payment plan, with $500 down and a 7% add-on charge. Repayment is to be made over 12 months. Find the true annual rate of interest, using Table 13.3.

Solution

Principal = Cash price − Down payment
= $5,500 − $500
= $5,000

Interest = P × R × T
= $5,000 × 7% × 12 months
= $5,000 × .07 × 1 year
= $350

I ÷ P = $350 ÷ $5,000 = .07 × 100 = 7.00
Table 13.3: 7.00, 12 payments = 12.75%

In Sample Problem 16, the *stated rate* is 7% but the *true rate* is $12\frac{3}{4}$%. There is a difference because the interest charge is calculated on a principal of $5,000—but *paid* on a principal of less than $5,000 each month after the first month. The principal keeps decreasing, but the payment stays constant.

Caution: Table 13.3 is used only for loans and accounts repaid in equal installments. Other tables apply to other situations, such as a single repayment of a loan.

The Rate on Actuarial Loans

Sample Problems 17, 18, and 19 show various situations in which loans are secured and repaid. In each case, the principal is the amount borrowed, and the interest is the excess of the repayments over the amount borrowed.

SAMPLE PROBLEM 17

A bank advertises that a loan of $600 can be repaid in 36 monthly installments of $20. Find the true annual rate of interest, using Table 13.3.

Solution

Amount to be repaid ($20.00 × 36)	$720.00
Principal	−600.00
Interest	$120.00

I ÷ P = $120.00 ÷ $600.00 = .2000 × 100 = 20.00
Table 13.3: 20.00, 36 months = 12.25%

SAMPLE PROBLEM 18

A credit union loan of $700 is to be repaid in 24 monthly installments of $32.95. Find the true annual rate of interest, using Table 13.3.

Solution

Amount to be repaid ($32.95 × 24)	$790.80
Principal	−700.00
Interest	$ 90.80

I ÷ P = $90.80 ÷ $700.00 = .1297 × 100 = 12.97
Table 13.3: 12.97, 24 payments = 12.00%

SAMPLE PROBLEM 19

A small loan company offers a $1,400 loan with repayment in 30 monthly installments of $59.87. Using Table 13.3, find the true annual rate of interest.

Solution

Amount to be repaid ($59.87 × 30)	$1,796.10
Principal	−1,400.00
Interest	$ 396.10

I ÷ P = $396.10 ÷ $1,400.00 = .2829 × 100 = 28.29
Table 13.3: 28.29, 30 periods = 20.25%

A Shortcut Sample Problem 20 sums up the use of Table 13.1, the preparation of a debt-reduction schedule, and the determination of the true annual rate of interest. Also, there may—or may not—be a shortcut. Can you find it?

SAMPLE PROBLEM 20

The Dune Savings Bank offers a loan of $500 to Fay Herbert at 1% a month on the unpaid balance.

(a) Find her monthly payment, if the debt is to be repaid in five months, using Table 13.1.
(b) Prepare a debt-reduction schedule.
(c) Use Table 13.3 to find the true annual rate of interest.

Solution

(a) Table 13.1: 5 months, 1% = $0.206040
Monthly payment = $0.206040 × 500 = $103.02

(b) **Schedule of Debt Reduction**

Loan of $500 at 1%
Repaid in Five Monthly Installments

Payment	Amount	Interest at 1%	Reduction of Principal	Principal Balance
				$500.00
1	$103.02	$5.00	$ 98.02	401.98
2	103.02	4.02	99.00	302.98
3	103.02	3.03	99.99	202.99
4	103.02	2.03	100.99	102.00
5	103.02	1.02	102.00	−0−

(c)

Amount repaid ($103.02 × 5)	$515.10
Principal	−500.00
Interest	$ 15.10

I ÷ P = $15.10 ÷ $500.00 = .0302 × 100 = 3.02
Table 13.3: 3.02, 5 months = 12.00%

A rate of 1% a month has yielded an annual rate of 12.00%. Therefore, when the *monthly rate is known,* multiply the monthly rate by 12. This shortcut applies in open-end credit, as we have already seen.

If the monthly rate is $1\frac{1}{2}\%$, what is the annual rate? The answer is $1\frac{1}{2}\% \times 12 = 18\%$. What about $\frac{8}{10}\%$ a month? The answer is 9.6%. This is the second method for finding the true annual rate of interest.

The Constant Ratio Formula

When tables are not available, the *constant ratio formula* can be used to approximate the true annual rate of interest:

$$R = \frac{24 \times I}{P \times (n + 1)}$$

Here, R = the true annual rate of interest, I = interest, P = principal, and n = the number of payments. The number 24 is a constant used for monthly payments. If the payments are weekly, 104 is used instead of 24.

Given I = $90, P = $300, and n = 23

$$R = \frac{24 \times 90}{300 \times (23 + 1)}$$

$$= \frac{\overset{1}{\cancel{24}} \times \overset{3}{\cancel{90}}}{\underset{10}{\cancel{300}} \times \underset{1}{\cancel{24}}}$$

$$= \frac{3}{10} = 30\%$$

Compare the calculation of the true annual rate of interest using this formula with two previous problems which used Table 13.3. For Sample Problem 15:

$$R = \frac{24 \times 38.80}{200 \times (10 + 1)}$$

$$= \frac{24 \times 38.80}{200 \times 11}$$

$$= 42.33\%$$

For Sample Problem 18:

$$R = \frac{24 \times 90.80}{700 \times (24 + 1)}$$

$$= \frac{24 \times 90.80}{700 \times 25}$$

$$= 12.45\%$$

The approximation is close but try $(I \div P) \times 100$ and Table 13.3 first. It can save time!

TEST YOUR ABILITY

1. Using Table 13.3, find the true annual rate of interest on each of the following loans:

	Interest	Principal	Time (Months)
(a)	$ 77.22	$ 200.00	20
(b)	$276.00	$1,000.00	48

	Interest	Principal	Time (Months)
(c)	$263.50	$ 500.00	54
(d)	$399.00	$ 700.00	96

2. A speedboat motor is purchased on the installment plan for $20 down and ten monthly payments of $105.80. The cash price would have been $1,020.00. Use Table 13.3 to find the true annual rate of interest.
3. The Warren Food Plan charges $500 for a four-month shipment of food, to be repaid in four installments, with interest of 8% added-on. Find the true annual rate of interest, using Table 13.3.
4. Using Table 13.3, find the true annual rate of interest on each of the following financial loans:

	Amount of Loan	Number of Payments	Amount per Payment
(a)	$ 200.00	10	$ 21.12
(b)	400.00	13	$ 38.42
(c)	600.00	6	$106.00
(d)	1,200.00	96	$ 20.00

5. Marge Landowska's loan of $100 is repaid in four months at a monthly rate of $\frac{7}{10}$% on the unpaid balance.
   (a) Find her monthly payment using Table 13.1.
   (b) Prepare a debt-reduction schedule. ($\frac{7}{10}\% = .007$)
   (c) What is the true annual rate of interest?
6. Use the constant ratio formula to find the true annual rate of interest, to the nearest tenth of a percent, in each situation:

	Down Payment	Number of Installments	Amount per Installment	Cash Price
(a)	$50.00	8	$25.00	$200.00
(b)	$20.00	15	$30.00	$420.00
(c)	$15.00	35	$20.00	$500.00
(d)	$ 8.00	3	$10.00	$ 30.00

SUMMARY Interest rates are important in our lives as consumers. It is the purpose of consumer legislation, such as Regulation Z, to protect us as consumers and to make us aware of our choices among sources of credit or money.

Let's summarize the three methods available to calculate or determine the true annual rate of interest.

Method	When Used
Monthly rate × 12	Monthly rate known
Table 13.3	Monthly rate not known; normal situations
Constant ratio formula	Table not available or I ÷ P entry not in table

Sample Problem 21 reviews the choice of credit sources.

SAMPLE PROBLEM 21

Johan and Jane Schwartz ask you to choose for them one of two sources from which to obtain a loan of $1,000. Source A offers terms of 36 monthly payments of $32.67. Source B offers a monthly rate of $\frac{3}{4}\%$, payments of $60.00, and an 18-month repayment term.

(a) What is the finance charge in each case?
(b) Find the true annual rate of interest in each case.
(c) Which loan would you choose, and why?

Solution

	A	B
(a) Repay	$32.67 × 36 = $1,176.12	$60.00 × 18 = $1,080.00
Principal	−1,000.00	−1,000.00
Interest	$ 176.12	$ 80.00

(b) Source A:

$I \div P = \$176.12 \div \$1{,}000 = .1761 \times 100 = 17.61$

Table 13.3: 17.61, 36 months is not available; use constant ratio formula

$$R = \frac{24 \times 176.12}{1{,}000 \times (36 + 1)} = 11.4\%$$

Source B:

Monthly rate known

$\frac{3}{4}\% = .0075 \times 12 = .09 = 9\%$

(c) Prepare to discuss your choice and reasons in class.

CHECK YOUR READING

1. What is the purpose of the Federal Reserve Board's *Regulation Z?*
2. Is a finance charge the same as interest?
3. Distinguish between *open-end credit* and *other credit* by giving an example of each.
4. How do payments on account affect the finance charge in the (a) previous balance method, (b) adjusted balance method, and (c) average daily balance method?
5. How do purchases during a month affect the finance charge in each of the three methods listed in Question 4?
6. How is the finance charge found in an installment purchase? What about in an add-on situation?
7. What method is used by lending institutions to determine the amount of monthly payments?
8. How is the finance charge found in an actuarial loan?
9. Use Table 13.3 to find the monthly payment, at $\frac{1}{2}\%$ interest a month, to repay a loan of $1 in 17 months.
10. Why are normal rounding rules overlooked in computing monthly payments on an actuarial loan?
11. What two factors determine the amount of each monthly payment figure in Table 13.2?
12. Explain how Table 13.2 can be used to find the size of a mortgage.
13. In addition to monthly payment and mortgage size, what other items can be found from an amortized mortgage table?

14. What three factors must be known in order to use Table 13.3?
15. What is the principal in (a) an installment purchase; (b) an add-on purchase; (c) an actuarial loan?
16. Find the true annual rate of interest if the monthly rate is (a) $\frac{3}{4}$%; (b) $\frac{2}{3}$%; (c) $\frac{9}{10}$%.
17. What statement of comparison can be made about answers obtained from the constant ratio formula and those obtained from Table 13.3?

EXERCISES *Group A*

1. A sofa costing $960 is purchased at an add-on charge of 8%, to be repaid over 30 months. Find (a) the finance charge; (b) the monthly payment.
2. Prepare a debt-reduction schedule for Lucy Sanchez's $3,000 loan, which is to be repaid in 12 months, at a rate of 1% a month on the unpaid balance.
3. Terrie Molleur wants to pay a maximum of $600.00 a month for a $50,000 amortized mortgage. If she is able to borrow money at the rate of 12%, what term mortgage should she seek to obtain?
4. Find the true annual rate of interest on each of the following loans, using Table 13.3:

	Interest	Principal	Time (Months)
(a)	$ 100.00	$ 600.00	30
(b)	$ 840.00	$2,000.00	72
(c)	$ 60.00	$ 400.00	16
(d)	$11,000.00	$5,500.00	84

5. A ten-speed racing bicycle is purchased for $20 down and 20 monthly payments of $11.10. Its cash price would have been $220. Use Table 13.3 to find the true annual rate of interest.
6. A car purchased at a cash price of $6,000, with an add-on charge of 7$\frac{1}{2}$%, is to be repaid in 36 months. Calculate the finance charge.
7. Find the monthly payment on each of the following installment purchases:

	Installment Price	Down Payment	Number of Payments
(a)	$ 750.00	$ 75.00	10
(b)	$ 95.00	$ 4.75	15
(c)	$ 875.00	$ 25.00	24
(d)	$1,875.00	$187.50	30

8. George Culver has just secured a 12$\frac{1}{2}$%, 25-year, $50,000 amortized mortgage loan from a local savings and loan association. (a) Calculate the monthly payment. (b) Calculate the annual cost. (c) What is the total cost for the life of the loan?
9. A loan of $500 is made to Michael Sanders for eight months at a monthly rate of $\frac{1}{2}$%. (a) Find the monthly payment, using Table 13.1. (b) Prepare a debt-reduction schedule. (c) What is the true annual rate of interest?

10. Use Table 13.1 to find the monthly payment for each of the following loans:

	Amount of Loan	Monthly Interest Rate	Number of Payments
(a)	$ 850	$\frac{7}{10}$%	7
(b)	$ 2,400	$\frac{2}{3}$%	31
(c)	$ 350	$\frac{7}{10}$%	6
(d)	$15,000	$\frac{6}{10}$%	48
(e)	$ 4,500	$\frac{3}{4}$%	40

11. For each part of Exercise 10, find the finance charge.
12. Barbara Rousseau can afford a maximum of $475.00 a month for a PI payment on a $40,000 mortgage that she needs. If she is able to secure a 30-year mortgage, what is the maximum rate she can afford?
13. Find the total cost of the following installment purchases:

	Down Payment	Number of Payments	Amount per Payment
(a)	$ 24.50	12	$18.25
(b)	$ 38.75	27	$38.75
(c)	$125.00	15	$95.50
(d)	$ 40.95	18	$32.20

14. Using the figures from Exercise 13, determine the finance charge in each case, if the corresponding cash prices are (a) $210; (b) $1,020; (c) $1,485; (d) $580.
15. A piano is purchased for $2,500, with an add-on charge of $6\frac{1}{2}$% stated for 24 months. Find the true annual rate of interest, using Table 13.3.
16. A loan of $3,000 is made to Jessica Corbet at a rate of $\frac{8}{10}$% a month, to be repaid in 19 months. Find (a) the monthly payment; (b) the finance charge.
17. Using Table 13.3, find the true annual rate of interest for each of the following loans:

	Amount of Loan	Number of Payments	Amount per Payment
(a)	$ 200.00	24	$10.20
(b)	$ 300.00	5	$61.92
(c)	$ 500.00	15	$36.07
(d)	$2,000.00	84	$71.43

18. Using the constant ratio formula, find the true annual rate of interest in each case. Round to the nearest tenth of a percent.

	Down Payment	Number of Installments	Amount per Installment	Cash Price
(a)	$20	15	$15	$200
(b)	$30	10	$40	$350
(c)	$10	39	$ 5	$180
(d)	$ 5	30	$ 5	$125

19. Walter Jenkins's credit card account shows a $350 balance on September 1. He charges purchases of $110 on the 18th and pays $200 on the 27th. At the annual rate of 18%, find his finance charge for September under (a) the previous balance method; (b) the adjusted balance method; (c) the average daily balance method.
20. Brian Fox can afford a monthly PI payment of $682.00. If he can obtain a $12\frac{1}{2}$%, 20-year mortgage, what is the maximum mortgage that he will be able to pay, to the nearest thousand?

Group B

1. Sally Hartford's charge account at Acme Department Store shows the following activity for September and October of the current year:

Balance, September 1	$250
Purchase, September 10	100
Payment, September 26	50
Purchase, October 13	100
Payment, October 27	50

Calculate her finance charge for each month, at an annual rate of 18%, using (a) the previous balance method; (b) the adjusted balance method. (*Hint:* the September finance charge becomes part of the October 1 balance.)

2. A wall unit is purchased on the installment plan for $735, with 20% down and the balance in ten equal monthly installments. Find the cost of each installment.
3. A sailboat with a cash price of $2,750 is purchased on the following terms: 10% down, 8% add-on interest, and the balance paid in 20 equal monthly installments. (a) Find the amount of each monthly payment. (b) If repayment were instead over 30 months, calculate the monthly payment.
4. How much interest can Henry Wallace save in a year if his 12-month, $800 loan is secured at $\frac{8}{10}$% a month instead of at 1% a month?
5. Molly Hirshfield borrows $600 from the Washington Savings Bank for ten months at $\frac{2}{3}$% a month.
   (a) Find her monthly payment.
   (b) Find the finance charge.
   (c) What is the true annual rate of interest? (*Hint:* the monthly rate is known.)
   (d) Calculate the true annual rate of interest by the constant ratio formula (nearest tenth of a percent).
   (e) Prepare a debt-reduction schedule.
6. Calculate the missing figures, using Table 13.2. Use the nearest rates and years for your answers.

	Principal	Annual Rate	Term (Years)	Monthly Payment
(a)	$ 60,000	$12\frac{1}{2}$%	10	____
(b)	90,000	15%	25	____
(c)	40,000	____	30	$ 322
(d)	70,000	____	20	771

(e)	120,000	$12\frac{1}{4}\%$	______	1,740
(f)	80,000	15%	______	1,120

7. The Charlotte Pillow Company is offered a choice of two repayment plans on a $4,000 loan. The Stevenson National Bank offers a 60-month loan, with $88.88 paid a month. The James Savings Bank offers a 36-month loan, with monthly payments of $132.86. Use either Table 13.3 or the constant ratio formula to determine which loan offers the lower true annual rate of interest.
8. Chet Ruiz's loan of $550 is to be repaid in 40 weekly installments of $15. Calculate the true annual rate of interest to the nearest tenth of a percent.

UNIT 4 REVIEW

Terms You Should Know

Unit 4 had had many terms. Can you define them?

actuarial loan	interest
adjusted balance	maker
amortization schedule	maturity date
amount	maturity value
annuity	mortgage
annuity schedule	negotiable
average daily balance	NOW account
bank discount	open-end credit
bankers' interest	other credit
cancellation method	payee
carrying charge	present value
compound discount	previous balance
compound interest	principal
constant ratio formula	proceeds
debt-reduction schedule	promissory note
discounting a note	rate
down payment	Regulation Z
due date	simple interest
endorsement	sinking fund
exact interest	sinking fund schedule
face	term of discount
finance charge	time
installment buying	true annual rate of interest

UNIT 4 SELF-TEST

The following items cover the highlights of Unit 4, with one notable exception: no schedules of accumulation of annuities or sinking funds, or for reductions of debts, are included. However, you can always review

Items 1–13 review the basic concepts of Chapter 11. Use Tables 11.1, 11.2, and 11.3 where needed.

1. The simple interest on $1,000 for $2\frac{1}{2}$ years at 8% is ________.
2. The time from April 14 to September 11 is ________ days.
3. Exact interest is based on a ________-day year, while bankers' interest is based on a ________-day year.

4. If P = \$1,200 and R = 6%, then I for 30 days = __________; 90 days = __________; 12 days = __________; 15 days = __________; 80 days = __________; 6 days = __________; 10 days = __________.
5. If P = \$1,200 and T = 60 days, then I at 5% = __________; 4% = __________; 9% = __________; 10% = __________; $1\frac{1}{2}$% = __________.
6. The exact interest on \$100 at $9\frac{1}{2}$% for 16 days is __________, while the bankers' interest is __________.
7. Smith gives a note to Henrik. The payee is __________; the maker is __________.
8. The due date of a three-month note dated June 16 is __________.
9. The due date of a 90-day note dated October 5 is __________.
10. On a \$100 loan for 60 days at 6%, the interest is \$1. At simple interest, __________ will be repaid. At bank discount, the proceeds are __________.
11. If the note in Question 8 is discounted on July 5, the term of discount is __________.
12. If the note in Question 9 is discounted on November 11, the term of discount is __________.
13. Consider a 90-day, 8%, \$600 note. Its maturity value is __________. If discounted for 80 days at 9%, the proceeds are __________.

Items 14–19 cover Chapter 12. For Items 14–18, use Tables 12.1–12.5 as needed and assume a rate of 6%.

14. Investing \$1,000 a year for four years, compounded annually, will yield Arthur Washburn a total of __________.
15. The sum for Erika Jacobs to deposit today in order to amount to \$1,000 in four years, compounded quarterly, is __________.
16. To withdraw \$1,000 a year for the next four years, compounded annually, you should invest __________ today.
17. Vincent Platt's investment of \$1,000 for four years will amount to __________ at annual compounding; __________ at semiannual compounding; __________ at quarterly compounding.
18. To reach \$1,000 in four years, the annual sinking fund deposit, compounded annually, is __________.
19. The first-quarter interest on Louisa Norton's savings account balance of \$385.99 at 8%, compounded quarterly, is __________.

Items 20–31 review Chapter 13.

20. The open-end credit method that considers both purchases and payments in the month being charged is the __________.
21. An annual rate of 9.6% is equivalent to a monthly rate of __________.
22. A coat purchased for \$10 down and nine payments of \$15, with a cash price of \$110, has a total cost of __________ and a finance charge of __________.

23. With $50 down and a total cost of $900, the amount of each of ten equal payments is __________.
24. A jeep costing $2,400 with an add-on charge of 8% for 24 months has a finance charge of __________ and a monthly payment of __________.
25. In the __________ method, payments are applied first to interest on the unpaid balance and then to reduction of __________.
26. The monthly payment on a $100 loan for six months at $\frac{8}{10}$% monthly interest is __________.
27. The monthly payment on a $100.000, 20-year, $10\frac{1}{2}$% mortgage loan is __________.
28. Using the data in Question 26, the finance charge is __________.
29. If the monthly rate of interest is $\frac{7}{10}$%, the annual rate is __________.
30. If $I = \$70$ and $P = \$200$, the true annual rate of interest for 18 periods from Table 13.3 is __________.
31. If $I = \$40$, $P = \$240$, and $n = 39$ periods, the constant ratio formula shows a true annual interest rate of __________.

KEY TO UNIT 4 SELF-TEST

1. $200
2. 150 days
3. 365; 360
4. $6; $18; $2.40; $3; $16; $1.20; $2
5. $10; $8; $18; $20; $3
6. $0.4165 (Table 11.3); $0.4222 (Table 11.2) (round to $0.42)
7. Henrik; Smith
8. September 16
9. January 3
10. $101; $99
11. 73 days
12. 53 days
13. $612; $599.76
14. $4,637.09 (Table 12.2)
15. $788.03 (Table 12.3)
16. $3,465.11 (Table 12.4)
17. $1,262.48; $1,266.77; $1,268.99 (Table 12.1)
18. $228.59 (Table 12.5)
19. $7.70
20. average daily balance method
21. .008 or $\frac{8}{10}$%
22. $145; $35
23. $85
24. $384; $116
25. actuarial; principal
26. $17.14
27. $999
28. $2.84
29. 8.4%
30. 40.50%
31. 10%

UNIT 5

The Mathematics of Ownership and Investment

Wise management uses its money to earn more money. It does this by the process of *investment*—by owning and loaning.

Unit 5 discusses corporate stocks and explores the math involved in determining investment cost, investment income, and the gain or loss from the sale of investments.

Chapter 14 is devoted to a specific discussion of investment in stocks, while Chapter 15 considers bonds, mutual funds, life insurance, and other forms of investment.

Read this unit as both a student and as a business owner. Knowledge of investments is useful to you in both these roles.

CHAPTER **14**

Corporation Stocks

In the early days of American business, two forms of ownership were common: the *sole proprietorship* (a business owned by one person), and the *partnership* (a business owned by two or more persons).

The bulk of American business activity today is conducted by the *corporation,* an artificial "person" created by state law and empowered to engage in commercial transactions. Ownership rights in a corporation are represented by shares of stock. The individual who buys stock hopes to gain by this investment in two ways: receiving quarterly dividends and later, selling the stock at a profit.

To be successful in stock investing, an individual must plan purchases and sales. This planning requires knowledge of both procedures and of the computations involved, as well as an understanding of how a corporation pays dividends.

Your Objectives

After completing Chapter 14, you will be able to

1. Discuss the unique legal features of the corporation.
2. Define the various terms related to stock.
3. Calculate the dividends earned by each share of stock in the corporation.
4. Read and interpret stock exchange listings in a newspaper.
5. Calculate the cost of stock, the amount received from selling stock, and the resulting gain or loss.

Legal Features of a Corporation

Let's look at six unique legal features that will help you understand the corporation.

(1) A corporation is a *separate legal entity* whose life begins when a formal document—a *charter*—is issued by the state. A corporation is empowered to enter into contracts, and it is subject to taxation.

(2) The life of a corporation, unless specifically arranged otherwise, is a *perpetual existence*. Its life is not limited to the life of the owners, as is the case in a sole proprietorship and a partnership.

(3) Because of the second feature, *ownership is transferable*. Thus, as stock is sold in the various exchanges across the country or from one individual to another, the firm is unaffected and is able to continue its day-to-day operations.

(4) The owner is relieved of the decision-making role in most corporations. Rather, the corporation is managed by *elected directors*, who may or may not be the owners. In this way, the corporate form benefits from the best of two worlds: the money of those who want to invest and the brainpower of those who want to manage.

The right of the owner to determine policy is usually limited to voting on policy or to electing the directors.

(5) An owner of a corporation has *limited liability*—in the event of a financial crisis, the owner can lose only what has been invested. This differs from the unlimited liability of sole proprietors and partners.

(6) In a sole proprietorship, the owner keeps *all* profits. In a partnership, the owners share *all* profits. In a corporation, it is the decision of the board of directors as to how much of the profits should be distributed. *Some* profits may be shared by the owners, while other profits may be retained by the firm for various reasons.

Stock Features and Terms

A second area you must master before doing actual calculations is the specialized language of corporation stocks. Several common terms are introduced in this section.

Some corporations issue a single class of stock, called *common stock*. Other corporations issue two classes: common and preferred. *Preferred stock* has a number of advantages when compared to common stock: it comes first in distributions of profit; it is even guaranteed a profit distribution in certain instances; and it is redeemed first if the corporation is dissolved. On the other hand, the holders of common stock usually have the voting power.

The charter specifies the maximum, or *authorized*, number of shares that a corporation can sell. As a share is sold, it is said to be *issued*. In certain situations, a corporation buys back some of its issued shares; those shares still in the hands of the stockholders are called the shares *outstanding*. If a firm has a maximum of 30,000 shares, sells 22,000 shares, and reaquires 1,000 shares, it has 30,000 shares authorized, 22,000 issued, and 21,000 outstanding. The arithmetic point to remember is that any distribution of profits would be made to only the *outstanding* shares, 21,000.

Stock is either *par* or *no-par*. Par stock has a given amount as its face value, just as a one-dollar bill has a one as its face value. If no value is stated on the stock certificate, it is no-par.

Certain preferred stock is guaranteed a distribution of profits. Should a year pass in which the board of directors decides not to distribute any profits, *cumulative* preferred stock has the right to these unpaid past dividends in the next year. *Noncumulative* preferred stock does not have this right, and common stock never has the cumulative right.

Finally, some preferred stock is *participating*. The normal sequence for profit distribution is first to preferred and then "the rest" to common. Participating preferred stock will share in "the rest"; *nonparticipating* will not.

Distribution of Profits

The corporation profits distributed to stockholders are known as *dividends*. In order to compute the dividends earned by a class or a share of stock, it is first necessary to know what kinds of stock have been issued, with what rights, and in what amounts.

Common Stock Only

When a corporation has only common stock outstanding, divide the amount of profits distributed by the number of shares outstanding to find the dividend per share.

SAMPLE PROBLEM 1

Bellows Fibers, Inc., distributes dividends of $54,000 to its 20,000 outstanding shares of stock. (a) Find the dividend per share. (b) Find the amount of dividends received by Mary Stuart, who owns 35 shares.

Solution

(a) $\text{Dividend per share} = \dfrac{\text{Total dividends declared}}{\text{Number of shares outstanding}}$

$$= \frac{\$54{,}000}{20{,}000 \text{ shares}} = \$2.70 \text{ per share}$$

(b) Stuart: 35 shares × $2.70 = $94.50

Note The problem did not state *common* stock, but since only one class of stock was involved, it has to be common.

Common and Preferred Stock

When both common and preferred stock are outstanding, and the preferred stock has no special rights:

(1) Give the preferred stock its dividend first.

(2) Give the common stock the remainder.

Preferred stock dividends are expressed in two ways:

(1) As a percent of par value. A 7%, $100 par preferred stock will earn an *annual* dividend of .07 × $100 = $7.00

(2) As dollars and cents. The board of directors might decide to pay $6.20 a share on its preferred stock.

SAMPLE PROBLEM 2

Snyder, Inc., has outstanding 3,000 shares of 8%, $100 par preferred stock and 20,000 shares of common stock. A dividend of $39,000 is declared. How much will be received by each share of each class of stock?

Solution

$$\text{Preferred, dividend per share} = 8\% \times \$100 = \$8.00$$
$$\text{Total preferred dividends} = \$8 \times 3{,}000 \text{ shares} = \$24{,}000$$
$$\text{Available for common} = \$39{,}000 - \$24{,}000 = \$15{,}000$$
$$\text{Common, dividend per share} = \frac{\$15{,}000}{20{,}000 \text{ shares}} = \$0.75$$

If Snyder, Inc., had a good year and declared a dividend of $100,000, then the total dividends available for common would have been $100,000 − $24,000 = $76,000. Each share of common would have received $76,000 ÷ 20,000 shares = $3.80. Thus, the holder of common stock can do either well or poorly. The holder of preferred stock receives a consistent dividend.

What if only $19,000 of dividends were declared? Common would receive nothing that year because all $19,000 would go to the preferred. But, is the preferred entitled to the "missing" $5,000 ($24,000 − $19,000)? Yes—if it is a *cumulative* stock.

Cumulative Preferred Stock

When the preferred stock has a legal claim to past dividends, the procedure for distributing dividends changes slightly.

(1) Give the preferred stock its unpaid past dividends (referred to as dividends in *arrears*).

(2) Give the preferred stock its normal current dividend.

(3) Give the common stock what is left over.

Look back to Sample Problem 2 again. If $5,000 were in arrears, then in the next dividend year, preferred would receive the $5,000 arrears plus the normal $24,000 current dividend before any distribution was made to common.

Sample Problem 3 presents a computation with cumulative preferred stock. Sample Problem 4 uses the same data for noncumulative preferred stock.

SAMPLE PROBLEM 3

The Lang Planking Corporation declares dividends as follows from 1985–1989:

1985	$51,000
1986	30,000
1987	75,000
1988	12,000
1989	67,000

Its outstanding stock consists of 5,000 shares of 9%, $100 par, cumulative preferred and 30,000 shares of common. No dividends are in arrears as of the end of 1984.

(a) What amount of dividends would be given to each class of stock each year?

(b) What are the total dividends for each class of stock for the five years?

Solution

(a) Normal preferred dividend: 9% × $100 × 5,000 shares = $45,000

Year	Total	Preferred Stock	Explanation	Common Stock
1985	$ 51,000	$ 45,000	Normal year	$ 6,000
1986	30,000	30,000	$15,000 arrears	-0-
1987	75,000	60,000	$15,000 arrears + $45,000 current	15,000
1988	12,000	12,000	$33,000 arrears	-0-
1989	67,000	67,000	$33,000 + $45,000 = $78,000; $11,000 arrears	-0-
(b)	$235,000	$214,000		$21,000

SAMPLE PROBLEM 4

Using the data given in Sample Problem 3, but assuming the preferred to be noncumulative, find (a) the amount given to each class of stock each year and (b) the total dividends per class of stock for the five years.

Solution

(a) Year	Total	Preferred Stock	Common Stock
1985	$ 51,000	$ 45,000	$ 6,000
1986	30,000	30,000	-0-
1987	75,000	45,000	30,000
1988	12,000	12,000	-0-
1989	67,000	45,000	22,000
(b)	$235,000	$177,000	$58,000

Sample Problem 5 shows calculation of dividends per share when outstanding stock includes cumulative preferred and common stock.

SAMPLE PROBLEM 5

Astrolite, Inc., has outstanding 3,500 shares of $8\frac{1}{2}$%, $100 par, cumulative preferred stock and 20,000 shares of common stock. Dividends on the preferred are currently $7,000 in arrears. If a dividend of $64,000 is declared, find the dividend earned by each share of each class of stock.

Solution

Preferred:

Arrears:	$ 7,000
Current dividend:	
($8\frac{1}{2}$% × $100 × 3,500 shares)	29,750
Total	$36,750

Per share: $36,750 ÷ 3,500 shares = $10.50

Common:
Available: $64,000 − $36,750 = $27,250

Per share: $27,250 ÷ 20,000 shares = $1.36

TEST YOUR ABILITY

1. The Alneas Loan Corporation is allowed to issue a maximum of 100,000 shares. During its first year, it sells 42,000 shares in January and 31,500 shares in March, reacquires 300 shares in July, and sells 15,300 in December. As of the end of the year, determine the number of shares (a) authorized; (b) issued; (c) outstanding.
2. The Multi-Method Company, Inc., acquires a special contract and declares a $26,000 dividend to its 6,500 outstanding shares of stock.
   (a) Find the dividend per share.
   (b) John Bowman owns 110 shares. How much of a dividend will he receive?
3. The Vogel Insurance Corporation declares a dividend to its 3,000 outstanding shares of 7%, $100 par, noncumulative preferred stock, and its 10,000 outstanding shares of common stock. Find the dividend per share for each class of stock if total dividends are (a) $31,000; (b) $27,000; (c) $9,000.
4. The Watertown Investigating Corporation has stock outstanding as follows: 2,000 shares of 8%, $50 par, cumulative preferred; 20,000 shares of common. At the end of 1984, no dividends are in arrears. From the data given, find the *total* amount of dividends received by each class of stock for the years 1985–1989.

Year	Dividends Declared
1985	$ 6,000
1986	17,000
1987	24,000
1988	3,000
1989	9,000

5. Solve Problem 4, assuming the preferred stock to be noncumulative.
6. The Mara Wheat Corporation's outstanding stock consists of 4,000 shares of 9%, $100 par, cumulative preferred and 14,000 shares of common. Dividends on the preferred stock are unpaid for the past two years. If a dividend of $136,000 is declared in the current year, find the dividend per share for (a) preferred stock and (b) common stock.

Participating Preferred Stock

Participating preferred stock receives an extra dividend after a normal distribution to preferred stock and a specified distribution to common stock have been made. In other words, the preferred stock *participates* in the distribution of the remaining profits.

Where there is participating preferred stock, dividends are distributed in this order:

(1) Give the preferred stock its normal dividend.
(2) Give the common stock a stated dividend.
(3) Divide the remaining distribution, as specified, between the common and the preferred stock.

Sample Problems 6 and 7 both assume a noncumulative preferred stock. In Sample Problem 6, the preferred stock is participating, while in Sample Problem 7, it is nonparticipating.

SAMPLE PROBLEM 6

Elena Dairy Products, Inc., acquires several important holdings and declares a dividend of \$78,000 in 1988 to its 4,000 outstanding shares of 7%, \$100 par, noncumulative, participating preferred stock, and its 10,000 outstanding shares of common stock. Preferred shareholders participate with the common shareholders, in the ratio of shares outstanding, after each common share has received \$1.50.

(a) Find the total dividends for each class of stock.
(b) Find the dividend per share for each class of stock.

Solution

(a) Preferred, normal dividend:

7% × \$100 × 4,000 shares =	\$28,000
Common, stated dividend:	
\$1.50 × 10,000 shares =	15,000
Remainder:	
\$78,000 − \$28,000 − \$15,000 = \$35,000	
Total shares outstanding:	
4,000 + 10,000 = 14,000	
Distribution by ratio:	
Preferred: $= \dfrac{4{,}000 \text{ shares}}{14{,}000 \text{ shares}} = \dfrac{2}{7} \times \$35{,}000 =$	10,000
Common: $\dfrac{10{,}000 \text{ shares}}{14{,}000 \text{ shares}} = \dfrac{5}{7} \times \$35{,}000 =$	25,000
	\$78,000

Total dividends per class of stock:

Preferred: \$28,000 + \$10,000 =	\$38,000
Common: \$15,000 + \$25,000 =	40,000
	\$78,000

(b) Per share:

Preferred: \$38,000 ÷ 4,000 shares = \$9.50
Common: \$40,000 ÷ 10,000 shares = \$4.00

The original dividends for preferred and common were \$7.00 and \$1.50 per share, respectively. Observe that each share of each class of stock received the same "extra" amount (\$2.50) from participation. Whenever participation is determined by the ratio of shares outstanding, the extra amount is the same.

SAMPLE PROBLEM 7

Using the data from Sample Problem 6, but assuming that the Elena Dairy's preferred stock is neither cumulative nor participating, find (a) the total dividends for each class of stock and (b) the dividend per share for each class of stock.

Solution

	Preferred	Common
(a) 7% × \$100 × 4,000 shares	\$28,000	
Remainder: \$78,000 − \$28,000		\$50,000
Total dividends	\$28,000	\$50,000
(b) Shares outstanding	4,000	10,000
Dividend per share	\$7.00	\$5.00

Common stockholders benefit most when the preferred stock is neither cumulative nor participating.

TEST YOUR ABILITY

1. The Blake Literary Agency, Inc., declares a dividend of \$29,000 to its holders of 5,000 outstanding shares of 8%, \$50 par, noncumulative, participating preferred and 10,000 outstanding shares of common stock. The preferred stock participates with the common stock in the ratio of shares outstanding, after each common share has received a dividend of \$1.00. Calculate the dividend per share for each class of stock.
2. Using the figures from Problem 1, but assuming the Board of Directors declares a dividend of \$42,000, calculate the dividend per share for each class of stock.
3. Using the data from Problem 1 and a \$42,000 distribution of dividends, calculate the dividend per share for each class of stock if the preferred is nonparticipating.
4. Alden Corporation has the following outstanding stock: 2,000 shares of 8%, \$100 par, cumulative, participating preferred and 14,000 shares of common. Money was tight for Alden last year; the preferred dividends were unpaid. After allowing common stock \$0.50 per share, the preferred and common stock participate in the remaining dividends in the ratio of shares outstanding. If dividends of \$55,000 are declared, find the dividend per share for each class of stock.

Purchasing and Selling Stock

A stock may be listed on an *exchange* or traded *over the counter*. The most stable corporations—those with a long history and reliability in paying dividends—are listed on an exchange, the most well-known of which is the New York Stock Exchange (NYSE). All our problems are concerned with only the NYSE. Stocks on an exchange are bought and sold at auction prices, which move up and down, depending on the number of shares available (supply) and the number of potential buyers (demand).

New stocks, generally more risky, are traded over the counter, as are the stocks of some established corporations that choose not to be traded on an exchange. Over-the-counter stocks may be purchased from a stockbroker or from the corporation itself. The price depends upon agreement between buyer and seller.

Since stocks traded on an exchange form the bulk of transactions, we will focus on the exchange. Stock prices flashed over a ticker tape increase the excitement of brokers and investors. The tape shows the latest price at which a

Table 14.1

Sample New York Stock Exchange Listings

Stock and Dividend	Sales in Hundreds	High	Low	Close	Net Change
Arkansas Gas 1.60	46	55	$54\frac{1}{2}$	$54\frac{3}{4}$	-0-
Brandywine Coal .80	116	$16\frac{1}{4}$	$15\frac{3}{4}$	$15\frac{3}{4}$	$-\frac{3}{8}$
CPS Corp.	15	26	$25\frac{3}{4}$	$25\frac{3}{4}$	$-\frac{1}{2}$
Flying Angel .76	305	$39\frac{1}{8}$	$38\frac{1}{4}$	$38\frac{7}{8}$	$+\frac{1}{4}$
Foster Industries .32	27	$10\frac{1}{8}$	10	10	-0-
Grainger R.R. 2.50	428	$83\frac{3}{4}$	82	$83\frac{3}{4}$	+2
Harris-Jones Elec. 1.25	386	$21\frac{1}{2}$	$21\frac{1}{8}$	$21\frac{3}{8}$	$-\frac{1}{8}$
Harris-Jones Elec. pf 4.50	3	51	51	51	+1
Jessup Industries 3.20	252	$63\frac{5}{8}$	$62\frac{3}{4}$	$63\frac{1}{2}$	$+\frac{3}{8}$
Lehigh Oil 10.70	12	$109\frac{1}{2}$	$108\frac{1}{2}$	$109\frac{1}{2}$	$+1\frac{1}{2}$
Pilmer Steel .20	9	$8\frac{1}{2}$	$8\frac{1}{2}$	$8\frac{1}{2}$	-0-
Reese Camera .60	318	$34\frac{7}{8}$	34	34	$-\frac{3}{4}$
Reese Camera pf 3.00	5	85	83	83	−4
Yates Co. 1.12	2240	$13\frac{3}{8}$	13	$13\frac{1}{4}$	-0-
Yates Co. pf 4.40	1	$55\frac{1}{2}$	$55\frac{1}{2}$	$55\frac{1}{2}$	$+\frac{1}{2}$

stock was purchased. At the end of every business day, a summary of the exchange transactions is given to the newspapers and appears on the financial page. An example of such a summary is Table 14.1. Let's see how it is interpreted.

(1) The left-hand column names the stock and its latest *annual* dividend. A common stock is not labeled; a preferred stock is followed by "pf." Reese Camera .60 is a common stock paying $0.60 a year dividend on each share. Reese Camera pf 3.00 is a preferred stock paying $3.00 a year per share. CPS Corp. is an unfortunate exception—no dividends were paid in the past year.

(2) The next column shows the sales for the day in hundreds. Pilmer Steel sold 900 for the day. What were the sales of Arkansas Gas? The answer of 4,600 shares is correct. Most NYSE stocks are sold in 100-share blocks.

(3) The next three columns show the stock's trading prices during the day. All the prices are expressed in dollars and fractions of dollars. Eighths are the smallest fractions of a dollar used. Take Flying Angel, for example. Its highest price during the day was $39.125 ($39\frac{1}{8}$); its lowest $38.25 ($38\frac{1}{4}$). It ended the day, when the exchange closed, at a final price of $38.875 ($38\frac{7}{8}$).

(4) *Net change* refers to the difference between *today's closing price and yesterday's closing price.* Harris-Jones Elec. pf shows a net change of +1, or +$1. If it closed today at $51, the company must have closed yesterday at $50. Check on this point: what did Foster Industries close at yesterday? The answer is 10, today's close, since the net change is −0−.

Round Lots The number of shares in a trading unit is a *round lot.* Round lots generally consist of 100 shares of stock, or multiples of 100 shares, with the exception of some inactive stocks.

The prices given in the stock listings are round-lot prices. The price paid or received per share in a stock transaction is called the *execution price.*

SAMPLE PROBLEM 8

Find the cost of 100 shares of Yates Co. if the stock was purchased at today's high price.

Solution

Execution price ($13\frac{3}{8}$)	\$ 13.375
Number of shares	× 100
Cost of shares	\$1,337.50

Odd Lots An amount of stock that is less than a round lot is called an *odd lot.* Thus, less than 100 shares of a 100-unit stock constitutes an odd lot.

Round lots are easily traded on a stock exchange and by a stockbroker, but odd lots present some difficulty. If you wanted 20 shares of a 100-unit stock, for example, your broker would probably have to buy a 100-unit block, sell you 20, and wait for other odd-lot purchasers to buy the other 80. To compensate for this extra service, the execution price per share on an odd-lot purchase differs from the price on a round-lot purchase. The price difference is called the *odd-lot differential* and is kept by the stockbroker. The amount of the odd-lot differential on 100-share units is negotiable, but is typically $\$\frac{1}{8}$.

For the *purchaser* of an odd lot:

Execution price = Round-lot price + Odd-lot differential

For the *seller* of an odd lot:

Execution price = Round-lot price − Odd-lot differential

SAMPLE PROBLEM 9

Find the execution price per share for each purchase.

(a) 100 shares of a 100-unit stock at 65.
(b) 90 shares of a 100-unit stock at 65.
(c) 80 shares of a 100-unit stock at 35.

Solution

(a) \$65 (round lot)
(b) $\$65\frac{1}{8}$ (odd lot)
(c) $\$35\frac{1}{8}$ (odd lot)

Had each transaction in Sample Problem 9 been a *sale* rather than a purchase, part (a) would be unchanged, since it is a round lot. Part (b) would have been $\$65 - \$\frac{1}{8} = \$64\frac{7}{8}$, and part (c), $\$35 - \$\frac{1}{8} = \$34\frac{7}{8}$.

SAMPLE PROBLEM 10

Find the cost of (a) 90 shares of Pilmer Steel at today's high; (b) 300 shares of Lehigh Oil at today's low.

Solution

(a) Pilmer Steel

Execution price is $8\frac{1}{2}$ (high) + $\frac{1}{8}$ (odd lot)	$ 8.625
Number of shares	× 90
Cost of shares	$776.25

(b) Lehigh Oil

Execution price is $108\frac{1}{2}$ (low); round lot	$ 108.50
Number of shares	× 300
Cost of shares	$32,550.00

On a transaction involving a round lot and an odd lot, the differential applies to the odd-lot portion. For the *purchase* of 110 shares of a 100-unit stock at 36, observe the following:

100 shares × $36 =	$3,600.00
10 shares × 36\frac{1}{8}$ =	361.25
Cost of shares	$3,961.25

Sample Problem 11 concludes this discussion with a couple of points. It shows a *sale* of a round lot and an odd lot with the differential deducted, and introduces a new term—*gross proceeds,* the total sales price of the shares.

SAMPLE PROBLEM 11

Find the gross proceeds from a sale of 230 shares of Grainger R.R. stock at today's closing price.

Solution

Execution price, round lot, close	$83\frac{3}{4}$
Execution price, odd lot, $-\frac{1}{8}$	$83\frac{5}{8}$
Round lot (200 × $83.75)	$16,750.00
Odd lot (30 × $83.625)	2,508.75
Gross proceeds	$19,258.75

Always try to answer two questions when finding the execution price per share:

1. Is it round lot or odd lot?
2. If odd lot, is it a purchase or sale?

TEST YOUR ABILITY

The companies mentioned are found in Table 14.1.

1. Find the cost of 100 shares of Jessup Industries if purchased at (a) today's low price; (b) today's high; (c) yesterday's close.
2. Find the execution price per share and the cost of the shares for each of the following purchases:
   (a) 300 shares of CPS Corp. at today's low
   (b) 80 shares of Yates Co. pf at today's high
   (c) 60 shares of Reese Camera pf at today's close

3. Assuming that each transaction in Problem 2 was a sale, find the execution price per share and the gross proceeds in each case.
4. Find the cost of 350 shares of Flying Angel at today's closing price.
5. Find the gross proceeds from 310 shares of Foster Industries at today's high price.

Commission

A commission is paid to a stockbroker on both purchases and sales. For many years, a set rate system was used by brokers who traded on the NYSE, but currently, commissions are determined by individual brokers, giving the investor a chance to "shop."

Rates depend on such factors as the number of shares traded and the execution price of the shares. The rates range from $\frac{1}{2}$% on large transactions to 4% on small transactions, "large" and "small" being defined by the total dollar value of the transaction. Let's assume that no one will go to extremes; therefore, we will use in our problems a standard rate of 1%—but with a *$30 minimum.*

To the purchaser, commission represents an added cost.

Total cost = Cost of shares + Commission

SAMPLE PROBLEM 12

Find the total cost of 300 shares of Sweetie Candies at a listed price of 37.

Solution

Round-lot purchase	
Cost of shares (300 × $37.00)	$11,100
Commission (1% × $11,100)	111
Total cost	$11,211

To the seller, commission is one of the three charges that reduce gross proceeds. Others include an SEC charge, imposed by the Federal Securities and Exchange Commission, and a transfer tax, imposed by several states. Neither is included in our problems.

Sample Problems 13 and 14 conclude this section by showing the calculation of *net proceeds* to the seller of stock, using the formula:

Net proceeds = Gross proceeds − Commission

Watch for the minimum commission in Sample Problem 14.

SAMPLE PROBLEM 13

Find the net proceeds from a sale of 200 shares of Halston Mining at 47.

Solution

Round-lot sale	
Gross proceeds (200 × $47.00)	$9,400
Less: commission (1% × $9,400)	−94
Net proceeds	$9,306

SAMPLE PROBLEM 14

Find the net proceeds from a sale of 140 shares of Yates Co. at today's close.

Solution

Execution price, round lot, close $13\frac{1}{4}$		
Execution price, odd lot, $-\frac{1}{8}$	$13\frac{1}{8}$	
Gross proceeds:		
Round lot (100 × \$13.25)		\$1,325
Odd lot (40 × \$13.125)		525
Total gross proceeds		\$1,850
Less: commission (1% × \$1,850) = \$18.50 use minimum		−30
Net proceeds		\$1,820

TEST YOUR ABILITY

(Extra data needed for these companies are listed in Table 14.1.)

1. Find the total cost, including commission, of a purchase of 220 shares of Jessup Industries at today's low.
2. Find the net proceeds from a sale of 300 shares of Shark Missiles, Inc. at 16.
3. Find the net proceeds from a sale of 410 shares of Harris-Jones Elec. at yesterday's close.

Gain or Loss on the Trading of Stock

Stocks are bought for either dividends, sale at a gain, or both. Of course, you may never receive a single dividend, and a sale may lead to a net loss, but these are the risks of any investment.

Calculating gain or loss can be a simple process—compare the price at which a stock is sold with the price at which the stock was bought, ignoring commission, SEC charges, and transfer taxes. This *investor's approach* is a handy, quickly done technique, compared to the more accurate but involved *income tax method* in which all charges are considered. Sample Problems 15 and 16 show the investor's approach, while Sample Problem 18 shows the income tax method.

SAMPLE PROBLEM 15

Eleanor Russell buys 200 shares of Bayside Gas stock at $81\frac{3}{8}$ and sells them two months later at $84\frac{1}{2}$. Determine her gain on this sale, using the investor's approach.

Solution

Gain per share ($84\frac{1}{2} - 81\frac{3}{8}$)	\$ 3.125
Number of shares	× 200
Gain on sale	\$625.00

The \$625 gain is only part of the story, however. If Ms. Russell received dividends while holding the stock, her gain would be greater. Sample Problem 16 highlights this point, and shows as well the calculation of the *annual rate of gain or loss* based on cost.

SAMPLE PROBLEM 16

Four hundred shares of Grainger R.R. were purchased by Sherman Tourneur two years ago at $77\frac{1}{2}$ and are sold at today's high price.

(a) Determine his gain or loss on the share price.

(b) What total dividends did he receive for the two-year period, assuming that dividends have been constant?
(c) What is his total gain or loss?
(d) Calculate the annual rate of gain or loss based on cost.

Solution

(a)	Sales price per share	\$ 83.75
	Purchase price per share	−77.50
	Gain per share	\$ 6.25
	Number of shares	× 400
	Gain on share price	\$2,500.00
(b)	Dividends: \$2.50 × 400 shares × 2 years	2,000.00
(c)	Total gain	\$4,500.00

(d) Annual gain: \$4,500 ÷ 2 years = \$2,250
Cost: 400 shares × \$77.50 = \$31,000
Annual rate of gain based on cost:

$$R = \frac{P}{B} = \frac{\text{Gain}}{\text{Cost}} = \frac{\$2,250}{\$31,000} = 7.3\%$$

Notes
(1) The gain from the share price in (a) is taxed to the seller in the year of sale.
(2) The dividends in (b) are taxed in the year received.
(3) The rate of gain was put on an *annual* basis by dividing the gain by the number of years involved. Always do this before calculating the rate. Round the rate to the nearest tenth of a percent.

Rate of Yield The rate of gain or loss is very important to a seller of stock. Another measure of profit is the *rate of yield*—the rate that is earned by simply holding the investment. It is determined from the calculation:

$$\text{Rate of yield} = \frac{\text{Annual dividends per share}}{\text{Cost per share}}$$

For the Grainger R.R. stock held in Sample Problem 16, the dividend per share is \$2.50; the cost per share, \$77.50. Dividing \$2.50 by \$77.50 = 3.2%.

SAMPLE PROBLEM 17

If they are bought at today's low price, determine the rate of yield on (a) Yates Co. 1.12 and (b) Yates Co. pf 4.40.

Solution

(a) Annual dividend per share \$1.12
Cost per share \$13.00
Rate of yield = 1.12 ÷ 13.00 = 8.6%
(b) Annual dividend per share \$4.40
Cost per share \$55.50
Rate of yield = 4.40 ÷ 55.50 = 7.9%

Rate of yield is a main consideration for selecting an investment. We will return to this concept many times in the remainder of Unit 5.

TEST YOUR ABILITY

1. Arthur Wills buys 300 shares of Olson Corporation stock at $53\frac{1}{2}$ and sells them four months later at $51\frac{7}{8}$. Determine the gain or loss on the sale, using the investor's approach.
2. Four years ago, 300 shares of Reese Camera were bought at 30 by Denise Nelson. They are sold at today's closing price.
   (a) Determine the gain or loss on the share price.
   (b) Find her dividends received in four years.
   (c) What is her total gain or loss?
   (d) Calculate the annual rate of gain or loss based on cost, to the nearest tenth of a percent.
3. Determine the rate of yield for each of the following, to the nearest tenth of a percent: (a) Xebec Shipping Company, paying $4.80, cost $240. (b) Frederic's Yarn Company, 8%, $100 par, cost $110. (c) Foster Industries, bought at today's high.

Summary Problem on Stocks

Sample Problem 18 reviews the essential points of investing in stocks, showing the calculation of gain or loss for income tax purposes. Dividends are not included in the gain or loss, since they are taxed separately in the year received. The problem also shows the investor's approach.

SAMPLE PROBLEM 18

Two years ago, 220 shares of Harris-Jones Elec. 1.25 were purchased at 19 by Frank Wilburs. They are sold at today's high.

(a) Calculate the total cost of the shares.
(b) Calculate the net proceeds from the sale of the shares.
(c) What is the taxable gain or loss?
(d) Determine the total gain or loss, using the investor's approach.
(e) Calculate the annual rate of gain or loss, using the investor's approach.
(f) What is the rate of yield?

Solution

(a) Total cost:

Round lot (200 × $19.00)	$3,800.00
Odd lot (20 × $19.125)	382.50
Cost of shares	$4,182.50
Commission (1% × $4,182.50)	41.83
Total cost	$4,224.33

(b) Net proceeds:

Today's high is $21\frac{1}{2}$

Round lot (200 × $21.50)	$4,300.00
Odd lot (20 × $21.375)	427.50
Gross proceeds	$4,727.50
Less: commission (1% × $4,727.50)	−47.28
Net proceeds	$4,680.22

(c) Taxable gain: $4,680.22 − $4,224.33 = $455.89

(d) Investor's gain:

Gross proceeds	$4,727.50
Cost of shares	−4,182.50
Gain on share price	$ 545.00
Dividends: $1.25 × 220 shares × 2 years	550.00
Total gain	$1,095.00

(e) Annual gain: $1,095.00 ÷ 2 years = $547.50
Cost of shares $4,182.50
Annual rate of gain based on cost:
$547.50 ÷ $4,182.50 = 13.1%

(f) Rate of yield:
Annual dividend per share $1.25
Cost per share $19.00
Rate of yield:
$1.25 ÷ $19.00 = 6.6%

Notes (1) The taxable gain or loss is the difference between net proceeds and total cost.
(2) The investor's gain or loss considers gross proceeds and the cost of shares, excluding commissions, but adding in dividends.
(3) The rate of yield is based on the round-lot cost per share.

CHECK YOUR READING

1. What formal document is legally required by a person or persons establishing a corporation?
2. What is unique about the life of a corporation?
3. Jeff Allen, a stockholder of the Acme Corporation, sells stock to Sally Wilson. What effect does the transfer of ownership have on the life of the business?
4. Why are all corporation profits not distributed as dividends?
5. Distinguish between common and preferred stock.
6. A charter reads "30,000 shares allowed." The corporation sells 27,000 shares and later reacquires 500 shares. State the number of shares (a) authorized; (b) issued; (c) outstanding.
7. What is the difference between par and no-par stock?
8. Distinguish between cumulative and noncumulative preferred stock.
9. What is the difference between participating and nonparticipating preferred stock?
10. Name the two rights of preferred stockholders that affect the dividends received by common stockholders.
11. Convert each of the following stock quotations into dollars and cents: (a) $13\frac{1}{2}$; (b) $48\frac{1}{8}$; (c) $62\frac{3}{4}$; (d) $57\frac{5}{8}$; (e) $1037\frac{7}{8}$.
12. How are common and preferred stocks distinguished in a stock listing?
13. If Lenart Gadgets closes at 52 today and its net change is $-1\frac{1}{8}$, what was yesterday's close?
14. What is the *execution price* per share in a stock transaction?
15. Why is an additional charge per share paid on an odd-lot purchase?

16. Find the execution price per share on each of the following stock purchases:

	Number of Shares	Unit of Trading	Quotation
(a)	100	100 shares	50
(b)	200	100 shares	90
(c)	80	100 shares	50
(d)	10	100 shares	90

17. Assuming that each transaction in Question 16 was a *sale*, find the execution price per share.
18. If 140 shares of a 100-unit stock are purchased at 43, how would they be priced?
19. Distinguish between the investor's approach and the tax method in determining gain or loss on the sale of stock.
20. If a stock costing $100 gives Jackie Ardrushka a $7.70 dividend, what is the rate of yield?

EXERCISES

The companies named that need additional data are found in Table 14.1. A commission rate of 1% is assumed whenever applicable.

Group A

1. Find the net proceeds from a sale of 410 shares of Arkansas Gas at yesterday's closing price.
2. The outstanding stock of the Cutler Corporation consists of 3,500 shares of 8%, $100 par, cumulative, nonparticipating preferred and 20,000 shares of common stock. At the end of 1984, no dividends are in arrears. Find the total dividends *per class* of stock for the five-year period 1985–1989. Dividends declared each year are as follows:

1985	$29,000
1986	32,000
1987	7,000
1988	14,000
1989	52,000

3. Using the Culter Corporation data from Exercise 2, but assuming the preferred stock to be noncumulative, find the total dividends *per class* of stock for the five-year period.
4. Determine the rate of yield in each of the following, rounding to the nearest tenth of a percent:
   (a) Minivehicle Co., paying $7.20, costing $350
   (b) Nielson Lighting, 7%, $100 par, costing $125
   (c) Brandywine Coal, purchased at today's high
5. Find the execution price per share and the cost of the shares for each of the following purchases made today, excluding commission:
   (a) 600 shares of Harris-Jones Elec. pf at the low
   (b) 90 shares of Lehigh Oil at the high
   (c) 50 shares of Brandywine Coal at the high

6. Assuming that each transaction in Exercise 6 was a sale, find the execution price per share and the gross proceeds in each case.
7. The Wilder Silver Corporation's charter allows for a maximum of 50,000 shares. Transactions during 1988 are as follows:

January:	Sold 17,500 shares
March:	Sold 16,210 shares
July:	Sold 15,030 shares
October:	Reacquired 830 shares
December:	Sold 1,105 shares

As of year-end, determine the number of shares (a) authorized: (b) issued; (c) outstanding.
8. The Baumeister Insurance Corporation's outstanding stock consists of 6,000 shares of 7%, $50 par, cumulative, nonparticipating preferred and 50,000 shares of common. Dividends on the preferred stock have not been paid for the past two years. If a dividend of $135,000 is declared, find the dividend *per share* for (a) the preferred stock and (b) the common stock.
9. Find the net proceeds from a sale of 600 shares at 53 to Carlos Girri.
10. Find the cost of a purchase of 200 shares of CPS Corp., ignoring commission, if bought at (a) today's high price; (b) today's closing price; (c) yesterday's closing price.
11. Find the cost, excluding commission, of 180 shares of Arkansas Gas at today's low.
12. Faris Wagons, Inc., declares a dividend to its 5,000 outstanding shares of 8%, $100 par, noncumulative, nonparticipating preferred stock and its 40,000 outstanding shares of common stock. Find the dividend *per share* for each class of stock if dividends declared are (a) $35,000; (b) $44,000; (c) $120,000.
13. The Dunglass Furniture Corporation declares a dividend of $70,400 to its 8,000 outstanding shares of 7%, $100 par, noncumulative preferred and its 24,000 outstanding shares of common stock. Each share of common stock receives a dividend of $0.75. Determine the dividend *per share* for each class of stock if the preferred stock participates in the remainder with the common stock in proportion to the number of shares outstanding.
14. Using the data from Exercise 13, but assuming the dividend declared to be $109,200, calculate the dividend *per share* for each class of stock.
15. Using the data from Exercise 13 and a $109,200 dividend declaration, determine the dividend *per share* for each class of stock if the preferred is nonparticipating. Round to the nearest cent.
16. Ralph Carlson buys 500 shares of Foresight Planners' stock at 59 and sells them two years later at $72\frac{1}{8}$. Find his gain or loss on the sale, using the investor's approach.
17. Find the gross proceeds from a sale of 47 shares of Brandywine Coal at today's low.
18. Five years ago, 300 shares of Arkansas Gas were purchased at $61\frac{1}{2}$. They are sold at today's high. (a) Determine, by the investor's approach, the gain or loss on the share price. (b) Determine the dividends earned for the five-year period. (c) What is the total gain or loss? (d) What is the annual rate of gain or loss based on cost, to the nearest tenth of a percent?

19. Matsuda Motors, Inc., declares a dividend of $66,000 for its shareholders. The number of shares outstanding is 24,000.
    (a) Find the dividend per share.
    (b) Find the dividends received by Marty Stein, holder of 410 shares.
20. Corbin Importers, Inc., has outstanding stock consisting of the following: 3,000 shares of 9%, $100 par, cumulative, participating preferred; 15,000 shares common. Dividends on the preferred are two years in arrears. After $0.80 per share for the common stock, preferred participates in remaining dividends in the ratio of shares outstanding. If dividends of $129,000 are declared, find the dividend *per share* for each class of stock.
21. Find the total cost, including commission, of a purchase of 640 shares of Reese Camera at today's low price.

Group B

1. The preferred stock of the Salvador Camera Corporation consists of 4,150 outstanding shares of $7\frac{1}{2}$%, $100 par, cumulative nonparticipating stock. Dividends are unpaid for the past $3\frac{1}{2}$ years. What amount of dividends must be declared this year before the common stock will receive any dividends?
2. The Carlisle Bus Corporation has outstanding 2,000 shares of 7%, $50 par, cumulative, nonparticipating preferred stock and 10,000 shares of common stock. The preferred stock is $5,000 in arrears at the end of 1984. Determine the total dividend per share for the five-year period 1985–1989 for each class of stock, if dividends declared are as follows:

1985	$ 6,000
1986	9,000
1987	14,000
1988	2,000
1989	16,000

3. Argo Vacation Inns, Inc., has outstanding common stock (8,000 shares) and noncumulative, participating preferred stock (7%, $100 par, 2,000 shares). After the normal dividends to preferred and common stock ($0.50 per share), any remaining dividends are shared in the ratio of shares outstanding. Determine to the nearest cent, the *average annual dividend* per share for each class of stock for the five-year period 1985–1989, if dividends are declared as follows:

1985	$28,000
1986	22,000
1987	15,000
1988	6,000
1989	33,000

4. Use the figures in Exercise 3 and assume the preferred to be cumulative as well as participating. No dividends are in arrears at the end of 1984. Find the average annual dividend per share for each class of stock for the period 1985–1989.
5. The outstanding stock of the Manfred Exploring Corporation consists of 5,000 shares of 7%, $100 par preferred and 20,000 shares of common stock.

Dividends of $90,000 are declared in 1988. Determine the dividend per share for each class of stock under each of the following conditions:

(a) The preferred stock is noncumulative and nonparticipating.
(b) The preferred stock is cumulative and nonparticipating. Dividends are unpaid for two years.
(c) The preferred stock is noncumulative and participates, in the ratio of shares outstanding, after a $2.00 per share common dividend has been paid.
(d) The preferred stock is cumulative and participating, as outlined in (b) and (c).

6. Andrea Passos buys 300 shares of Brandywine Coal at the day's low and sells them two years later at $19\frac{3}{8}$. Find (a) her total cost; (b) the net proceeds; (c) the taxable gain or loss; (d) the investor's gain or loss; (e) the annual rate of gain or loss; (f) the rate of yield.
7. If 120 shares of Harris-Jones Elec. 1.25 were purchased four years ago by Harry Burke at $19\frac{1}{2}$ and are sold at today's closing price, find (a) the taxable gain or loss; (b) the investor's annual rate of gain or loss; (c) the rate of yield.

CHAPTER 15

Bonds and Other Investments

Another major area of investment for both individuals and business firms is the purchase of bonds. A *bond* is a long-term obligation of a corporation or government, given in return for a loan of money by the investor. The investor is paid interest during the term of the bond, and at the end of the term, the loan itself is repaid.

Mutual funds, real estate, life insurance, or savings certificates can also be used as investments. In this chapter, you will look at these methods of investment, and some day, perhaps, apply one as a means of making money.

Your Objectives

After completing Chapter 15, you will be able to

1. Understand the language of bonds.
2. Calculate the costs, proceeds, gains or losses, and yield from bond investments.
3. Calculate the costs and returns from investments in mutual funds.
4. Calculate the appreciation on a real estate investment.
5. Calculate the net cost or net return on an investment in life insurance .
6. Develop your ability to compare investments.

Bonds

Investing in bonds differs from investing in stocks. The bondholder generally seeks a regular income from the interest, while the stockholder seeks a gain on the sale of a rising stock, in addition to varying dividends. A bondholder is a

Table 15.1

New York Exchange Bond Listings

Bonds	Interest	Maturity	Yield	Sales*	High	Low	Close	Net Change
A-One	8's	93	8.3	45	96	$95\frac{5}{8}$	96	$-\frac{1}{4}$
Brown	$6\frac{1}{2}$'s	89	8.9	62	$73\frac{1}{8}$	$72\frac{7}{8}$	$72\frac{7}{8}$	$-\frac{1}{4}$
Eason	$7\frac{3}{4}$'s	94	8.3	15	$92\frac{7}{8}$	$92\frac{3}{4}$	$92\frac{7}{8}$	$+\frac{1}{2}$
Freeway	$10\frac{3}{4}$'s	97	11.0	266	$93\frac{3}{4}$	$93\frac{1}{2}$	$93\frac{1}{2}$	$+\frac{1}{8}$
HSL	10's	88	9.3	10	107	107	107	−2
Jackson	$12\frac{1}{2}$'s	90	11.0	14	110	110	110	-0-
LMN	10's	14	9.1	17	$108\frac{7}{8}$	$108\frac{7}{8}$	$108\frac{7}{8}$	$-1\frac{1}{2}$
Manorhaven	6's	07	—	3	55	55	55	$-4\frac{1}{2}$
Par-com	$2\frac{3}{4}$'s	89	3.7	6	73	73	73	$+3\frac{1}{8}$
Pyrovan	$8\frac{1}{2}$'s	05	8.4	10	101	101	101	$+\frac{1}{2}$
Steffens	9's	95	8.9	9	$101\frac{1}{2}$	101	101	$-\frac{3}{4}$
Tenga	$8\frac{1}{2}$'s	92	8.5	5	$100\frac{1}{8}$	100	100	-0-
Wyoming	6's	90	9.5	31	$63\frac{1}{8}$	$62\frac{7}{8}$	$62\frac{7}{8}$	$-\frac{1}{8}$
Yale	8.2's	92	8.0	48	$102\frac{1}{2}$	$101\frac{3}{8}$	102	$+\frac{1}{2}$
Zeta-com	$4\frac{3}{4}$'s	90	6.3	51	$74\frac{7}{8}$	$74\frac{7}{8}$	$74\frac{7}{8}$	$+\frac{7}{8}$

* Number of $1,000 bonds

creditor of a corporation and is paid, in the event of the corporation's failure, before the stockholder, who is an owner.

Bonds, like stocks, are priced on the New York Stock Exchange. The information in this chapter reflects exchange practices. We are limiting our discussion to *corporate bonds*—those issued by a private firm—and will not include *government bonds*.

Bond Listings

Table 15.1 is an example of bond trading on the NYSE.

1. The number of bonds sold is given in $1,000 units, because $1,000 is the standard *face value* of a corporation bond. On this date, 14 Jackson bonds were sold, with a total face value of $14,000.
2. Three items of information are given for each bond, aside from its name, number sold, and prices:
   (a) The interest rate. Jackson is a $12\frac{1}{2}$% bond ($12\frac{1}{2}$'s). A holder of one Jackson bond will receive interest at $12\frac{1}{2}$% of face ($1,000), or $125 *a year*. However, since bond interest is normally paid semiannually, two $62.50 payments would be received.
   (b) The rate of yield. Jackson has a rate of yield of 11.0%. This is the *true rate that the investor will earn* by buying Jackson bonds. New bond issues generally do not print a rate of yield. Manorhaven bonds were just issued this year, so no yield is indicated. Additional discussion of this idea appears later in this chapter.

(c) The maturity date. Jackson falls due in 90, or 1990. The "19" is quickly understood except, perhaps, in cases such as Manorhaven, where an answer of 1907 is not logical. When do you think Manorhaven is due? The only answer is the year 2007. On the due date, each bondholder will be paid $1,000 per bond.

3. Three prices are also reported on the bond market. Highs, lows, and closing prices are given in *percents* of $1,000 to the nearest eighth of a percent. Jackson sold at 110 today at each price—110% of $1,000, or $1.10 \times \$1{,}000 = \$1{,}100$. Manorhaven sold at 55% of $1,000, or $550. A bond listing at 100 sells at face of $1,000. You can see that bond prices vary less than stock prices, if you compare Table 15.1 to Table 14.1. The reason is understandable: Bonds are much more stable than stocks.
4. As in the stock market, *net change* is the difference between today's closing price and yesterday's. Jackson shows a net change of –0–, so yesterday's close was 110. Manorhaven dropped by $4\frac{1}{2}$, since yesterday, to today's 55 close. What was yesterday's? The answer is $59\frac{1}{2}$.

Buying Bonds

The stock purchaser pays the execution price plus a commission when buying stock. The *bond* purchaser pays three amounts when buying bonds: market price (*bond quotation*), *commission,* and *accrued interest.* However, accrued interest is paid only if the purchase is made between interest dates. Let's examine each part of bond cost.

Sample Problems 1 and 2 show the calculation of *quotation cost* on bond purchases, at prices below and above face.

SAMPLE PROBLEM 1

Calculate the quotation cost of five Wyoming bonds at today's high price, using Table 15.1.

Solution

High price per bond $= 63\frac{1}{8} = 63\frac{1}{8}\% =$
$.63125 \times \$1{,}000 = \$631.25$ per bond

Number of bonds $= 5$

Quotation cost $= 5 \times \$631.25 = \$3{,}156.25$

SAMPLE PROBLEM 2

Find the quotation cost of 16 Yale bonds at today's low price. Use Table 15.1.

Solution

Low price per bond $= 101\frac{3}{8} = 101\frac{3}{8}\% =$
$1.01375 \times \$1{,}000 = \$1{,}013.75$ per bond

Number of bonds $= 16$

Quotation cost $= 16 \times \$1{,}013.75 = \$16{,}220.00$

Commission

The buyer of bonds pays a commission to the broker on the purchase. As with stock commission, what was once fixed is now negotiable. Let's use a $10.00 per bond fee for all transactions.

The purchaser in Sample Problem 1 will pay a $10 commission per bond for the five bonds, or $50. What commission per bond is paid in Sample Problem 2? What total commission? The answers are $10 and 16 × $10 = $160.

Accrued Interest

As we have seen, bonds pay interest every six months. Whoever holds the bond on an interest date will receive six months' interest. But what if the bond is sold *just before* the interest date? Will the seller lose the interest? The answer is no, because of the following standard practice: When a bond is sold, the buyer pays the seller the interest earned (*accrued*) from the last interest payment date to the date of purchase. Thus, if a bond—paying interest on June 1 and December 1—is sold on October 10, the buyer pays the seller the accrued interest from June 1 to October 10. On December 1, the buyer then receives six months' interest.

Interest is computed by the usual formula, $I = P \times R \times T$, where

P = *face* of all the bonds purchased
R = rate of bond interest
T = time from last interest payment date to date of purchase

The difference in bond interest calculation is the determination of time. Exact number of days is *not* used. Instead, *bond time* (also called *30-day month time*) is used to count the days. From a specific date in one month to the same date in another month, count the number of months and multiply by 30. For example, from January 1 to March 1, there are two months, or 60 days. A year is considered to be 360 days.

SAMPLE PROBLEM 3

Five bonds with interest dates of January 1 and July 1 are sold on June 1 to Simon Burns. If the bonds pay $7\frac{1}{2}\%$, calculate the accrued interest to be paid by Mr. Burns.

Solution

$P = 5 \text{ bonds} \times \$1{,}000 = \$5{,}000$

$R = 7\frac{1}{2}\%$

$T = \text{January 1 to June 1} = 5 \text{ months} \times 30 \text{ days} = 150 \text{ days}$

$$I = \$5{,}000 \times 7\tfrac{1}{2}\% \times \tfrac{150}{360}$$

$$= \frac{\$5{,}000}{1} \times \frac{15}{200} \times \frac{150}{360}$$

$$= \$156.25 \text{ paid by Mr. Burns}$$

When the dates of interest payment and of purchase are not the same, bond time is calculated as follows: (Assume that the purchase in Sample Problem 3 was made on the 18th of June.)

January 1 to June 1 = 5 months	= 150 days
June 1 to June 18 = 18 − 1	= 17 days
Bond time	167 days

This procedure is seen again in Sample Problem 4.

SAMPLE PROBLEM 4

Pierre Ferreira purchases 80 Freeway bonds on April 10 at today's low. Calculate the accrued interest paid by Mr. Ferreira if the interest dates are March 1 and September 1. (Use Table 15.1.)

Solution

P = 80 bonds × $1,000 = $80,000

R = $10\frac{3}{4}\% = \frac{43}{400}$

T = March 1 to April 1 = 30 days
April 1 to April 10 = 9 days
Bond time 39 days

$$I = \frac{\$80{,}000}{1} \times \frac{43}{400} \times \frac{39}{360} = \$931.67$$

Notice, in Sample Problem 4, that the bond interest is based on the *face* of the bonds. The quotation of $93\frac{1}{2}$ in Table 15.1 did not influence the math.

Sample Problem 5 highlights the computation of the cost of a bond purchase by using the formula:

Total cost = Quotation cost + Commission + Accrued interest

SAMPLE PROBLEM 5

Find Nancy Green's total cost of her purchase of three Steffens bonds on July 8 at today's high price, if the interest dates are May 1 and November 1.

Solution

High price per bond = $101\frac{1}{2} = 101\frac{1}{2}\%$ = 1.015 × $1,000 = $1,015.00 per bond
Number of bonds = 3
Quotation cost = 3 × $1,015.00 = $3,045.00
Commission on bonds is $10.00 each

Accrued interest:
P = 3 × $1,000 = $3,000
R = 9%
T = May 1 to July 8 = 2 months, 7 days = 67 days

$$I = \frac{\$3{,}000}{1} \times \frac{9}{100} \times \frac{67}{360} = \$50.25$$

Total cost:

Quotation cost	$3,045.00
Commission	30.00
Accrued interest	50.25
Total cost	$3,125.25

Interest will be accrued in all bond transactions *except* when a sale takes place *on an interest date*. In this case, the seller will receive the full six months' interest *from the issuing corporation*—the buyer will owe no interest to the seller.

TEST YOUR ABILITY

1. Look at the the A-One bond listing in Table 15.1.
   (a) How many bonds were sold?
   (b) What is the face value of the bonds sold?
   (c) What annual interest would the holder of one bond receive?
   (d) What is the rate of yield?
   (e) In what year is the bond due?
   (f) What is the low price in dollars and cents?
   (g) What was yesterday's closing price in dollars and cents?
2. Calculate the quotation cost of the following bond purchases:

	Number of Bonds	Quotation
(a)	12	100
(b)	6	$93\frac{1}{2}$
(c)	10	$114\frac{1}{4}$
(d)	20	$78\frac{7}{8}$

3. Find the quotation cost of 15 Yale bonds at today's high price, using Table 15.1.
4. Determine the accrued interest on each of the following bond purchases:

	Number of Bonds	Interest Dates	Date of Purchase	Interest Rate
(a)	9	April 1, October 1	September 1	4%
(b)	36	May 1, November 1	December 1	5%
(c)	21	January 1, July 1	October 19	6%
(d)	6	February 1, August 1	July 15	7%
(e)	18	March 1, September 1	June 22	8%

5. Celina Austen purchases 20 Tenga bonds on August 7 at today's high price. If the interest dates are January 1 and July 1, find the accrued interest paid by Ms. Austen, using Table 15.1.
6. Determine the total cost for Mark Granger of each of the following purchases on May 21: (a) five Wyoming bonds at today's closing price (interest dates, January 1 and July 1); (b) 120 Freeway bonds at today's high (interest dates, March 1 and September 1).

Selling Bonds

To find the *gross proceeds* from a sale of bonds, add the quotation price to the accrued interest.

Gross proceeds = Quotation price + Accrued interest

For the seller in Sample Problem 5, gross proceeds are \$3,045.00 + \$50.25 = \$3,095.25.

To compute *net proceeds* from a sale of bonds, deduct the broker's commission from the gross proceeds:

Net proceeds = Gross proceeds − Commission

For our seller in Sample Problem 5, the commission is the same as the buyer's—\$30.00. The net proceeds are then \$3,095.25 − \$30.00 = \$3,065.25.

An SEC charge is also deducted in practice, but is not shown here.

Review the complete procedure for finding net proceeds in Sample Problem 6.

SAMPLE PROBLEM 6

Calculate the net proceeds from Jack McCormack's sale of ten Pyrovan bonds at today's closing price on January 17, if the interest dates are January 1 and July 1. Use Table 15.1.

Solution

Closing price per bond = 101 = 101% = 1.01 × \$1,000 = \$1,010.00 per bond

Number of bonds = 10

Quotation price = 10 × \$1,010.00 = \$10,100.00

Accrued interest:

$P = 10 \times \$1{,}000 = \$10{,}000$

$R = 8\frac{1}{2}\% = \frac{17}{200}$

$T = \text{January 1 to 17} = 16 \text{ days}$

$$I = \frac{\$10{,}000}{1} \times \frac{17}{200} \times \frac{16}{360} = \$37.78$$

Gross proceeds = \$10,100.00 + \$37.78	\$10,137.78
Commission (10 × \$10.00)	−100.00
Net proceeds	\$10,037.78

Gain or Loss on Bond Transactions

As we learned in our study of stocks, there are two ways of calculating gain or loss on bond transactions. The *tax method* compares *total cost* (excluding accrued interest) with *net proceeds* (excluding accrued interest) to find the gain or loss. The *investor's method* compares *quotations* and adds *interest received.* Both methods are shown in Sample Problem 7.

SAMPLE PROBLEM 7

Julie Heathrow purchases five 4% bonds at 93 and sells them two years later at 101. Both the purchase and the sale take place on interest dates. Determine the gain or loss by (a) the tax method and (b) the investor's method.

Solution

(a) Tax method:

Gain or loss = Net proceeds − Total cost

Quotation price (101 = \$1,010 per bond × 5)	\$5,050
Commission (5 × \$10)	− 50
Net proceeds	\$5,000
Quotation cost (93 = \$930 per bond × 5)	\$4,650
Commission (5 × \$10)	50
Total cost	\$4,700

Taxable gain = \$5,000 − \$4,700 = \$300

(b) Investor's method:

Quotation price on sale	$5,050
Quotation price on purchase	−4,650
Gain on quotation	$ 400
Interest received ($5,000 × 4% × 2 years)	400
Investor's gain	$ 800

For tax purposes, the gain reported in the year of sale is $300. The interest is reported separately, in the year earned. Notice that there was no accrued interest in the problem; both the sale and purchase took place on interest dates. *Had there been accrued interest,*

1. the total cost would have been higher;
2. the net proceeds would have been higher;
3. the gain would still have been $300, since accrued interest does not affect taxable gain or loss. (You will use this concept in the *Group B* exercises only.)

The *investor's gain* answers the question: "How much have I gained overall?" The answer is $800. The clever investor will go one step further and seek the *rate of gain.*

In Sample Problem 7, Julie Heathrow gained $800 in two years on an investment of $4,650. The *average annual gain* is $400 ($800 ÷ 2). The *rate of gain* is:

$$\frac{400}{4650} = 8.6\%$$

Bond Yield Table 15.1 includes a list of yield rates for each bond, except one new issue. Calculating these rates is a complex process that we will explore just a bit. However, here's a practical formula:

$$\text{Rate of yield} = \frac{\text{Annual interest}}{\text{Quotation cost of one bond}}$$

The formula is similar to one that we used for stock. Let's see if we can arrive at A-One's yield of 8.3%:

$$\text{Annual interest is } 8\% \times \$1{,}000 = \$80$$
$$\text{Cost is } 96 = \$960$$
$$80 \div 960 = 8.3\%$$

Let's also calculate Jackson's yield. Dividing $125 by $1,100, we get 11.4%—not 11.0%. In other words, a simple formula does not always give an exact bond yield. We can use the formula, but remember that it is only a rough answer. If you intend to invest, you will need to know the actual rate of yield. You should therefore research the more complex method on your own. (It is beyond the level of this text.)

Sample Problem 8 reviews the calculation of two rates for the investor by using the formulas:

$$\text{Rate of gain or loss} = \frac{\text{Investor's gain or loss}}{\text{Total quotation cost}}$$

$$\text{Rate of yield} = \frac{\text{Annual interest}}{\text{Quotation cost of one bond}}$$

SAMPLE PROBLEM 8

Fred Cranshaw buys ten bonds at 102 and two years later sells them at 105. Interest of 5% is earned. Find (a) the rate of gain or loss and (b) the rate of yield.

Solution

(a) Rate of gain or loss:

Quotation price on sale (10 × $1,050)	$10,500
Quotation cost on purchase (10 × $1,020)	−10,200
Gain on quotation	$ 300
Interest received ($10,000 × 5% × 2 years)	1,000
Total gain	$ 1,300
Average annual gain ($1,300 ÷ 2)	$ 650

$$\text{Rate of gain} = \frac{650}{10{,}200} = 6.4\%$$

(b) Rate of yield:

Interest on one bond = 5% × $1,000 = $50

Quotation cost of one bond = $1,020

$$\text{Rate of yield} = \frac{50}{1{,}020} = 4.9\%$$

TEST YOUR ABILITY

1. Determine the net proceeds from the sale of six 10% bonds on August 1 at 102, if the interest dates are May 1 and November 1.
2. Determine the net proceeds from the sale of five HSL bonds on June 1 at today's low price, if the interest dates are June 1 and December 1. Use Table 15.1.
3. If the HSL bonds in Problem 2 were sold on July 19, calculate the net proceeds.
4. Eight 5% bonds are purchased at 94 by Florence Pliny and sold five years later at 102. Both the purchase and the sale take place on interest dates. Calculate the gain or loss by the tax method.
5. Use the data in Problem 4 and find the gain or loss by the investor's method.
6. Nine bonds are bought at 101 by Seagroves International and sold three years later at 104. Interest of 6% is received. Calculate (a) the gain or loss by the investor's method; (b) the annual rate of gain or loss; (c) the rate of yield. Round rates to the nearest tenth of a percent.
7. Find the rate of yield to the nearest tenth of a percent on each problem.

	Interest Rate	Quotation
(a)	$6\frac{1}{2}\%$	98
(b)	7%	102
(c)	$4\frac{1}{4}\%$	$80\frac{1}{8}$
(d)	$5\frac{3}{4}\%$	$73\frac{3}{4}$

Mutual Funds

Many investors today prefer to pool their money, rather than risk individual investment. They join in a company known as an *investment trust*. The individuals buy shares in the trust, which in turn invests the pooled capital. The choice of investments is decided by a manager or management group, depending on the size of the trust.

If the number of shares in the fund is fixed, like the authorized stock of a corporation (Chapter 14), the trust is a *closed-end fund.* Closed-end funds may be listed on an exchange and have their shares bought and sold just as any other stock.

If the number of shares is unlimited, the trust is an *open-end fund* or, more commonly, a *mutual fund.* Mutual funds are not listed on an exchange. Shares are purchased directly from the fund or through a broker representing the fund.

A mutual fund may be either of two types. If it has a sales force, the investor will pay a fee to cover sales representatives' commissions. This is called a *load fund.* If it has no sales force and does not charge a fee, it is called a *no-load fund.*

Many mutual funds are reported daily in the financial pages of the newspaper with highs, lows, closing prices, and net changes. The price listed for the fund is the *offering price per share,* which is the *net asset value per share* plus the load fee, if there is one.

Net Asset Value per Share

A firm's net worth is the difference between what it owns (assets) and what it owes (liabilities). (We will examine this point in Chapter 16.) Thus, if assets are \$50,000 and liabilities are \$30,000, net worth is \$20,000 (\$50,000 – \$30,000).

If the net worth of a corporation is divided by the number of shares outstanding, we arrive at a *net asset value per share.* If a mutual fund with 10,000 shares outstanding has a net worth of \$100,000, each share is worth \$10 (\$100,000 ÷ 10,000 shares). Each share has a net asset value of \$10.

SAMPLE PROBLEM 9

A mutual fund with 1,800,000 shares outstanding has assets of \$25,000,000 and liabilities of \$5,000,000. Calculate the net asset value per share to the nearest penny.

Solution

$$\text{Net worth} = \$25{,}000{,}000 - \$5{,}000{,}000 = \$20{,}000{,}000$$

$$\text{Net asset value per share} = \frac{\$20{,}000{,}000}{1{,}800{,}000 \text{ shares}} = \$11.11 \text{ per share}$$

Investing in a Mutual Fund

In a no-load fund, the cost of each share is its net asset value. If you purchase 100 shares of a no-load fund at the market price (net asset value) of \$13.71 per share, you pay 100 × \$13.71 = \$1,371.

In a load fund, you pay the net asset value plus the load fee. Table 15.2 has a typical schedule of load fees. Notice that the rate decreases as the dollar value increases. The published offering price for the fund always includes the maximum load fee of $8\frac{1}{2}\%$.

Assume that you have a share with a net asset value of \$18.30 and you want to price it for sale to the public at an offering price. As mentioned above, the offering price *includes* the load fee at $8\frac{1}{2}\%$. But the $8\frac{1}{2}\%$ is not part of the \$18.30 net asset value—it is part of the offering price. What is the offering price? We have here nothing less than a base, rate, and percentage problem in which the

Table 15.2

Schedule of Load Fees

Investment	Load Fee
Under $25,000	$8\frac{1}{2}$%
$25,000 but under $50,000	6%
$50,000 but under $100,000	4%
$100,000 but under $250,000	3%
$250,000 but under $500,000	2%
$500,000 and over	1%

offering price is the base—100%. The net asset value is 100% − $8\frac{1}{2}$% = $91\frac{1}{2}$% of the offering price. To solve the problem, find the base as shown below:

Given:	$	%
Net asset value	18.30	
Load fee		$8\frac{1}{2}$
Offering price		100

Find: Offering price

Solution:

Net asset value = 100% − $8\frac{1}{2}$% = $91\frac{1}{2}$%

$18.30 = $91\frac{1}{2}$% of base

$$B = \frac{P}{R} \qquad \frac{\$18.30}{.915} = \$20.00$$

Check Is load fee $8\frac{1}{2}$% of offering price?

Load fee = $20.00 − $18.30 = $1.70

$8\frac{1}{2}$% of $20.00 = $1.70

What you have seen is that one share of the mutual fund costs $20.00. From this amount, $1.70 will cover the fund's expenses, and $18.30 is your investment.

Sample Problem 10 begins with an offering price and shows the deduction of the $8\frac{1}{2}$% load fee in determining the amount with which the investor is credited.

SAMPLE PROBLEM 10

Joan Moore wishes to purchase 100 shares of a load fund listed at an offering price of $32. (a) How much will she pay? (b) What fee per share will she pay? (c) What total load fee will she pay? (d) With how much will she be credited?

Solution

(a) 100 × $32.00 = $3,200.00 (the offering price includes the load)
(b) $8\frac{1}{2}$% × $32.00 = $2.72 (Table 15.2)
(c) 100 × $2.72 = $272.00
(d) $3,200.00 − $272.00 = $2,928.00

Investing in a mutal fund can be done in either of two ways. In Sample Problem 10, Ms. Moore bought a number of shares. The alternate method is simply to invest money. If you decide, for example, to invest $50,000 in a mutual fund, the load will be deducted from your investment and the remainder will be used to buy the shares. Sample Problem 11 illustrates this procedure.

SAMPLE PROBLEM 11

Bill Esson invests $50,000 in a load fund. The net asset value per share is $12.50. With how many shares will he be credited?

Solution

Load fee on $50,000 from Table 15.2 is 4%
$50,000 × 4% = $2,000 fee
Amount credited = $50,000 − $2,000 = $48,000
Net asset value per share is $12.50

$$\text{Number of shares purchased} = \frac{\$48{,}000}{\$12.50} = 3{,}840 \text{ shares}$$

If the number of shares worked out unevenly, the investor would be credited with a fractional share, to three decimal places.

TEST YOUR ABILITY

1. In each of the following problems, calculate the net asset value per share to the nearest cent:

	Assets	Liabilities	Number of Oustanding Shares
(a)	$ 250,000	$ 120,000	13,000
(b)	$ 219,500	$ 82,650	17,000
(c)	$ 4,154,800	$ 1,058,000	280,000
(d)	$38,250,000	$14,576,000	1,400,000

2. Walter Bowen buys 250 shares of a no-load fund with a net asset value per share of $16.21. Calculate his total cost.
3. Bob Ryan purchases 100 shares of a load fund at an offering price of $28.00.
   (a) How much will he pay?
   (b) What is the load fee per share?
   (c) What is the total load fee?
   (d) With how much will he be credited?
4. In each of the following load fund problems, use Table 15.2 and find the number of shares credited to the investor, to three decimal places, if necessary.

	Amount Invested	Net Asset Value per Share
(a)	$ 75,000	$24.00
(b)	$100,000	$19.40
(c)	$260,000	$25.48
(d)	$600,000	$11.88

Income from a Mutual Fund

A mutual fund earns income in two ways and distributes this income to the individual investors:

1. It receives interest on bonds held and dividends on stocks held.
2. It gains (or loses) on the sale of stocks and bonds.

Each investor receives a prorated share of both incomes. The fund is required by law to distribute 90% of all *dividends and interest.*

SAMPLE PROBLEM 12

A fund earns $315,000 in dividends and interest, all of which is distributed to the fund's shareholders. Bruce Moore owns 75 of the fund's 150,000 shares. How much will he receive?

Solution

$$\text{Earnings per share} = \frac{\$315{,}000}{150{,}000 \text{ shares}} = \$2.10$$

$$75 \times \$2.10 = \$157.50$$

Alternate Solution

$$\text{Moore holds } \frac{75}{150{,}000} \text{ or } \frac{1}{2{,}000} \text{ of the shares}$$

$$\frac{1}{2{,}000} \times \$315{,}000 = \$157.50$$

Gains from the sales of stocks and bonds by the fund can be distributed in cash, distributed as additional shares, or reinvested in the fund. This is decided by the fund manager.

Rates of Gain or Loss and Yield

The same rates that were calculated for stock investments can be calculated for investments in mutual funds. Sample Problem 13 shows the *investor's approach* to the calculation of gain or loss and the *annual rate of gain or loss* from a mutual fund.

SAMPLE PROBLEM 13

Laura Adams invests $25,000 in a load fund for 2,500 shares at $10 per share. Over a two-year period, she receives distributions of $0.70 per share and then sells the shares at $12 each. Determine by the investor's approach (a) the gain or loss on the fund holdings; (b) the annual rate of gain or loss.

Solution

(a)	Sales price (2,500 × $12.00)	$30,000
	Cost	−25,000
	Gain on share price	$ 5,000
	Distributions received (2,500 × $0.70)	1,750
	Total gain	$ 6,750

(b) Average annual gain: $\$6{,}750 \div 2 = \$3{,}375$

$$\text{Annual rate of gain} = \frac{\$3{,}375}{\$25{,}000} = 13.5\%$$

Notice that in the investor's approach, as with stocks, the load fees (commission) were not included in the calculation. For tax purposes, they would have been.

Rate of yield is calculated by dividing the *annual* distribution by the cost per share, ignoring the load fees. For Sample Problem 13, the annual distribution is $\$0.70 \div 2 \text{ years} = \$0.35$. The cost per share is $10. The rate of yield is $\$0.35 \div \$10.00 = 3.5\%$.

TEST YOUR ABILITY

1. A fund gains $40,000 in income, of which 90% is distributed among 72,000 outstanding shares. What will each share receive?
2. Helen Sudermann, an investor in the fund in Problem 1, holds 4,200 shares. How much will be distributed to her?
3. Brenda Romano holds 400 shares in a mutual fund that she bought at $9.50 each. Two years later, after receiving distributions of $0.90 a share in total, she sells the shares for $8 each.
   (a) Use the investor's approach to calculate her gain or loss.
   (b) What is the annual rate of gain or loss?
   (c) Calculate the rate of yield.

Real Estate as an Investment

Scarlett O'Hara, the heroine of *Gone With the Wind,* knew the value of land—as do many investors who devote their efforts to speculating in real estate. An investor will purchase a home or plot of land and hold it during a period of *appreciation,* which is a gradual increase in value over a period of time. When the property is sold at a later date, the investor realizes a gain (or a loss, though a real-estate loss is not likely in today's economy). Sample Problem 14 illustrates the calculation of appreciation.

SAMPLE PROBLEM 14

Five years ago, Barbara Cushion purchased a home for $50,000. When she sold it this year, she realized a 60% appreciation in the value of her purchase. What was her profit on the sale of her home?

Solution

This is a base, rate, percentage problem—given B and R.

$P = B \times R$

$P = \$50{,}000 \times .60 = \$30{,}000$ gain

Sample Problem 14 is an oversimplification of the actual process of real estate investment, since it does not take into consideration such items as rental of the property to produce additional revenue, costs of maintaining the property, taxes, or legal fees. It makes the point that real estate is a valid area for investment. Sample Problem 15 illustrates one additional area of related arithmetic.

SAMPLE PROBLEM 15

Hank Northridge purchased a home five years ago for $56,000 and sold it this year for $84,000. (a) What percent of appreciation did he realize? (b) What was his annual rate of appreciation?

Solution

(a) This is a B, R, P problem—given B and P.

$$R = \frac{P}{B}$$

$$R = \frac{\$84{,}000 - \$56{,}000}{\$56{,}000} = \frac{\$28{,}000}{\$56{,}000} = 50\%$$

(b) $50\% \div 5$ years $= 10\%$ a year

Life Insurance as an Investment

In Chapter 9, we studied life insurance as part of our discussion of the various types of business insurance. We talked about two of life insurance's benefits—extended term and paid-up insurance—and we mentioned a third—*cash surrender value*. It is this third benefit that intrigues the investor.

All life insurance policies, except term insurance, have a money value to the insured who wishes to cash in the policy. Most cash surrender values start after three years of paying premiums. The amount of the cash value depends on such factors as the time span of paying premiums, the age of the insured at issue, the type of policy, and the face of the policy.

Table 15.3 shows typical cash surrender values for ordinary whole life, 20-payment life, and 20-year endowment policies, listed according to the age

Table 15.3

Cash Surrender Values per $1,000 Face for Males

	Years Premiums Paid				
Age at Issue	**5 Years**	**10 Years**	**15 Years**	**20 Years**	**Until Age 65**
		ORDINARY WHOLE LIFE			
18	$ 21	$ 68	$122	$ 184	$595
19	22	71	127	191	592
20	24	74	132	198	585
25	32	93	161	236	570
30	43	115	194	279	547
35	55	140	230	325	517
40	68	167	269	374	478
45	83	197	312	427	427
50	100	227	353	472	353
		20-PAYMENT LIFE			
18	59	152	261	387	690
19	60	156	268	397	690
20	62	161	275	407	690
25	73	184	313	459	690
30	85	211	353	515	690
35	99	238	395	574	690
40	112	266	437	632	690
45	126	293	478	690	690
50	140	318	514	744	514
		20-YEAR ENDOWMENT			
18	175	408	681	1,000	
19	175	408	681	1,000	
20	175	408	681	1,000	
25	175	408	681	1,000	
30	175	409	680	1,000	
35	176	409	679	1,000	
40	177	408	677	1,000	
45	177	406	673	1,000	
50	177	404	667	1,000	

of the insured. Values are listed for males and per $1,000 of face. Notice that the 20-year endowment policy, after 20 years, pays $1,000 per $1,000 of face—an understandable amount since the use of an endowment policy is very much a form of investment.

A man who took out a $10,000 ordinary whole life policy at age 18 would receive $68 × 10 thousands = $680, if he cashed it in after ten years. If he were to wait ten more years, he would then receive $184 × 10 = $1,840, more than double. Cash value increases at a rapid rate as time elapses.

SAMPLE PROBLEM 16

Marcia Angell took out a $20,000, 20-payment life policy when she was 23. What cash surrender value can she get at age 38? Use Table 15.3. (*Hint:* Remember the age difference for females.)

Solution

Subtract three years for females; use 20 at issue, 15 years later.
Table 15.3: $275 per $1,000
Cash surrender value = $275 × 20 = $5,500

Gain or Loss on Insurance Investments

To find the gain or loss on an investment in life insurance, compare the *premiums paid* with the *cash surrender value.* When the premiums paid are greater than the cash surrender value, there is a loss, or *net cost,* in insurance terms. When the cash surrender value is greater, there is a gain, or *net return.* Annual premiums are calculated from Table 9.8. Sample Problem 17 shows the calculation of net cost or net return.

SAMPLE PROBLEM 17

Bill Mitchell, age 25, takes out a $20,000, 20-year endowment policy. Use Tables 9.8 and 15.3 to calculate his net cost, or net return, after ten years.

Solution

Premiums paid (Table 9.8)	
$42.45 × 20 thousands = $849 × 10 years =	$8,490
Cash value (Table 15.3)	
$408 × 20 thousands =	−8,160
Net cost	$ 330

We see that the annual premium must be multiplied by ten (years), but the cash value does *not* have to be multiplied, since Table 15.3 includes time.

Sample Problem 18 concludes this section by calculating a rate of return.

SAMPLE PROBLEM 18

Susan Wilson, at age 21, is insured for $12,000 on a 20-payment life insurance plan. After 47 years, she has a change of heart and cancels the policy. Find the (a) net cost, or net return, and (b) rate of cost, or rate of return.

Solution

Use ages 18 and 65.
(a) Premiums paid (Table 9.8)

$18.18 × 12 thousands = $218.16 × 20 years =	−$4,363.20
Cash value (Table 15.3)	
$690 × 12 =	8,280.00
Net return	$3,916.80

(b) Average annual return = $3,916.80 ÷ 47 years = $83.34
Rate of return = $83.34 ÷ $4,363.20 = 1.9%

A return of 1.9% is small, *but* Ms. Wilson has also been insured for a long time.

TEST YOUR ABILITY

1. Nine years ago, Charles Mitchell purchased a home for $35,000. When he sold it this year, he realized a 95% appreciation. What was his profit on the sale of the home?
2. Marianne Russell purchased a home six years ago for $50,000 and sold it this year for $85,000. (a) What percent of appreciation did she realize? (b) What was her annual rate of appreciation on her home?
3. Find the cash surrender value of each of the following policies:

	Age at Issue	Sex	Type of Policy	Face	Years Premiums Paid
(a)	30	Male	Ordinary whole	$ 7,000	to age 65
(b)	35	Male	20-year endowment	$12,000	10
(c)	28	Female	20-payment life	$17,000	15
(d)	45	Male	Ordinary whole	$ 8,500	10
(e)	53	Female	20-payment life	$24,500	5

4. Find the net cost or return to men with the following policies:

	Age at Issue	Type of Policy	Face	Time in Years
(a)	35	Ordinary whole	$10,000	10
(b)	18	20-year endowment	$ 6,000	15
(c)	19	20-payment life	$ 9,000	20
(d)	50	20-year endowment	$ 8,000	5
(e)	20	Ordinary whole	$20,000	at 65

5. Arthur Cross, age 30, takes out an ordinary whole life policy for $25,000. After 20 years, he cancels the policy. Determine (a) total premiums paid; (b) cash surrender value; (c) net cost or return; (d) rate of net cost or return.

Comparing Investment Choices

Let's compare our investment choices. In addition to stocks, mutual funds, bonds, and life insurance, there are such areas as *coins, savings accounts,* and *savings certificates.* Some firms invest in race horses (a lot of people do, too)! The possibilities are endless. The choice should be based on sound judgment.

The savings certificate is a long-term, single-sum deposit that earns higher than usual interest. Savings certificates require a minimum investment—such as $1,000—that is tied up for the term of the item; in addition, a substantial penalty is incurred for early withdrawal of the amount invested. Items yielding even higher returns are the *money market certificate* and the *U. S. Treasury bill* (which is issued weekly at extremely high rates); both required an even larger investment, such as a $10,000 minimum.

Two points conclude this unit. The following section presents a caution to the investor; the second offers a summary problem and discussion.

Nominal and Effective Rates

The primary caution that should be heeded by all investors is that of distinguishing between stated and actual rates of interest. We have seen this distinction several times already, but let's go over the difference between *nominal rates* (stated) and *effective rates* (actual) of interest in two major areas.

(1) In savings accounts, or other funds accumulating at compound interest, the nominal rate given is usually lower than the effective rate. For example, if interest is compounded *daily,*

A Nominal Rate of	Is an Effective Rate of
5%	5.12%
$5\frac{1}{2}$%	5.64%
6%	6.18%
$7\frac{1}{2}$%	7.79%
$7\frac{3}{4}$%	8.06%
$11\frac{3}{4}$%	12.47%
$14\frac{1}{2}$%	15.22%

(2) On bond investments, the yield quoted is the effective rate, while the interest quoted is the nominal rate. Notice in Table 15.1 that a bond selling below face (under 100) has an effective rate *greater than* its nominal rate. Why? Because you are being paid interest on $1,000 but investing less than $1,000. A bond selling above face has a higher nominal rate than effective rate.

The terms *yield* and *effective rate* are similar in all types of investment. It is the effective rate that the investor must consider when comparing investment choices.

SUMMARY PROBLEM

Three investments are made by Richard Wilson:

(a) $10,000 in 80 shares of Springer preferred stock, costing $125 a share, paying 8%, par $100
(b) $10,000 in 10 Ballin bonds, 10's, at 100
(c) $10,000 in Wright Fund, no-load, 800 shares, net asset value per share, $12.50

Two years later, he sells the stock at $135, the bonds at 105, and the fund at $13.00. During this period, he has received normal stock and bond distributions plus $1.20 a share annually on the fund. Which has been the best investment?

Solution

(a) Stock:

Gross proceeds (80 × $135)	$10,800
Cost of shares (80 × $125)	−10,000
Gain on share price	$ 800
Dividends received ($100 × .08 = $8 a share × 80 × 2 years)	1,280
Total gain	$ 2,080
Average annual gain ($2,080 ÷ 2)	$ 1,040
Rate of gain = $1,040 ÷ $10,000 =	10.4%

(b) Bonds:

Quotation price (105% × $10,000)	$10,500
Quotation cost (100% × $10,000)	−10,000
Gain on quotation	$ 500
Interest received ($10,000 × 10% × 2 years)	2,000
Total gain	$ 2,500
Average annual gain ($2,500 ÷ 2)	$ 1,250
Rate of gain = $1,250 ÷ $10,000 =	12.5%

(c) Mutual fund:

Sales price (800 × $13.00)	$10,400
Cost (80 × $12.50)	−10,000
Gain on shares	$ 400
Distributions received (800 × 1.20 × 2 years)	1,920
Total gain	$ 2,320
Average annual gain ($2,320 ÷ 2)	$ 1,160
Rate of gain = $1,160 ÷ $10,000 =	11.6%

The bond investment shows the highest rate of gain, but this is hindsight. You can look back and ask: "What *did* I gain?" An investor should look ahead and use the *rate of yield* as a better basis for comparison of where to invest. Let's calculate yield rates for each investment.

On the stock:

$$\frac{\text{Dividend per share}}{\text{Cost per share}} = \frac{\$8.00}{\$125.00} = 6.4\%$$

On the bonds, the yield is 10%, since the bonds were purchased at face.

On the fund:

$$\frac{\text{Distribution per share}}{\text{Cost per share}} = \frac{\$1.20}{\$12.50} = 9.6\%$$

Would he be best off investing in a long-term savings certificate? If he did, would he have had as much fun? After all, it is safe but rather dull, putting everything in a savings certificate. Perhaps on a different investment, a yield of 15% or more might be possible.

CHECK YOUR READING

1. What is meant by $4\frac{3}{4}$'s bonds?
2. How much interest will the holder of a 7% bond receive on each semiannual interest date?
3. When is an 89 bond due? An 04 bond?
4. If a bond of B. F. Motors closes at $83\frac{1}{8}$ today, with a net change of $+1\frac{1}{4}$, what was yesterday's closing quotation?
5. What is the quotation cost of Timonthy Quarry Company's bond listed at (a) 96; (b) 102; (c) $95\frac{1}{2}$; (d) $103\frac{1}{8}$?
6. Find the quotation cost of three bonds at 94.
7. What total commission is paid on a purchase of 13 bonds?
8. Find the bond time from (a) August 1 to December 1; (b) March 1 to June17.
9. How much principal is represented by 36 bonds held by Selma Thomson?
10. Given a quotation price of $960, accrued interest of $40, and commission of $10, find the (a) gross proceeds and (b) net proceeds.
11. What two amounts are compared in determining gain or loss on bonds in (a) the tax method; (b) the investor's method?
12. Distinguish between an open-end fund and a closed-end fund.
13. Distinguish between a load fund and a no-load fund.
14. Compare the terms *net asset value* and *offering price.*
15. If the net asset value is $9.15 and the offering price is $10, compute and label the difference.
16. What amount of fee is charged to Farley Morgan on an $80,000 investment in a mutual fund?
17. What two types of income are earned by a mutual fund?
18. What is meant by the term "appreciation"?
19. On a 20-payment life policy issued to Larry Mellows at age 30 for $1,000, what cash will be received for the policy at (a) age 50; (b) age 65?
20. If the premiums paid total $861 and the cash value is $1,000, is there a net cost or a net return? Of how much?
21. What is a savings certificate?
22. A 5% bond is purchased at 97 by Juan Rioseco. (a) What is the nominal rate of interest? (b) Will the effective rate be higher or lower than the nominal rate?
23. On what time basis must rates of gain or loss be compared?

EXERCISES *Group A*

1. Calculate the quotation cost of the following bond purchases:

	Number of Bonds	Quotation
(a)	30	$83\frac{3}{8}$
(b)	110	$104\frac{1}{2}$
(c)	18	111
(d)	25	$96\frac{1}{4}$

2. Determine the net proceeds from a sale of ten LMN bonds on July 1 at today's low price, if interest dates are January 1 and July 1. Use Table 15.1.

3. If the LMN bonds in Exercise 2 are sold on October 19, calculate the net proceeds from the sale.
4. In each of the following problems, find the net asset value per share to the nearest cent:

	Assets	Liabilities	Number of Outstanding Shares
(a)	$ 906,240	$ 625,180	26,000
(b)	$ 1,649,010	$ 720,210	108,000
(c)	$ 1,507,160	$ 936,560	60,000
(d)	$74,267,910	$37,225,410	2,750,000

5. Phil Atkins buys 350 shares in a mutual fund at $13.00 each. Four years later, he sells the shares for $10.50 each. During the four-year period, he has received distributions of $0.30 per share *annually*.
   (a) Find his gain or loss, using the investor's approach.
   (b) Calculate the annual rate of gain or loss.
   (c) Calculate the rate of yield.
6. Elise Hood purchased a home five years ago for $42,000 and sold it this year for $54,600.
   (a) What percent of profit did she realize?
   (b) What is her annual rate of profit on the house?
7. In each of the following load fund problems, find the number of shares credited to the investor. Use Table 15.2. Carry to three places if necessary.

	Amount Invested	Net Asset per Share
(a)	$400,000	$24.50
(b)	$120,000	$24.25
(c)	$ 35,000	$11.45
(d)	$650,000	$25.00

8. Find the cash surrender value of each of the following policies, using Table 15.3.

	Age at Issue	Sex	Type of Policy	Face	Years Premiums Paid
(a)	28	F	Ordinary whole	$ 22,000	15
(b)	45	M	20-year endowment	$ 45,000	5
(c)	53	F	20-payment life	$150,000	10
(d)	25	M	Ordinary whole	$ 6,000	to age 65
(e)	35	M	20-year endowment	$ 2,500	20

9. Using Table 15.1, determine Ann O'Hara's total cost of purchases on July 16 of (a) six Steffens bonds at today's low price, with interest dates of March 1 and September 1; (b) three Manorhaven bonds at yesterday's close, with interest dates of April 1 and October 1.
10. Look at the Eason bonds in Table 15.1.
    (a) How many bonds were sold?
    (b) What is the face value of the bonds sold?

(c) What annual interest would the holder of one bond receive?
(d) What is the rate of yield?
(e) In what year are the bonds due?
(f) What is the closing price in dollars and cents?
(g) What was yesterday's closing price in dollars and cents?

11. Seven bonds are purchased at 103 by John Aldrich and sold four years later at $104\frac{1}{2}$. The interest is 5%. Find (a) the gain or loss by the investor's method; (b) the annual rate of gain or loss, to the nearest tenth of a percent; (c) the rate of yield, to the nearest tenth of a percent.
12. Rudorf Hays purchases 350 shares of a no-load fund. If the net asset value per share is $24.35, find his total cost.
13. Determine the accrued interest in each of the following bond transactions:

	Number of Bonds	Interest Dates	Date of Purchase	Rate of Interest
(a)	18	January 1, July 1	December 21	9%
(b)	62	March 1, September 1	July 1	6%
(c)	5	May 1, November 1	August 19	7%
(d)	9	April 1, October 1	March 29	5%
(e)	12	June 1, December 1	May 13	8%

14. Sidney Gordon buys eleven 6% bonds at 77 and sells them three years later at 82. Both transactions occur on interest dates. Find the gain or loss by the tax method.
15. Use the data in Exercise 14 to find the gain or loss by the investor's method.
16. Use Table 15.1 to find the quotation cost of six Par-com bonds at yesterday's closing price.
17. Forest International purchases 24 A-One bonds on September 21 at today's low price. If the interest dates are June 1 and December 1, calculate the interest accrued. Use Table 15.1.
18. Bill Kawecki, age 35, takes out an ordinary whole life policy for $50,000. After 30 years, he can no longer afford the policy, so he cancels it. Calculate the (a) total premiums paid; (b) cash surrender value; (c) net cost or return; (d) rate of cost of return.
19. A fund gains $36,000 in income, of which 90% is distributed. If 60,000 shares are outstanding, what will each share receive?
20. Stella Zangwill, an investor in the fund in Exercise 19, holds 3,721 shares. How much will she receive?
21. Mary Oates purchases 90 shares of a load fund at an offering price per share of $36.
    (a) How much will she pay?
    (b) What is the load fee per share?
    (c) What is the total load fee?
    (d) With how much will she be credited?
22. Seven years ago Hank Wallace bought a home for $27,000. When he sold it this year, he realized a 75% appreciation. What was the amount of appreciation on the house?
23. Find the net proceeds from a sale of eighteen $7\frac{1}{2}$% bonds on July 9 at 103, if the interest dates are February 1 and August 1.

24. Find the rate of yield to the nearest tenth of a percent in each case:

	Interest Rate	Quotation
(a)	$4\frac{1}{2}\%$	95
(b)	6%	103
(c)	$7\frac{3}{4}\%$	84
(d)	$6\frac{1}{4}\%$	$62\frac{1}{2}$

25. Find the net cost, or net return, to each of the men holding the following insurance policies: (Use the appropriate tables.)

	Age at Issue	Type of Policy	Face	Time in Years
(a)	25	Ordinary whole	$ 4,000	to age 65
(b)	35	20-year endowment	$ 9,000	5
(c)	40	20-payment life	$ 25,000	15
(d)	19	Ordinary whole	$ 6,500	10
(e)	50	20-year endowment	$100,000	20

Group B

1. Twelve Zeta-com bonds were purchased on May 25, 1988 at 51 by Myra Goldstein and sold at today's low price on October 25, 1989. The interest dates are January 1 and July 1. Find the (a) total cost; (b) net proceeds; (c) taxable gain or loss. *Hint:* Accrued interest is considered in (a) and (b), but not in (c). Use Table 15.1 again.
2. Forty Brown bonds (Table 15.1) were purchased on February 1, 1988, at today's high price and are held to maturity. Interest is paid on February 1 and August 1.
   (a) Find the total cost of the bonds.
   (b) What will the holder receive at maturity?
   (c) Determine the taxable gain or loss.
   (d) What total interest will be received on these bonds?
   (e) What is the investor's gain or loss? (*Hint:* Omit commission)
   (f) Calculate the rate of gain or loss, to the nearest tenth of a percent.
3. By including the maximum commission of $8\frac{1}{2}\%$, find the offering price for the following net asset values. (Round to the nearest penny.) (a) $27.45; (b) $36.60; (c) $40.00; (d) $6.50.
4. Cynthia Robinson invests $75,000 in a load fund. The net asset value of each share is $11.06. With how many shares, to three decimal places, will she be credited?
5. Harold Godwin, age 20, takes out an ordinary whole life policy for $50,000. According to Table 15.3, after how many years will he *begin* to have a net return?

UNIT 5 REVIEW

Terms You Should Know

accrued interest
appreciation
authorized stock
bond
bond quotation
bond time
cash surrender value
charter
closed-end fund
common stock
corporation
cumulative stock
dividend
effective rate
execution price
gross proceeds
issued stock
load fee
load fund
mutual fund
net asset value
net cost
net proceeds
net return
no load fund
no-par
nominal rate
odd lot
odd-lot differential
offering price
open-end fund
outstanding stock
over the counter
par
participating stock
partnership
preferred stock
rate of yield
round lot
savings certificate
sole proprietorship

UNIT 5 SELF-TEST

Items 1–12 review the highlights of Chapter 14.

1. If 35,000 shares are outstanding and 1,000 have been reacquired, ________ were issued.
2. A $7\frac{1}{2}$%, $50 par preferred stock pays ________ per share.
3. The normal dividend for 10,000 shares of $9\frac{1}{2}$%, $100 par preferred stock is ________.
4. If the stock in Item 3 is in arrears for three years, ________ must be paid before the common stock receives anything.
5. If outstanding stock consists of 5,000 shares preferred and 35,000 shares common, and $20,000 of dividends are available for participation, the common will receive ________ of the $20,000.
6. A stock with a closing price of $17\frac{1}{2}$ and a net change of $-1\frac{1}{8}$ closed at ________ yesterday.

7. Twenty shares of a 100-unit stock form a(n) __________ lot; 200 shares form a(n) __________ lot.
8. A round-lot price on a 100-unit stock is $57\frac{1}{4}$. If ten shares are bought, the execution price per share is __________; if ten shares are sold, the execution price per share is __________.
9. If 200 shares of Delta Paper are purchased at $15\frac{1}{2}$, the cost of the shares is __________; the total cost is __________.
10. If 200 shares of B. F. Petrosian, Inc., are sold at $49\frac{1}{2}$, gross proceeds are __________; the commission is __________; the net proceeds are __________.
11. Three hundred shares of stock are bought at 10 by Betty Ross and sold at 15. The investor's gain or loss on share price is __________.
12. If the cost per share of Daniel Wagons, Inc., is $32 and the annual dividend is $2.56, the yield is __________.

Items 13–22 review Chapter 15.

13. The quotation cost of a bond listed at $93\frac{7}{8}$ is __________.
14. A $7\frac{1}{2}$% bond pays __________ per interest period.
15. The total commission charge on a purchase of nine bonds by Harry James is __________.
16. A bond bought at $98\frac{1}{2}$ and sold at $96\frac{3}{4}$ is sold at a loss in dollars of __________.
17. The rate of yield on an 8% bond bought by Gloria Smith at 80 is __________.
18. A(n) __________ fund charges a fee; a(n) __________ fund has an unlimited number of shares.
19. If assets are $20,000, liabilities are $5,000, and 30,000 shares of the fund are outstanding, the net asset value per share is __________.
20. An investment of $30,000 in a mutual fund with a fee is subject to a commission amount of __________.
21. The Wingate Memorial Fund distributes 90% of its $80,000 earnings. If 20,000 shares are outstanding, each share earns __________.
22. After ten years, the cash surrender value of Jason Polo's $10,000 ordinary whole life policy that was taken out at age 50 is __________.

Problem 23 is an interesting review situation, similar to the summary problem in Chapter 15.

23. Margaret Williams invests $60,000 as follows:
    $20,000 in 160 shares of Orvax $5.00 stock, costing $125.00 a share
    $20,000 in 20 Stone bonds, 4's, at 100
    $20,000 in Voorhees Fund, no load, 1,600 shares, $12.50 per share net asset value
    Three years later, after receiving normal stock and bond distributions and $0.50 per share a year on the fund, she sells the stock at $122, the bonds at 103, and the fund at $14.00. Find:
    (a) the investor's gain or loss on each security
    (b) the annual rate of gain or loss on each security

KEY TO UNIT 5 SELF-TEST

1. 36,000
2. $3.75
3. $95,000
4. $380,000
5. $17,500
6. $18\frac{5}{8}$
7. odd; round
8. $57\frac{3}{8}$, $57\frac{1}{8}$
9. $3, 100; $3, 131
10. $9,900; $99; $9,801
11. $1,500
12. 8.0%
13. $938.75
14. $37.50
15. $90
16. $17.50
17. 10.0%
18. load; mutual, or open-end
19. $0.50
20. $1,800
21. $3.60
22. $2,270
23. (a) stocks $1,920 gain
    bonds $3,000 gain
    fund $4,800 gain
    (b) stocks 3.2%
    bonds 5.0%
    fund 8.0%

UNIT 6

The Mathematics of Summary and Analysis

Our last two chapters explore the end results of business mathematics—the summary and analysis of data.

The two tools of summary and analysis are accounting and statistics, which you will study in other courses. The summarizing function of the accountant is expressed in financial statements, while that of the statistician is expressed in averages. Chapter 16 is involved with both types of summaries.

The ratios used by the accountant to analyze data and the index numbers used by the statistician for the same purpose are the main methods discussed in Chapter 17.

CHAPTER **16**

Summary Techniques

Summary is the gathering of related bits of information into a meaningful form. One of the jobs of both the accountant and the statistician is to provide business with such summaries.

Your Objectives

After completing Chapter 16, you will be able to

1. Prepare basic balance sheets and income statements.
2. Compute a series of averages from ungrouped data.
3. Prepare a frequency distribution.
4. Compute a series of averages from grouped data.

Financial Statements

A *financial statement* is a formal document summarizing the results of business activity. We will investigate two types of financial statements, the balance sheet and the income statement. The *balance sheet* lists the *assets, liabilities,* and *capital* (or net worth) of a firm as of a specific date. The *income statement* shows the results of the operations for the period just ended.

The Balance Sheet

Let's first explore the balance sheet. But before doing so, it is important to know the fundamental equation of accounting:

$$\text{Assets} = \text{Liabilities} + \text{Capital}$$

An *asset* is any property, owned by a firm, that has money value. One type of asset is *cash*—bills, coins, money orders, and checks. If a firm is the payee on a note, it will have the asset *notes receivable.* A firm selling on credit will show the claims against its customers as the asset *accounts receivable.* The goods in a firm's stockroom are its *inventory. Supplies,* on hand in the office or store, are also assets. Finally, services paid for in advance, such as insurance policies, are called *prepaid expenses.*

All of these assets will have value within the next time period—usually a year—and are collectively called *current assets.* Other possible current assets are stocks and bonds that are expected to be sold within the next time period. The term *marketable securities* is used for this current asset.

Assets that will last beyond a year fall into a second group—*fixed assets,* such as *land, buildings, machinery,* and *furniture and fixtures.*

Other types of assets are *investments* (long-term stocks and bonds) and *intangible assets* (patents, copyrights, trademarks). All of these asset values are summarized in the asset section of the balance sheet.

A *liability* is the claim of a creditor against the firm—a debt that the firm owes. If the firm bought goods on credit, it has *accounts payable* as the name of the liability. If the firm signed a note as maker, the debt is listed under *notes payable. Income taxes payable* is a liability to the government. All liabilities, such as these, that are due in the next time period are *current liabilities.*

Long-term liabilities include *bonds payable* and *mortgages payable.* All liabilities are listed on the balance sheet at their values on the date specified.

Thus, the balance sheet is a listing of the firm's assets and liabilities—each divided into current and long-term sections—and the firm's capital. If the total assets equal total liabilities plus total capital, the equation is balanced.

Capital

The *capital* (or net worth) of a firm represents the owner's claim to the firm's assets. In a proprietorship, the sole owner claims all the capital. In a partnership, the partners share the capital. In a corporation, the stockholders own the capital.

Thus, on a proprietor's balance sheet, we will see a line reading:

John Brown, capital	$20,000

On a partnership's balance sheet, we will see:

Jones, capital	$10,000
Smith, capital	10,000
Total capital	$20,000

In a corporation, it is common to see another line in the capital section called *retained earnings*—the profits kept in the firm. Also, individual owners' names are not listed. Therefore, we will see:

Common stock	$15,000
Retained earnings	5,000
Total capital	$20,000

Sample Problem 1 shows the preparation of a balance sheet for a proprietorship.

SAMPLE PROBLEM 1

From the data presented below, prepare a balance sheet as of December 31, 1988 for Stanley Wilkes, proprietor of the Wilkes Cleaning Store.

Cash, $17,200; notes receivable, $2,600; accounts receivable, $5,020; inventory, $17,500; supplies, $500; equipment, $6,000; furniture and fixtures, $4,000; notes payable, $3,120; accounts payable, $11,200; mortgage payable (on equipment and furniture), $6,000; and capital, $32,500.

Solution

Wilkes Cleaning Store
Balance Sheet
December 31, 1988

Assets			
Current Assets			
Cash	$17,200		
Notes receivable	2,600		
Accounts receivable	5,020		
Inventory	17,500		
Supplies	500		
Total current assets		$42,820	
Fixed Assets			
Equipment	$ 6,000		
Furniture & fixtures	4,000		
Total fixed assets		10,000	
Total assets			$52,820
Liabilities			
Current Liabilities			
Notes payable	$ 3,120		
Accounts payable	11,200		
Total current liabilities		$14,320	
Long-term Liabilities			
Mortgage payable		6,000	
Total liabilities			$20,320
Capital			
Stanley Wilkes, capital			32,500
Total liabilities and capital			$52,820

Notes (1) Every financial statement heading consists of three lines, answering, in order, the questions: Who? What? When? In this case, Wilkes Cleaning Store is "who," the balance sheet is "what," and December 31, 1988 is "when."

(2) The order of the items and sections listed is typical.

(3) The balance sheet shows that *assets* ($52,820) = *liabilities* ($20,320) + *capital* ($32,500).

The Income Statement

The income statement shows the results of a period's operations by listing net sales, cost of goods sold, gross profit, operating expenses, and net operating profit. These are the major divisions we referred to in Chapter 10 in our discussion of profit calculation for a trading business.

Two other sections commonly found on income statements are *other income or expense* and *income taxes.* "Other income or expense" is income from sources other than operations, such as income from investments.

For a corporation, the last three items on its income statement are *net profit before taxes, income taxes,* and *net profit after taxes,* since the corporation is taxed on its profits. Tax rates are indicated for each problem.

SAMPLE PROBLEM 2

Prepare an income statement for Stanley Wilkes's store (Sample Problem 1), using the following operating data for the year ended December 31, 1988:

Net sales	$19,400
Beginning inventory (1/1/88)	11,820
Net purchases	16,450
End-of-year inventory (12/31/88)	17,500
Operating expenses	6,120
Other income	500

Solution

Wilkes Cleaning Store
Income Statement
For the Year Ended December 31, 1988

Net sales		$19,400
Cost of goods sold:		
Inventory, 1/1/88	$11,820	
+ Net purchases	16,450	
Total	$28,270	
− Inventory, 12/31/88	−17,500	
Cost of goods sold		−10,770
Gross profit		$ 8,630
Operating expenses		−6,120
Net operating profit		$ 2,510
Other income		500
Net profit		$ 3,010

Notes (1) An income statement covers the period of time preceding the date of preparation, so its heading reads "For the Year (or other period) Ended. . . ."

(2) The statement presented here is in outline form. A full statement would include, for example, an itemized list of operating expenses. We are using a simple form to present the basic idea.

TEST YOUR ABILITY

1. From the data listed below, prepare a balance sheet for John Conroy's Delicatessen as of December 31, 1988.
Cash, $6,800; notes receivable, $800; accounts receivable, $1,250; inventory,

$4,100; supplies, $75; equipment, $8,000; furniture and fixtures, $5,800; notes payable, $500; accounts payable, $4,900; capital, $21,425.

2. From the data presented below, prepare a balance sheet for the Michigan Development Corporation as of December 31, 1988.
   Cash, $59,550; marketable securities, $5,000; notes receivable, $2,700; accounts receivable, $31,200; inventory, $39,800; supplies, $550; prepaid expenses, $920; land, $12,000; building, $100,000; equipment, $58,000; furniture and fixtures, $40,000; notes payable, $5,000; accounts payable, $15,600; mortgage payable, $30,000; bonds payable, $140,000; common stock, $100,000; retained earnings, $59,120.
3. The Barker Corporation uses the following data to determine net profit. Prepare an income statement for the year ended December 31, 1988.

Net sales	$420,000
Inventory, 1/1	61,000
Inventory, 12/31	65,000
Net purchases	290,000
Operating expenses	120,000
Other income	4,000
Income taxes	3,960

4. Prepare an income statement for Wallace Plastics, Inc., for the year ended December 31, 1988, based on the following data:

Net sales	$62,000
Inventory, 1/1	5,700
Inventory, 12/31	5,100
Net purchases	49,500
Operating expenses	10,400
Other income	200
Income taxes	22% of net profit before taxes

Basic Statistical Summaries

The statistician enjoys summarizing numerical data into a single representative number. We call this single number an *average.*

Data can be classified into a series of single numbers or into a natural group. In this section, we will study three different "averages" for both grouped and ungrouped data.

Averages

The number that is the focal point of a series of numbers can be said to yield three different averages: the *mean,* the *median,* and the *mode.*

The Mean

The *arithmetic mean*—or more simply, "the mean"—is the most familiar average. Add a series of numbers, divide the total by the number of numbers, and you have the mean. Sample Problem 3 shows the calculation in action. It is not essential—but it is easier—to arrange the numbers in decreasing order.

SAMPLE PROBLEM 3

The Winston Company's sales for the second quarter of the year are as follows: April, $2,750; May, $1,875; June, $2,425. Find the mean of the monthly sales.

Solution

$2,750 + $2,425 + $1,875 = $7,050 total sales
$7,050 ÷ 3 months = $2,350 mean monthly sales

For a short list of numbers, the procedure used in Sample Problem 3 is fine. But suppose you wanted to know your weekly mean income for a year. You would have 52 numbers to add, a process filled with the danger of mistakes. When many of the individual numbers are the same, a *weighted average* procedure is good.

First, arrange the data into a table, listing how many times an amount appears and then listing the amount, in decreasing value, next to the times. A firm's weekly payroll may appear as follows:

Number of Weeks	Payroll
10	$10,000
12	9,500
16	9,000
6	8,500
8	7,500

Next, multiply the number of weeks by its payroll figure, add the weeks, and then add the products.

Number of Weeks	×	Payroll	=	Product
10		$10,000		$100,000
12		9,500		114,000
16		9,000		144,000
6		8,500		51,000
8		7,500		60,000
52				$469,000

Divide $469,000 by 52, and we get a weighted mean average of $9,019.23.

How reliable is the mean? We had better do Sample Problem 4 before deciding.

SAMPLE PROBLEM 4

Find the mean weekly pay in a store's optical department in which one person earns $6,200 and 29 others earn $200 each.

Solution

Number of Workers	×	Pay	=	Product
1		$6,200		$ 6,200
29		200		5,800
30				$12,000

Mean weekly pay = $12,000 ÷ 30 = $400

You could interpret this answer to mean that the "average" worker earns $400 a week, which is far from the truth. The mean is the wrong average for this problem, since it distorts the facts. This is why we have three averages available. Let's turn to the median.

The Median

The *median* is the *middle number* of a series of numbers arranged in *increasing* or *decreasing order.* Exactly half of the remaining numbers fall above the median and half fall below.

In Sample Problem 3, the median is $2,425. In Sample Problem 4, it is $200.

To find the median, add 1 to the number of numbers, and then divide by 2. The result is the *position* that you are seeking. For a list of five numbers: $5 + 1 = 6 \div 2 = 3$. The third number in the ordered list is the median.

For a list of six numbers: $6 + 1 = 7 \div 2 = 3\frac{1}{2}$. The median is located halfway between the third and fourth numbers. The solution to this situation is to add the third and fourth numbers, and then divide by 2. The result is the median.

SAMPLE PROBLEM 5

Find the median of each list.

(a) 270, 312, 690, 410, 78, 621, 519
(b) 304, 264, 801, 191, 206, 625

Solution

(a) Rearrange high to low.

690, 621, 519, 410, 312, 270, 78

Median $= 7 + 1 = 8 \div 2 =$ 4th number or 410

(b) 801, 625, 304, 264, 206, 191

Median $= 6 + 1 = 7 \div 2 = 3\frac{1}{2}$ number

3rd number	304
4th number	264
Median	$568 \div 2 = 284$

When the data are arranged in a weighted average form, the median is found in the same manner. Sample Problem 6 demonstrates this point and compares the mean and the median.

SAMPLE PROBLEM 6

From the following test data, find (a) the median and (b) the weighted mean:

17 persons scored 19	12 persons scored 77
19 persons scored 76	1 person scored 6

Solution

(a) Rearrange high to low.

12 scores of 77
19 scores of 76
17 scores of 19
1 score of 6
49

Median $= 49 + 1 = 50 \div 2 =$ 25th number. The 25th number appears in the second highest group ($12 + 19 = 31$ numbers), so the median is 76.

(b)

12 × 77 =		924
19 × 76 =		1,444
17 × 19 =		323
1 × 6 =		6
49		2,697

Mean = 2,697 ÷ 49 = 55.0 (rounded)

If you were to choose the "best" average for this test series, is 76 or 55 more reliable? The median can also be misleading.

The Mode

The *mode* is the number in a series that appears *most often*. In Sample Problem 4, the mode is $200, since it appears 29 times.

In Sample Problem 5(a), there is no mode, nor is there one in Sample Problem 5(b), since no number appears more than once. In Sample Problem 6, the mode is 76.

A distribution may have no mode or, on the other hand, have two modes. In the series 12, 12, 10, 6, 6, 5, we have a *bimodal* situation since both 12 and 6 appear most often.

TEST YOUR ABILITY

1. The Adams Fruit Company shows the following monthly sales for 1988: January, $21,200; February, $18,600; March, $19,300; April, $19,700; May, $18,500; June, $22,300; July, $20,400; August, $17,300; September, $25,800; October, $26,700; November, $29,400; December, $27,200.
   Find the mean monthly sales to the nearest dollar.
2. The Edo Fish Market's staff of employees during 1988 is as follows:

First 9 weeks, 195	Next 6 weeks, 140
Next 11 weeks, 206	Next 10 weeks, 190
Next 8 weeks, 170	Last 8 weeks, 225

   Using a weighted average, find the mean to the nearest whole number.
3. Find the mean average weekly earnings for Carlson Motors' 780 factory workers if 247 earned $230 a week, 267 earned $240 a week, 184 earned $270 a week, and the rest earned $320 a week. Compute to the nearest penny.
4. Find the median of the following numbers:
   (a) 420, 610, 94, 870, 530, 529, 210
   (b) 580, 445, 312, 630, 804, 711
5. Scores on Bowden Computer Corporation's aptitude exam are as follows: 94, 87, 56, 73, 85, 94, 75, 63, 68, 99, 85, 94, 80, 71, 74
   Find, to the nearest tenth, the (a) mean; (b) median; (c) mode.
6. From the following grades scored on Morris TV Repair's in-service program, prepare a weighted average and find, to the nearest tenth, the mean, the median, and the mode.

Grade	Number
95	8
85	23
75	39
65	5

Frequency Distributions

When the data becomes overwhelming, it is not practical to list individual numbers. We have already seen that if the same numbers reappear, a weighted average can be used. But what if there are many numbers and no repeats?

In such a case, numbers are arbitrarily grouped into *intervals*. Each interval includes a low point and a high point. Intervals may be small—with lower and upper *limits* of 0 and 9, 10 and 19, 20 and 29—or large—such as 0–49,999; 50,000–99,999; and 100,000–149,999.

Given a list of numbers, an appropriate set of intervals is chosen, and the numbers are then tallied by intervals. For instance, here are the numbers: 79; 887; 2,640; 729; 556; 1,121; 1,863; 1,947; 2,012; 847; 621; 309; 515; 1,375; 1,163; 1,285; 2,175; 1,830.

If we choose intervals of 500, then our groups would be 0–499, 500–999, and so on. The table appears below:

Interval	Tally	Frequency
0–499	||	2
500–999	~~||||~~ |	6
1,000–1,499	||||	4
1,500–1,999	|||	3
2,000–2,499	||	2
2,500–2,999	|	1
		18

The choice of interval is your own. Generally, from 5 to 15 intervals are chosen. The final format of the data is a *frequency table*, or *frequency distribution*. "Frequency" means the number of times each type of number appears.

Both the mean and the median can be calculated from the frequency distribution. To find the *mean*, assume that all the scores in each interval fall at its *midpoint*. For the interval 0–499, the midpoint is 250, obtained by adding two successive lower limits (0 + 500), and then dividing by 2. What is the midpoint of 500–999? The answer is 500 + 1,000 = 1,500 ÷ 2 = 750.

Once the midpoints are established, a procedure similar to the weighted mean is followed: frequency × midpoint = product. The completed solution follows:

Interval	Midpoint	×	Frequency	=	Product
0– 499	250		2		500
500– 999	750		6		4,500
1,000–1,499	1,250		4		5,000
1,500–1,999	1,750		3		5,250
2,000–2,499	2,250		2		4,500
2,500–2,999	2,750		1		2,750
			18		22,500

Mean = 22,500 ÷ 18 = 1,250

To find the *median*, you will use a very different procedure:

(1) Divide the number of numbers by 2. *Do not add 1* (as with ungrouped data). 18 ÷ 2 = ninth number.

(2) Locate the group of the middle number. We are looking for the ninth number. After the 0–499 interval, there are 2 scores; after the 500–999, there are

a total of $2 + 6 = 8$ scores. The ninth number, therefore, is the first number in the 1,000–1,499 interval. We could look up its exact value, since we know the original scores, but that isn't any fun.

(3) The real lower limit of the group 1,000–1,499 is 999.5. That is because the real lower limit of any group is .5 below its printed limit.

(4) The group 1,000–1,499 has four numbers in it. We are interested in the first of the four, or $\frac{1}{4}$ of the group. The group's interval is 500, so in order for the median to be in its proper place within the group, $\frac{1}{4} \times 500 = 125$.

(5) The median is $999.5 + 125 = 1{,}124.5$.

If the median falls between two groups, add the upper limit of the first group to the lower limit of the second group and divide by 2.

The *mode* cannot be determined from a frequency distribution, but the *modal class,* the most common class, can. In our problem, 500–999 is the modal class.

Sample Problem 7 reviews the calculation of all three averages from a frequency table.

SAMPLE PROBLEM 7

From the following frequency distribution, calculate the mean, median, and modal class.

Interval	Frequency
\$ 0–\$ 399	3
400– 799	9
800– 1,199	10
1,200– 1,599	11
1,600– 1,999	5
2,000– 2,399	2
	40

Solution

Interval	Midpoint	×	Frequency	=	Product
\$ 0–\$ 399	\$ 200		3		\$ 600
400– 799	600		9		5,400
800– 1,199	1,000		10		10,000
1,200– 1,599	1,400		11		15,400
1,600– 1,999	1,800		5		9,000
2,000– 2,399	2,200		2		4,400
			40		\$44,800

Mean = \$44,800 ÷ 40 = \$1,120

Median = 40 ÷ 2 = 20th number
located in \$800–\$1,199 interval
$20 - (3 + 9) = 8$ needed to make 20, out of 10 in interval
= \$799.50 + ($\frac{8}{10}$ × \$400) = \$799.50 + \$320 = \$1,119.50

Modal class is \$1,200 – \$1,599 (most frequent)

Had the distribution contained 41 numbers, the median would be the $20\frac{1}{2}$ score and would have been calculated in the same way.

TEST YOUR ABILITY

1. From the following frequency distribution of bonuses at Bruce Lowcourt, Inc., calculate the mean, the median, and the modal class, all to the nearest cent.

Interval	Frequency
$ 0–$ 49	2
50– 99	5
100– 149	6
150– 199	5
200– 249	2
	20

2. The following data show the number of units produced by the Stern Company's newest computerized equipment during a 30-day test period. Prepare a frequency distribution and find the mean, median, and modal class to the nearest tenth. Use intervals of 10 units (0–9, 10–19, etc.).

Day	Units	Day	Units
1	12	16	7
2	17	17	53
3	26	18	47
4	27	19	29
5	31	20	18
6	33	21	27
7	29	22	36
8	46	23	41
9	39	24	44
10	41	25	47
11	52	26	50
12	47	27	55
13	45	28	57
14	39	29	58
15	16	30	59

CHECK YOUR READING

1. Write the fundamental equation of accounting.
2. If L = $26,000 and C = $19,000, find A.
3. What is the difference between a current asset and a fixed asset?
4. Distinguish between a current liability and a long-term liability.
5. Name the owner or owners of the capital in (a) a proprietorship; (b) a partnership; (c) a corporation.
6. What heading would appear on your own balance sheet, prepared today? How about on your own income statement?
7. Given the numbers 10, 9, 6, 5, 5, what is (a) the mean; (b) the median; (c) the mode?
8. In a series of numbers, which average has the highest value?
9. When is a weighted average used? And a frequency distribution?
10. What is the median number in an *ungrouped* list of (a) 41 numbers; (b) 30 numbers?
11. What is the mode of (a) the numbers 6, 5, 4; (b) the numbers 8, 8, 7, 5, 5?
12. If one interval is 300–599 and the next is 600–899, what is the midpoint of each interval?

13. For median calculation, what is the real lower limit of the intervals (a) 10–19; (b) $500–$999?
14. If a median falls two numbers into the interval 1,000–1,499, which has a frequency of eight numbers, what is the median?

EXERCISES *Group A*

1. Health insurance contributions for the Larington Sheet Metal Company's 940 workers are as follows: 305 paid $21.10; 264 paid $15.40; 194 paid $11.20; and the rest paid $9.50 a month. Find the mean payment to the nearest penny.
2. The following frequency table shows the number of cars sold by 30 sales representatives in a given month. Find, to the nearest tenth, (a) the mean; (b) the median; (c) the modal class.

Interval	Frequency
0–3	4
4–7	3
8–11	6
12–15	12
16–19	2
20–23	2
24–27	1
	30

3. The Delmar Rubber Corporation shows the following income data for the year ended December 31, 1988. Prepare an income statement for them.

Net sales	$410,000
Inventory, 1/1	59,000
Inventory, 12/31	47,000
Net purchases	342,000
Operating expenses	39,000
Other income	6,000
Income taxes	4,180

4. The Hunt Market's daily sales for the week are listed below. Find the mean average daily sales to the nearest cent.
   Monday, $946.18; Tuesday, $847.44; Wednesday, $719.90; Thursday, $1,104.17; Friday, $656.20; Saturday, $834.19; Sunday, $511.42.
5. Scores on Henderson's Multitude Talent aptitude test are as follows: 68, 74, 71, 93, 72, 85, 91, 80, 65, 50, 64, 77, 83, 88, 79. Find, to the nearest tenth, (a) the mean; (b) the median; (c) the mode.
6. From the data presented below, prepare a balance sheet for the Whitman Corporation as of December 31, 1988.
   Cash, $19,470; marketable securities, $5,210; notes receivable, $3,160; accounts receivable, $14,245; inventory, $37,540; supplies, $950; prepaid expenses, $475; land, $25,000; building, $75,000; equipment, $80,000; furniture and fixtures, $32,500; notes payable, $5,100; accounts payable, $34,240; mortgage payable, $45,500; bonds payable, $125,000; common stock, $50,000; retained earnings, ?

7. From the following data, find, to the nearest cent, (a) the weighted mean; (b) the median; (c) the mode.

Weekly commissions of 150 Part-Time Office Workers

Commission	Number of Workers
$200	14
180	37
160	51
140	30
120	13
100	5
	150

8. Find the median of the following numbers: (a) 637, 426, 914, 83, 450, 374, 97; (b) 512, 804, 701, 99, 110, 144.
9. Prepare an income statement from the following data for Witt Advertising, Inc., for the year ended December 31, 1988.

Net sales	$77,500
Inventory, 1/1	4,750
Inventory, 12/31	6,210
Net purchases	61,800
Operating expenses	12,460
Other income	350
Income taxes	22% of net profit before taxes

10. A cement mixer was tested over a 40-day span. The machine's production figures were:

    for 6 days, 57 units a day
    for 9 days, 71 units a day
    for 13 days, 46 units a day
    for 5 days, 75 units a day
    for 7 days, 62 units a day

    Find the weighted mean average, to the nearest unit.
11. Set up a frequency distribution with intervals of $60 for the following data. Find—to the nearest cent—the mean, median, and modal class.

Sales Representative Number	Commission	Sales Representative Number	Commission
1	$ 58.50	14	$164.71
2	131.36	15	163.85
3	211.26	16	171.74
4	94.22	17	261.85
5	149.95	18	301.19
6	155.41	19	86.30
7	79.25	20	298.40
8	104.11	21	127.16
9	209.50	22	122.09

Sales Representative Number	*Commission*	*Sales Representative Number*	*Commission*
10	$179.99	23	$125.04
11	91.18	24	57.25
12	182.54	25	86.95
13	195.01	26	79.50

12. From the information given below, prepare a balance sheet for Al Weber's Variety Store as of December 31, 1988.
    Cash, $7,400; notes receivable, $3,210; accounts receivable, $11,240; inventory, $17,580; supplies, $770; equipment, $12,000; furniture and fixtures, $8,400; notes payable, $2,490; accounts payable, $23,918; capital, ?

Group B

1. From the following data, construct an income statement for the Brown Dairy Association for the month of January, 1988.

Net profit *after* taxes of 22%	$ 14,820
Gross profit	52,250
Cost of goods sold	137,750
Other expenses	2,660
Inventory, 1/1	14,200
Inventory, 1/31	23,500

2. From the following information, recording the smoking habits of 100 individuals per day, determine, to the nearest tenth, the mean and median number of cigarettes smoked per day. Use intervals of 0–9, 10–19, and so forth.

3, 6, 0, 9, 18, 12, 11, 17, 21, 2
5, 12, 10, 11, 15, 51, 40, 38, 1, 14
16, 19, 5, 2, 8, 8, 11, 14, 26, 28
1, 17, 36, 23, 2, 14, 37, 24, 8, 57
38, 18, 7, 25, 19, 25, 39, 28, 3, 15
0, 19, 12, 27, 31, 1, 0, 26, 12, 11
37, 14, 39, 10, 22, 4, 5, 28, 6, 41
3, 20, 32, 33, 9, 13, 36, 15, 5, 29
21, 30, 16, 31, 20, 5, 6, 10, 35, 34
3, 18, 21, 35, 31, 2, 5, 7, 11, 14

CHAPTER 17

Analytical Procedures

In Chapter 16, we inspected some of the accounting and statistical summaries used in business. A summary is more than just a collection of data; it is the essential prerequisite to the process of *analysis*. Analysis brings out the crucial facts that determine whether or not a firm has met its goals.

In Chapter 17, we will apply some of the basic procedures of accounting and statistical analysis to the summaries from Chapter 16. This, however, is only a brief introduction to an enormous field.

Your Objective

After finishing Chapter 17, you will be able to

1. Carry out certain vertical, horizontal, and ratio analyses on financial statements.
2. Work accurately with index numbers.

Analyzing Financial Statements

Of what value is the statement, "Inventory is $20,000 today"? It has *absolute* value—you know what the inventory is but, is $20,000 a "good" or a "bad" figure?

A number becomes far more important with *comparison*. Comparison with what? An inventory can be compared with:

1. The total assets today—to see if the percentage of inventory is too high, too low, or just right.
2. Last year's inventory, to see if inventory has increased or decreased.

3. Standard figures of other firms in the area, or in the same line of business, or of accounting practice in general.

All these comparisons, plus many others, form the basis of *financial statement analysis*—the search of financial statements to determine the firm's degree of success.

Analysis may be *vertical, horizontal,* or by *ratios* and *measures.* Let's examine each in turn.

Vertical Analysis

Vertical analysis is the process of determining what percent of a total is represented by each individual amount.

In the vertical analysis of assets on a balance sheet, each individual asset will be the *percentage,* while total assets will be the *base.* We are finding the rate by our old friend:

$$R = \frac{P}{B}$$

If the cash is $30,000 and total assets amount to $500,000, divide $30,000 by $500,000 and get .06. Cash is 6% of the total assets.

SAMPLE PROBLEM 1

Summarize the following data of the Delta Corporation by using vertical analysis:

Cash	$ 30,000
Accounts receivable	40,000
Inventory	20,000
Land	100,000
Building	230,000
Furniture and fixtures	80,000
Total assets	$500,000

Solution

		%
Cash	30,000/500,000	6.0
Accounts receivable	40,000/500,000	8.0
Inventory	20,000/500,000	4.0
Land	100,000/500,000	20.0
Building	230,000/500,000	46.0
Furniture and fixtures	80,000/500,000	16.0
Total Assets		100.0

Notes
(1) The base in this problem is *total assets.* The total of the individual percents should come to 100.0%. The total may be off 0.1% or 0.2% due to rounding.
(2) A percent should always be rounded to the nearest tenth.

One goal of vertical analysis is to judge the "goodness" of a figure. Is 6% cash good? Is 46% in buildings too high? By specifying our questions, we have taken one step toward the answers.

In vertically analyzing the liabilities and capital section of a balance sheet, *total liabilities and capital* is the base. In vertically analyzing an income statement, *net sales* is the base. Each figure is expressed as a percent of net sales, as shown in Sample Problem 2.

SAMPLE PROBLEM 2

Interpret the following rough income statement of Getty Stores, Inc., by using vertical analysis.

Net sales	$200,000
Cost of goods sold	−120,000
Gross profit	$ 80,000
Operating expenses	−36,500
Net operating profit	$ 43,500
Other income	400
Net profit before taxes	$ 43,900
Income taxes	−14,572
Net profit after taxes	$ 29,328

Solution

		%
Net sales		100.0 (base)
Cost of goods sold	120,000/200,000	−60.0
Gross profit	80,000/200,000	40.0 (100 − 60 = 40)
Operating expenses	36,500/200,000	−18.3 (rounded)
Net operating profit	43,500/200,000	21.8 (0.1 rounding error)
Other income	400/200,000	0.2
Net profit before taxes	43,900/200,000	22.0 (21.8 + 0.2 = 22.0)
Income taxes	14,572/200,000	−7.3
Net profit after taxes	29,328/200,000	14.7 (22.0 − 7.3 = 14.7)

Each calculation above can be checked by addition or subtraction, as indicated in parentheses. *Do not* add or subtract to find answers, as you may carry an error through, but do use these processes to check your division.

Let's return to our earlier question: Is 14.7% a good profit? We are not, however, quite ready to answer it.

Horizontal Analysis

Horizontal analysis is the comparison of two years. This is done for either comparative percents on the same item each year—or for the rate of change from one year to the next. Let's examine the first use. What if the 1988 profit is, as shown, $29,328 after taxes, and the 1989 profit is $30,515? Use vertical analysis to find percents and horizontal analysis to compare the percents.

In 1988, the net profit after taxes is 14.7% of the net sales. Suppose the 1989 sales are $210,000. Then, the 1989 net profit after taxes is:

$$30{,}515/210{,}000 = 14.5\% \text{ of net sales}$$

By comparing the two percents, we note that 1988 was a better year. As you can see, dollars alone can be misleading. There was, in this case, a *dollar increase* in 1989, but a *percent decrease*. Figures take on a fuller meaning when they are compared.

SAMPLE PROBLEM 3

Interpret the figures of H.F. Smithers, Inc., by using vertical analysis. Then compare your answers horizontally to determine in which year Smithers had a smaller percentage tied up in inventory.

Inventory, 12/31/89	$ 21,870
Inventory, 12/31/88	22,880
Total assets, 12/31/89	243,000
Total assets, 12/31/88	286,000

Solution

1988 $22{,}880/286{,}000 = .08 = 8.0\%$
1989 $21{,}870/243{,}000 = .09 = 9.0\%$

Since a smaller percent of 1988's total assets was inventory, the former was a better year.

Note The conclusion above is based on the thought that most firms prefer assets in the form of cash and receivables rather than inventory, which is supposed to be sold.

The second type of horizontal analysis is the calculation of *rates of change* from one year to the next. (Does this appear familiar? In Chapter 4, this topic was covered under the heading "Rates of Increase and Decrease.")

To find a rate of change, divide the dollar change by the base, which is always the earlier year. The inventory in Sample Problem 3 went from $22,880 to $21,870, a drop of $1,010. The rate of change is

$$\begin{aligned}\text{Rate of change} &= \frac{\text{Dollar change}}{\text{Base amount}}\\ &= \frac{1{,}010}{22{,}880}\\ &= 4.4\%\end{aligned}$$

SAMPLE PROBLEM 4

From the following data, find the rates of change for inventory and sales from 1988 to 1989:

	1989	1988
Inventory	$ 9,020	$ 8,200
Sales	121,900	115,000

Solution

Inventory: $9,020 – $8,200 = $820 increase
Base: 1988 = $8,200
Rate of change: 820/8,200 = 10%

Sales: $121,900 – $115,000 = $6,900 increase
Base: 1988 = $115,000
Rate of change: 6,900/115,000 = 6%

Horizontal analysis allows us to compare rates of increase or decrease. "Why," a company might ask, "has our inventory increased by 10% but our sales by only 6%? Are we overstocking goods?" Our purpose is not to answer that difficult question, but to show how to obtain the information so that the company can make a valid decision.

TEST YOUR ABILITY

1. Using vertical analysis, summarize the following assets of the Newton Telescope Company:

Cash	$13,500
Accounts receivable	4,500
Inventory	18,000
Land	9,000
Building	27,900
Furniture and fixtures	17,100
Total assets	$90,000

2. Using vertical analysis, summarize the following liabilities and capital of the Green Textiles Company:

Notes payable	$ 5,125
Accounts payable	20,500
Income taxes payable	12,500
Mortgage payable	61,375
J. Green, capital	25,500
Total liabilities and capital	$125,000

3. Interpret the following data by vertical analysis:

Net sales	$60,000
Cost of goods sold	–42,000
Gross profit	$18,000
Operating expenses	–19,800
Net operating loss	($ 1,800)
Other income	3,000
Net profit before taxes	$ 1,200
Income taxes	–264
Net profit after taxes	$ 936

4. First vertically analyze the data below; then apply horizontal analysis to determine the firm's better year.

	1989	1988
Cash	$ 28,050	$ 28,560
Total assets	255,000	272,000

5. Find the rate of change from 1988 to 1989 for each of the following items:

	1989	1988
(a) Cash	$ 65,100	$ 62,000
(b) Inventory	$ 18,612	$ 19,800
(c) Capital	$ 69,000	$ 46,000
(d) Sales	$240,000	$300,000
(e) Net profit after taxes	$ 31,875	$ 37,500

Ratios and Other Measures

A third procedure for analyzing financial statements is to determine the *ratios* between various individual items on the statements and then compare these with standard ratios.

One very common ratio is the *current ratio*—the ratio of *current assets* to *current liabilities*. If current assets = $20,000 and current liabilities = $10,000, the current ratio is calculated as follows:

$$\text{Current ratio} = \frac{\text{Current assets}}{\text{Current liabilities}}$$
$$= \frac{20{,}000}{10{,}000}$$
$$= 2{:}1$$

A ratio is always expressed as "something to 1" when analyzing statements.

The purpose of the current ratio is to see if the assets in the current period will be enough to pay the liabilities due in the current period. Banks (and other firms extending credit) generally look for a minimum of 2:1 in the current ratio.

Sample Problem 5 shows this calculation. Ratios are carried to the nearest tenth.

SAMPLE PROBLEM 5

Compute the current ratios of the J.F.D. Company and the K.F.M. Company from the data below:

	J.F.D	K.F.M.
Current assets	$40,000	$50,000
Current liabilities	25,000	20,000

Solution

J.F.D. Company: 40,000/25,000 = 1.6:1
K.F.M. Company: 50,000/20,000 = 2.5:1

On the surface, K.F.M. seems to be in the better position. J.F.D. does not meet the 2:1 standard.

But let's analyze in a bit more depth. The tabulation below spells out each firm's current assets:

	J.F.D.	K.F.M.
Cash	$12,000	$ 2,000
Marketable securities	6,000	1,000
Notes receivable	2,000	1,000
Accounts receivable	5,000	4,000
Inventory	11,000	37,000
Supplies	4,000	5,000
Total current assets	$40,000	$50,000

If all of J.F.D.'s current liabilities are due in 30 days, how could J.F.D. pay them? It could use its cash, sell its marketable securities, discount or collect its notes receivable, and sell or collect its accounts receivable. It could raise, with relative ease, $12,000 + $6,000 + $2,000 + $5,000 = $25,000 to pay its current liabilities.

The K.F.M. Company could raise $2,000 + $1,000 + $1,000 + $4,000 = $8,000 in the same manner. Which firm can better pay its debts? The *acid-test* ratio, the ratio of *liquid assets* (cash and those easily converted into cash) to current liabilities, is our second important ratio.

$$\text{Acid-test ratio} = \frac{\text{Cash} + \text{Marketable securities} + \text{Receivables}}{\text{Current liabilities}}$$

J.F.D.'s ratio is 25,000/25,000 or 1:1. K.F.M.'s is 8,000/20,000 or 0.4:1. Thus, J.F.D. is in a better position when put to the "acid-test."

Let's consider three other measures used to analyze financial statements.

The owner's *rate of return* on an investment is found by dividing the year's net profit after taxes by the owner's capital investment.

SAMPLE PROBLEM 6

The Sandow Corporation's capital is $120,000. Its net profit after taxes is $8,400. Calculate the rate of return on investment.

Solution

$$\text{Rate of return} = \frac{\text{Net profit after taxes}}{\text{Capital invested}}$$

$$= \frac{8{,}400}{120{,}000}$$

$$= .07 = 7.0\%$$

Is 7% a good return? Compare it with last year's return, or compare it with yields from other investments, such as stocks or bonds.

Inventory turnover is another good measure. It is the ratio of the goods sold during a period to the average inventory held during the period.

$$\text{Inventory turnover} = \frac{\text{Cost of goods sold}}{\text{Average inventory}}$$

The cost of goods sold is determined from the income statement. Average inventory is the sum of the beginning and ending inventories, divided by 2.

SAMPLE PROBLEM 7

Given cost of goods sold of $26,000, beginning inventory of $4,000, and ending inventory of $4,800, find the inventory turnover of the Berry Hill Corporation.

Solution

Average inventory = $4,000 + $4,800 = $8,800 ÷ 2 = $4,400
Inventory turnover = $26,000 ÷ $4,400 = 5.9 times

A turnover of 5.9 times means that Berry Hill "sold out" roughly six times this year. Is this good? Again, it is necessary to compare to last year, other firms, and/or other items of the firm's own inventory.

Accounts receivable turnover is the final measure that we will inspect. As a measure of its credit granting policy, a firm wants to know how often it collects its accounts during the year. The calculation is:

$$\text{Accounts receivable turnover} = \frac{\text{Net credit sales}}{\text{Average accounts receivable}}$$

Cash sales are not considered. For our purposes, assume all sales are on credit. The average accounts receivable is computed the same way as average inventory. Sample Problem 8 shows the method in action.

SAMPLE PROBLEM 8

Net sales for Ease-All Cleaners, Inc., are $400,000. Beginning accounts receivable are $36,000, and ending accounts receivable are $44,000.

(a) Calculate the accounts receivable turnover.
(b) If credit terms are 30 days, are the accounts being collected on time?

Solution

(a) Average accounts receivable = $36,000 + $44,000 = $80,000 ÷ 2 = $40,000
Accounts receivable turnover = $400,000 ÷ $40,000 = 10 times
(b) If collected ten times in 365 days, average time is 365 ÷ 10 = 36.5 days. Since 36.5 days is longer than the 30 days allowed, the accounts are not being collected on time.

There are many other measures used in analyzing financial statements. You have seen a few but will probably get to know many more in later studies.

TEST YOUR ABILITY

1. Find the current ratio in each of the following cases:

	Current Assets	Current Liabilities
(a)	$50,000	$25,000
(b)	$60,000	$20,000
(c)	$25,000	$10,000
(d)	$18,000	$36,000
(e)	$16,000	$40,000

2. The current assets of Firms C and D are as follows:

	Firm C	Firm D
Cash	$19,000	$13,000
Marketable securities	5,000	-0-
Receivables	8,000	7,000
Other current assets	32,000	60,000
Total current assets	$64,000	$80,000

Current liabilities are $32,000 for C and $40,000 for D. Find each firm's (a) current ratio and (b) acid-test ratio.

3. Find the rate of return on an investment of $36,000 if the net profit after taxes is (a) $7,200; (b) $10,800; (c) $540; (d) ($1,800).
4. In each case, find the rate of inventory turnover:

	Beginning Inventory	Ending Inventory	Cost of Goods Sold
(a)	$57,000	$61,000	$354,000
(b)	$21,800	$26,200	$141,600
(c)	$ 9,700	$11,200	$ 37,620
(d)	$18,500	$26,300	$219,520

5. If, for Helen Products, Inc., net purchases are $82,000, beginning inventory is $25,000, and ending inventory is $19,000, find (a) the cost of goods sold; (b) the rate of inventory turnover.
6. The Terrific Department Store has net sales of $343,100, beginning accounts receivable of $46,000, and ending accounts receivable of $48,000.
   (a) Find the rate of accounts receivable turnover.
   (b) Are accounts being collected on time if the credit terms are 60 days?

Index Numbers

Are things expensive today? Do they cost "double what they used to"? To answer these questions, we can use a statistical method of analysis—calculating index numbers. An *index number* is an expression of a current price in terms of the price of another year, called the *base year,* which is the "normal" year in the judgment of the chooser.

For example, let's use 1982 as a base year for comparing prices. If a radio that cost $20 in 1982 costs $40 today:

$$\text{Index number of current year} = \frac{\text{Current price}}{\text{Base year price}} \times 100$$

$$= \frac{40}{20} \times 100$$

$$= 200$$

This year's *price index* is twice as much, or 200.

SAMPLE PROBLEM 9

Using 1982 as the base year, calculate the price index for (a) a table costing $50 in 1982 and $95 today; (b) a wheel costing $80 in 1982 and $70 today.

Solution

(a) $95/50 \times 100 = 190$ (b) $70/80 \times 100 = 87.5$

These answers are interpreted as follows: The table has increased in cost by 90% since 1982; and the wheel has decreased in cost to $87\frac{1}{2}$% of its 1982 price.

The procedure may also be carried out in reverse, as shown in Sample Problem 10.

SAMPLE PROBLEM 10

A lawn mower costs $72 today, a year in which the price index is 180 compared to 1982 as the base year. What did it cost in 1982?

Solution

Today's price is 180% of 1982 price.

$$B = \frac{P}{R} = \frac{\$72}{1.80} = \$40$$

Note We were given the rate and percentage and were looking for the base.

Many workers get salary and wage raises to keep up with the times. Consequently, to keep a check on this, there is the "cost-of-living" index based on consumer prices. Sample Problem 11 uses this fact in posing a problem of analysis: Has John Adamson kept up?

SAMPLE PROBLEM 11

John Adamson earned $12,000 in 1977. Today, a year in which prices are generally 60% higher than in 1977, he earns $18,000. Has his income kept up with the cost of living?

Solution

Current index is $100 + 60 = 160$
Adamson's salary index is $18{,}000/12{,}000 \times 100 = 150$
No, unfortunately his salary has not kept pace.

Many wage agreements contain a cost-of-living clause that guarantees an automatic increase in salary or wage whenever the general price index rises.

SAMPLE PROBLEM 12

Francine Yamato earns $256 a week in 1987, when prices are 160% of what they were in base year 1977. During 1988, the price index goes to 165. Her salary agreement provides for an automatic increase to meet the rising price index. Find her 1988 salary.

Solution

There are two ways to solve this problem:

(1) Increase: $165 - 160 = 5$

$\frac{5}{160} \times \$256 = \$8$

1988 salary $= \$256 + \$8 = \$264$

(2) Find the base year salary:

$$B = \frac{P}{R} = \frac{\$256}{1.60} = \$160$$

1988 salary = $160 × 165% = $264

In the first method, we found the increase as a fraction of the current salary and then added. In the second method, we went back to the base year and applied the new index to the base year salary.

Index numbers are very handy measures for comparing today's data with similar data for past years.

TEST YOUR ABILITY

1. Using 1982 as the base year, determine the price index for (a) a shirt costing $12 in 1982 and $18 today; (b) a dictionary costing $25 in 1982 and $35 today; (c) a mug that sold for $40 in 1982 and costs $28 today.
2. A fan costs $40 today. It cost $30 in 1977 and $32 in 1982. Determine today's price index if the base year is (a) 1977; (b) 1982.
3. A chair costs $78 today, a year in which the price index is 130 base on 1984 prices. Find the 1984 price.
4. Peter Mott earned a $120 a week in 1982. He earns $160 a week today. If today's price index is 135, with 1982 as the base year, has his salary kept up with the rising prices?

CHECK YOUR READING

1. Define *vertical analysis.*
2. What is the base for vertical analysis in (a) a balance sheet; (b) an income statement?
3. Define *horizontal analysis.*
4. In finding a rate of change between two years, which is the base year?
5. What does the current ratio show?
6. How is it possible for P. L. Stern, Inc., with a 3:1 current ratio to have an 0.5:1 acid-test ratio?
7. What purpose is served by calculating (a) inventory turnover; (b) accounts receivable turnover?
8. What is the purpose of calculating the rate of return on an investment?
9. How is a base year chosen for index number calculation?
10. What is the purpose of an index number?

EXERCISES

Group A

1. From the figures presented below, use vertical analysis to determine the percent of total assets represented by cash each year, and then state in which year the firm had the highest percentage of its assets in cash:

	1989	1988	1987
Cash	$ 71,595	$ 75,000	$ 66,600
Total assets	645,000	625,000	600,000

2. In each of the following cases, find the amount of inventory turnover:

	Beginning Inventory	Ending Inventory	Cost of Goods Sold
(a)	$42,500	$46,700	$267,600
(b)	$54,400	$51,800	$589,410
(c)	$ 8,450	$ 7,590	$ 75,388
(d)	$ 4,100	$ 4,600	$ 13,485

3. A pillow costing $56 today cost $40 in 1977 and $50 in 1982. Determine today's price index if the base year is (a) 1977; (b) 1982.
4. Vertically analyze the following company in terms of its total liabilities and capital:

Notes payable	$ 10,080
Accounts payable	35,460
Income tax payable	18,360
Mortgage payable	73,800
M. Carter, capital	13,500
T. Willis, capital	28,800
Total liabilities and capital	$180,000

5. Porter Chemicals has net sales of $512,000; beginning accounts receivable of $62,000; and ending accounts receivable of $66,000. (a) Find the rate of accounts receivable turnover. (b) If credit terms are 60 days, are the accounts being collected on time?
6. From the following data, calculate the rate of change for each item from 1987 to 1988:

	1988	1987
(a) Cash	$ 43,424	$ 47,200
(b) Accounts receivable	$ 10,695	$ 9,300
(c) Capital	$124,800	$104,000
(d) Sales	$324,500	$275,000
(e) Operating expenses	$ 58,446	$ 57,300

7. Find the rate of return on Anken Electricity's investment of $52,000 if the net profit after taxes is (a) $7,800; (b) $15,600; (c) $1,560; (d) ($2,080).
8. Using 1977 as the base year, determine the price index for (a) a shelf costing $15 in 1977 and $24 today; (b) a dress costing $60 in 1977 and $150 today; (c) a sled costing $75 in 1977 and $45 today.
9. Vertically analyze the following items in terms of net sales:

Net sales	$95,000
Cost of goods sold	−68,875
Gross profit	$26,125
Operating expenses	−15,295
Net operating profit	$10,830
Other expenses	−1,330
Net profit before taxes	$ 9,500
Income taxes	−2,090
Net profit after taxes	$ 7,410

10. Find the current ratio in each case below:

	Current Assets	Current Liabilities
(a)	$108,000	$54,000
(b)	$ 70,000	$17,500
(c)	$ 44,800	$64,000
(d)	$ 42,000	$30,000
(e)	$ 50,400	$56,000

11. Julie McAusland earned $180 a week in 1980. She earns $300 a week today. If today's price index is 165, with 1980 as the base year, has McAusland's salary kept up with the prices?
12. If purchases of Ferdinand International are $77,000; beginning inventory, $29,000; and ending inventory, $25,000, find (a) the cost of goods sold; (b) the rate of inventory turnover.
13. The current assets of Firms M and N are given below. If their current liabilities are, respectively, $50,000 and $120,000, find each firm's (a) current ratio; (b) acid-test ratio.

	Firm M	Firm N
Cash	$ 20,000	$100,000
Marketable securities	2,000	20,000
Receivables	18,000	60,000
Other current assets	110,000	60,000
Total current assets	$150,000	$240,000

14. Using vertical analysis, interpret the following data:

Cash	$ 18,000
Notes receivable	40,950
Accounts receivable	49,500
Inventory	106,650
Building	162,900
Equipment	72,000
Total assets	$450,000

15. A chess set costs $97.50 today, a year in which the price index is 125 based on 1980 prices. What did it cost in 1980?

Group B

1. The Argus Corporation's balance sheets and income statements for 1988 and 1989 follow on pages 338–39. (a) Copy the statements and fill in the horizontal and vertical figures. The first line of each has already been filled in. (b) Find the current ratio, acid-test ratio, rate of return on investment, and inventory turnover for each year. (c) What was the accounts receivable turnover for 1989? How efficient was the 15-day credit policy?
2. Shirley Botteril earns $166.40 a week in 1987, when prices are 130% of the base year 1980. In 1988, the price index rises to 135%. She is granted an automatic cost-of-living raise. Find her 1988 salary.

Argus Corporation
Comparative Balance Sheet
December 31, 1988 and 1989

	12/31/89	12/31/88	Percent of Total Assets 1989	1988	Increase (Decrease) $	%
Assets						
Current Assets						
Cash	$ 31,200	$ 37,080	19.5	20.6	(5,880)	15.9
Marketable securities	5,600	6,840				
Notes receivable	4,320	4,500				
Accounts receivable	15,520	14,940				
Inventory	16,000	18,000				
Supplies	6,560	8,100				
Prepaid expenses	4,160	5,400				
Total current assets	$ 83,360	$ 94,860				
Fixed Assets						
Land	$ 8,000	$ 9,000				
Building	19,200	21,960				
Equipment	26,400	28,440				
Furniture & fixtures	23,040	25,740				
Total fixed assets	$ 76,640	$ 85,140				
Total assets	$160,000	$180,000	100.0	100.0		
Liabilities						
Current Liabilities						
Notes payable	$ 5,760	$ 7,020				
Accounts payable	21,920	28,440				
Income taxes payable	8,000	16,560				
Total current liabilities	$ 35,680	$ 52,020				
Long-term Liabilities						
Mortgage payable	19,200	21,960				
Total liabilities	$ 54,880	$ 73,980				
Capital						
Common stock	$ 72,000	$ 72,000				
Retained earnings	33,120	34,020				
Total capital	$105,120	$106,020				
Total liabilites and capital	$160,000	$180,000	100.0	100.0		

Argus Corporation
Comparative Income Statement
For the Years Ended December 31, 1988 and 1989

	1989	1988	Percent of Net Sales 1989	Percent of Net Sales 1988	Increase (Decrease) $	Increase (Decrease) %
Sales (net)	$400,000	$360,000	100.0	100.0	40,000	11.1
Cost of goods sold						
Inventory, 1/1	$ 18,000	$ 15,120				
+ Purchases (net)	258,000	243,360				
Total	$276,000	$258,480				
– Inventory, 12/31	– 16,000	– 18,000				
Cost of goods sold	– $260,000	– $240,480				
Gross profit	$140,000	$119,520				
Operating expenses	– 124,800	– 87,840				
Net operating profit	$ 15,200	$ 31,680				
Other income	800	1,440				
Net profit before taxes	$ 16,000	$ 33,120				
Income taxes	– 3,520	– 7,300				
Net profit after taxes	$ 12,480	$ 25,820				

UNIT 6 REVIEW

Terms You Should Know

acid-test ratio	index number
asset	interval
average	liability
balance sheet	liquid assets
base year	long-term liability
capital	mean
current assets	median
current liabilities	modal class
current ratio	mode
financial statement	ratio
fixed asset	retained earnings
frequency distribution	turnover
horizontal analysis	vertical analysis
income statement	weighted average

UNIT 6 SELF-TEST

Only two problems appear for Unit 6. The first deals with a thorough review of financial statement classification and ratios. The second deals with other miscellaneous topics in the unit. What is not included in the review is a frequency distribution problem with averages; you have done several already.

1. From the various data presented below, find the value of (a) current assets; (b) fixed assets; (c) current liabilities; (d) long-term liabilities; (e) cost of goods sold; (f) gross profit; (g) net operating profit; (h) net profit before taxes; (i) net profit after taxes; (j) current ratio; (k) acid-test ratio; (l) rate of return on investment; (m) inventory turnover; (n) accounts receivable turnover, if the beginning balance was $25,000.

Notes receivable	$ 4,000	Net sales	$280,000
Land	12,000	Other income	1,000
Income taxes	3,740	Accounts payable	12,000
Capital	70,000	Accounts receivable	8,000
Cash	19,500	Supplies	3,000
Income taxes payable	3,740	Furniture & fixtures	10,000
Operating expenses	85,000	Inventory, 1/1	25,000
Mortgage payable	40,000	Equipment	20,000
Notes payable	3,000	Net purchases	175,000
Marketable securities	4,500	Building	40,000
Inventory, 12/31	21,000		

2. The sales and net profit after taxes figures from 1979 through 1989 are presented below for the Garcia Manufacturing Company:

Year	Net Sales	Net Profit after Taxes
1979	$22,000	$2,000
1980	24,000	3,000
1981	24,000	3,000
1982	28,000	4,000
1983	24,000	4,000
1984	32,000	5,000
1985	34,000	6,000
1986	36,000	6,000
1987	34,000	6,000
1988	32,000	8,000
1989	40,000	8,000

(a) Find the rates of change in net sales and net profit after taxes from 1988 to 1989.

(b) Use vertical analysis to find the percent of net sales represented by net profit in 1988 to 1989.

(c) Find the mean, median, and mode for both net sales and net profit after taxes for the 11-year period.

(d) If 1979 is the base year, express 1989 net sales and net profit after taxes in terms of 1979 prices, to the nearest tenth.

KEY TO UNIT 6 SELF-TEST

1. (a) $60,000
   (b) $82,000
   (c) $18,740
   (d) $40,000
   (e) $179,000
   (f) $101,000
   (g) $16,000
   (h) $17,000
   (i) $13,260
   (j) 3.2:1
   (k) 1.9:1
   (l) 18.9%
   (m) 7.8 times
   (n) 17 times

2. (a) Net sales 25%
   Net profit –0–
   (b) 1988 25%
   1989 20%
   (c)

	Net sales	Net profit
Mean:	$30,000	$5,000
Median:	32,000	5,000
Mode:	24,000	6,000

   (d) Net sales 181.8
   Net profit 400.0

Appendix

Operating a Pocket Calculator

The Electronic Calculator

You may have access to either the electronic printing calculator, which prints entries and results of calculations on a paper tape, or the electronic display calculator, which displays entries and results of calculations on a screen above the keyboard; or you may have a calculator with both the display and printing features, which can be used alternately as needed by flipping a switch. You probably have your own display calculator in a small, inexpensive pocket model. Although most of the newer pocket calculators have features in common and are becoming increasingly more standardized, there are some differences among the many models, especially in the arrangement of functional keys and switches.

Characteristics of the Pocket Calculator

1. The pocket calculator features the standard ten-key keyboard, which includes the digits 1 through 9, a 0, and a decimal point. This simple arrangement allows you to use the touch method of fingering and to keep your eyes on your work and thus handle papers efficiently.
2. Calculations and results are either printed on a tape or displayed on a screen above the keyboard.
3. The separate operational registers and the automatic repeat feature provide wide calculating versatility. Each group of operational keys represents a uniquely separate register including:
   (a) The *Keyboard Register.* All keyboard entries and results originate from this input/output register, which links together all the other registers. It automatically retains the last number printed; this number can be restored as a factor in any further immediate calculation by merely pressing the desired operational key.
   (b) The *Add/Subtract Register.* Addition, subtraction, and accumulation are performed in this register. It also holds an accumulated negative balance, as well as a positive balance.
   (c) The *Multiplication/Division Register.* Multiplication and division are performed in this register completely independent of the Add/Subtract Register. This makes individual extensions possible without interfering with accumulated totals.

(d) The *Memory/Accumulation Register.* Some machines include a Memory/Accumulation Register, which is useful in applications requiring two accumulations (such as sell/cost invoicing) or whenever two constants are used. Positive and negative entries may be made directly and accumulated with either the Memory Recall Key or the Memory Clear Key.

(e) The *Memory Register.* Some machines have a single Memory Register, which retains one constant which can be recalled for use in any calculation.

4. Several special function keys facilitate calculations:

   (a) **Clear Keys.** The Clear Key (C) clears the machine of all entries, except those in the Memory Register, which must be cleared by pressing the Memory Clear Key (MC). The Clear Entry Key (CE) is used to clear (1) an amount which has been entered into the keyboard prior to use of a function key or (2) an overflow error caused by an entry. Once an entry has been registered, it may be cleared by either subtracting out a positive amount or adding in a negative amount. The Add/Subtract Register may be cleared by pressing the Total Key. The Multiplication/Division Register clears automatically at the end of each simple operation. An entry into the single Memory Register clears the previous entry.

   (b) **Constant Control.** Although some machines will automatically retain as constants the first number in multiplication and the divisor in division, others permit the retention of these constants by a switch position, usually marked "K."

   (c) **Decimal Control.** This permits the operator to preset any number of decimal places desired in the result, depending on the range of settings listed. Some machines also feature the Add Mode (+) position, which places a decimal automatically between the tens and hundreds position without keyboard entry, or the Floating (F) position, which permits the placing of as many decimal places as the digit capacity of the machine allows, without the loss of whole number digits. Entries may include any combination of as many as 14 whole and decimal digits.

   (d) **Rounding Control.** Although some machines round automatically at the nearest selected decimal point, others include a Rounding Control (RO or 5/4), which is preset to round at a desired decimal place. If the next digit after the decimal setting is 5 or over, it rounds to the next higher number; if the next digit is 4 or under, it drops that digit and leaves the rest of the number as is.

   (e) The symbol "E" will either print or appear on the screen whenever (1) a whole number plus the decimal setting exceeds the digit capacity of the machine, or (2) a whole number exceeds the machine's digit capacity. The Keyboard Register clears automatically and the incorrect numbers are not printed or shown.

5. To repeat entries or to change last entries from one function to another, simply operate any of the desired function keys. Even after they have been printed or shown, entries or answers may be retained in the Keyboard Register for further calculation. Numbers flow easily from one function to another, and there are many shortcuts that increase speed and simplify calculations.

6. Common operational keys include the following:

On/Off Switch Switching from "Off" to "On" position automatically clears and readies the machine for use.

Numeric Entry Keys (0 through 9, and a decimal point) Enter numbers and decimal point into the calculator in the same order as read.

Decimal Point Key (.) Positions the decimal point in an entered amount.

Plus Key (+) Adds an entered amount to the contents of the Add Register. Repeated depression of the key repeats an addition.

Minus Key (−) Subtracts an entered amount from the contents of the Add Register. Repeated depression of the key repeats a subtraction.

Multiplication Key (×) Enters an amount as a multiplicand.

Division Key (÷) Enters an amount as a dividend.

Single Equals Key (=) Completes a simple operation of multiplication or division.

Equals/Plus Key (= +) Adds an entered amount to the contents of the Add Register. This key is also used to obtain the results in multiplication and division on some machines.

Equals/Minus Key (= −) Subtracts an entered amount from the contents of the Add Register. On some machines this key is also used to obtain the results in multiplication and division.

Percent Key (%) Permits entry and/or answers of actual percentages rather than their decimal equivalents.

Subtotal Key (ST or ◇) Displays or prints the contents of the Add Register. This amount may be used for further calculation.

Total Key (T or *) Displays or prints and clears contents of the Add Register. This amount may be used for further calculation.

Memory Plus Key (M+) Adds an entry or answer to the Memory. In multiplication or division, when this key is depressed instead of the Equals (=) or Equals Plus Key (= +), the products or quotients are added directly to the Memory.

Memory Minus Key (M−) Subtracts an entry from the Memory. In multiplication or division, when this key is depressed instead of the Equals (=) or Equals Minus Key (= −), the products or quotients are subtracted directly from the Memory.

Memory Recall Key (MR) Prints or displays contents stored in the Memory but functions as a subtotal, leaving the contents unchanged.

Memory Clear Key (MC) Prints or displays the contents of the Memory and clears the Memory Register.

Operating the Ten-Key Pocket Calculator Keyboard

Pocket calculators vary in size. Some are as large as transistor radios, others as small as business cards. Pocket calculator keys and keyboards vary with machine size. Operating technique will differ from person to person, depending on the size of the machine and of the user's fingers. You will probably find it comfortable to use all your fingers (the touch method) on some models but only one finger—and sometimes no fingers—on others. Follow this guide to find the method most comfortable for you:

1. Use all the fingers, if possible.
2. If not possible, then use the index finger only.
3. As a last resort, use the eraser at the end of a pencil.

CASIO CP-10

MACHINE PARTS

1 Tape
2 Number/Date/Time Display
3 Day-of-the-Week Indicator
4 Time/Date Set Key
5 Date Key
6 Time Key
7 Memory Minus Key
8 Memory Plus Key
9 Division Key
10 Multiplication Key
11 Minus Key
12 Plus Key
13 Equals Key
14 Decimal Point/P.M. Key
15 Number Keys
16 All-Clear Key
17 Clear Key
18 Percent Key
19 Print Key
20 Tape Advance Key
21 Memory Clear Key
22 Memory Recall Key
23 All Reset Key
24 Mode Selector Switch
25 A.M./P.M. Indicator

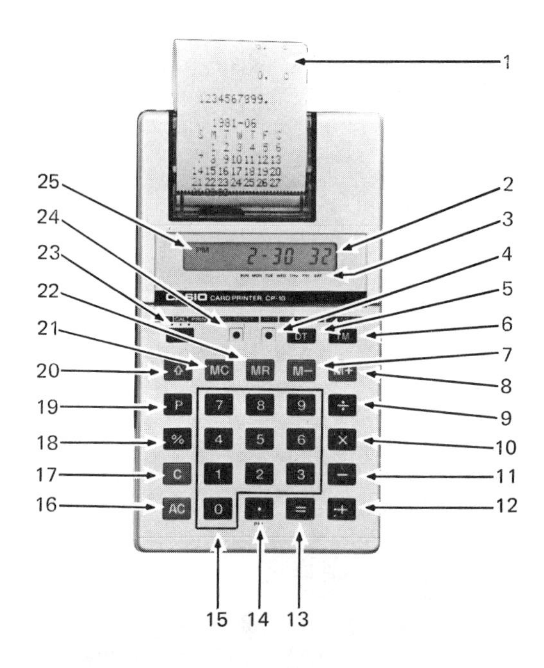

CASIO CARD PRINTER CP-10

TEXAS BUSINESS CARD

MACHINE PARTS

1 Number Display
2 Battery Content Indicator
3 Compute Key
4 Off Key
5 On/Clear Key
6 Square/Square Root Key
7 Memory/Annuity Due Key
8 Profit Margin Key
9 Division/Future Value Key
10 Multiplication/Present Value Key
11 Subtraction/Payment Key
12 Addition/Periodic Interest Rate Key
13 Equals/Number of Periods Key
14 Change Sign Key
15 Decimal Point Key
16 Number Keys
17 Cost Key
18 Selling Price Key
19 Percent/Percent Change Key
20 Reciprocal/Constant Key
21 2nd Function Key
22 Memory Content Indicator
23 2nd Function Indicator
24 Memory Use Indicator

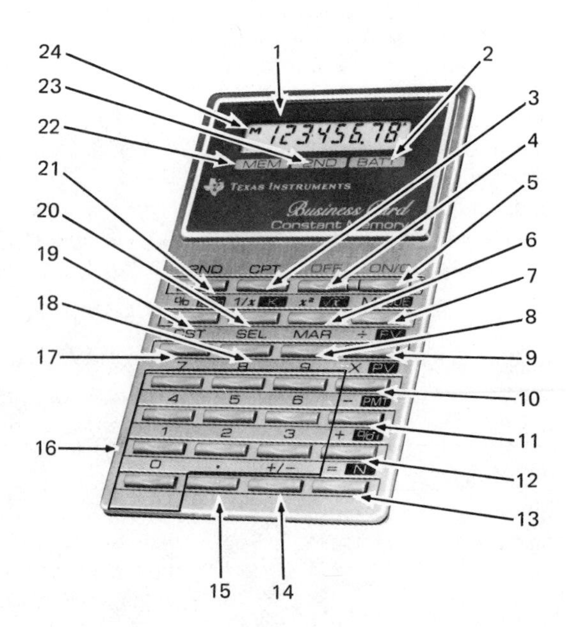

TEXAS INSTRUMENTS BUSINESS CARD

Machine Logic

Your calculator will function with either arithmetic or algebraic logic. Most pocket calculators have algebraic logic. One distinguishing feature of the machines with algebraic logic is that they require use of the Equals Key (=) to obtain results in all operations, including addition and subtraction.

Procedures for Machine Addition and Subtraction

Calculation: 40.15 − 5.10 + 15.20 = 50.25

Index	Touch	Result	Key Points
(Arithmetic instructions for display calculator showing a running total)			
40.15	+	40.15	Each amount is added or subtracted as it is entered,
5.10	−	35.05	giving a running total after each entry.
15.20	+	50.25	
(Algebraic instructions for machines using the Equals Key in addition and subtraction)			
40.15	−	40.15	After each entered amount, you touch the sign of the
5.10	+	35.05	operation you want the next amount to perform.
15.20	=	50.25	Touch the Equals Key after the last amount entered to find the answer.

Procedures for Other Basic Machine Operations

Following are the procedures for performing most of the basic machine operations on pocket calculators. By checking carefully the printed manual that comes with your particular machine, you will find that these operations can be adapted easily to it, whether it uses algebraic or arithmetic logic. Once you understand the mathematical principles involved, it is simple to use any calculator in order to solve any math problem quickly and accurately.

Multiplication

Calculation: 32.185 × 18.29 = 588.66365 (unrounded product)

Index	Touch	Result	Key Points
32.185	×	32.185	Set the Decimal Control at 5 to include the total of decimal places in each of the factors, or at F position to result in
18.29	=	18.29	all the whole digits plus as many decimal digits as the capacity of the machine will
		588.66365	allow.

Calculation: 48.276 × 62.159 = 3,000.787884 = 3,000.79 (Manually rounded to two decimal places)

Index	Touch	Result	Key Points
48.276	×	48.276	When the machine shows more decimal digits than are needed, you may round the results manually. If the third digit is 5 or above, raise the second digit by 1; if the third digit is 4 or less, drop it and leave the other two digits as they are.
62.159	=	62.159	
		3,000.787884	
		3,000.79 (Manually rounded)	

Calculation: 48.276 × 62.159 = 3,000.79 (Machine rounded to 2 decimal places)

Index	Touch	Result	Key Points
48.276	×	48.276	To round products at 2, set Decimal Control at 2 on machines with automatic rounding. On other machines, set the R/0, or 5/4 control (for rounding products) at the desired decimal setting.
62.159	=	62.159	
		3,000.79 (Machine rounded)	

Constant Multiplication

Calculation: 49 × 38 = 1,862.00
49 × 8.56 = 419.44
49 × 3.48 = 170.52

Index	Touch	Result	Key Points
49	×	49	On some machines, the multiplicand (the first number in multiplication) is retained automatically as the constant. On other machines, you may have to move the Constant Control switch to the K position. To check whether the first or second number is retained as a constant, multiply by 1.
38	=	38	
		1,862.00	
8.56	=	8.56	
		419.44	
3.48	=	3.48	
		170.52	

Multifactors or Chain Multiplication

Calculation: 13.4 × 27.6 × 39.9 = 14,756.616

Index	Touch	Result	Key Points
13.4	×	13.4	Chain multiplication can be accomplished easily by touching the
27.6	×	27.6 369.84	operational signs in the same order that they appear in the calculation.
39.9	=	39.9 14,756.616	

Accumulative Multiplication

Calculation: (34.56 × 22.19) + (14.72 × 39.87) = 1,353.78

Index	Touch	Result	Key Points
34.56	×	34.56	If the machine has a Memory Register, you may add the
22.19	M+	22.19 766.89	product directly into it by tapping M+ instead of the
14.72	×	14.72	Equals Key.
39.87	M+	39.87 586.89	To add the individual products in the Memory,
	MR	1,353.78	tap the MR Key.

Calculation: (157.86 × 385.97) − (52.17 × 47.38) = 58,457.41

Index	Touch	Result	Key Points
157.86	×	157.86	Add the first product directly into the Memory Register by
385.97	M+	385.97 60,929.22	tapping M+.
52.17	×	52.17	Subtract the second product directly from the Memory
47.38	M−	47.38 2,471.81	Register by tapping M−. Then tap MR to find the
	MR	58,457.41	contents of the Memory.

Division

Calculation: 58,794 ÷ 49 = 1,199.8776 (Machine rounded to four decimal places)

Index	Touch	Result	Key Points
58,794	÷	58,794	To round quotients to four decimal places, set Decimal Control at 4 on machines with automatic rounding. On other machines, set the R/O or 5/4 control (for rounding quotients) at the desired decimal setting.
49	=	49	
		1,199.8776	

Constant Divisor

Calculation: 849 ÷ 27 = 31.4444
1,247 ÷ 27 = 46.1852
3,899 ÷ 27 = 144.4074

Index	Touch	Result	Key Points
849	÷	849	On some machines, the divisor (the second number in division) is retained automatically as the constant. On other machines, you may have to move the Constant Control Switch or Lever to the K position.
27	=	27	
		31.4444	
1,247	=	1,247	
		46.1852	
3,899	=	3,899	
		144.4074	

Accumulative Division

Calculation: (2,595 ÷ 24.17) + (439 ÷ 4.45) = 206.01

Index	Touch	Result	Key Points
2,595	÷	2,595	If your machine has a Memory Register, you may
24.17	M+	24.17 107.36	add the quotient directly into it by tapping M+ instead of
439	÷	439	the Equals Key.
4.45	M+	4.45 98.65	To add the individual quotients in the Memory, tap the MR Key.
	MR	206.01	

Calculation: (3,875 ÷ 13.19) − (896 ÷ 4.77) = 105.94

Index	Touch	Result	Key Points
3,875	÷	3,875	Add the first quotient directly into the Memory
13.19	M+	13.19 293.78	Register by tapping M+.
896	÷	896	Subtract the second quotient directly from the Memory by tapping M−. Then tap MR to
4.77	M−	4.77 187.84	find the contents of the Memory.
	MR	105.94	

Percentages

Calculation: 40% of 3,000 = 1,200

Index	Touch	Result	Key Points
(With a Percent Key)			You may enter the percent without converting it to its
3,000	×	3,000	decimal equivalent when you
40	%	40 1,200	use the Percent Key instead of the Equals Key.
(Without a Percent Key)			Without the Percent Key, you have to convert the percent
3,000	×	3,000	to its decimal equivalent by
.40	=	.40 1,200	moving the decimal two places to the left (40% = .40).

Calculation: $500 is what percent of $1,200?

(With a Percent Key)			When the desired result is a rate, you may use the Percent Key instead of the Equals Key after entering the divisor.
500	÷	500	
1,200	%	1,200 41.67(%)	
(Without a Percent Key)			Without the Percent Key, you will have to convert your answer to the correct rate by moving the decimal two places to the right (.4167 = 41.67%).
500	÷	500	
1,200	=	1,200 .4167 (Decimal Equivalent) 41.67%	

Calculation: 300 is 5% of what number?

(With a Percent Key)			When an amount and its rate are known, 100% of the number can be found by using the Percent Key after the divisor.
300	÷	300	
5	%	5 6,000	
(Without a Percent Key)			Without the Percent Key, the divisor (the rate) must be converted to its decimal equivalent (5% = .05).
300	÷	300	
.05	=	.05 6,000	

Glossary

accounting The process of recording the financial activities of a business, summarizing these activities, and analyzing the results.

accrued interest Interest earned on a bond from the most recent interest payment date to the date of purchase, paid by the buyer to the seller.

accumulated depreciation The total depreciation on a plant asset from its purchase to any future date.

acid-test ratio The comparison of liquid assets (such as cash, marketable securities, or receivables) to current liabilities. The standard is 1:1.

ACRS method A new depreciation method in which predetermined rates are used for different classes of assets.

actuarial loan A loan on which monthly payment amounts are calculated using compound interest tables; used by financial institutions.

adjusted balance In open-end credit, the previous balance less any payments on account.

aliquot part A fractional part of a whole, usually based on 1, $1.00, or 100%.

allocation The process of distributing a total to various items or areas; used in overhead calculations.

amortization The gradual reduction of a debt, such as a loan or a mortgage; the gradual writeoff of an intangible asset.

amortization schedule A table showing the reduction of a debt.

amount The sum of principal plus interest; that is, the figure to which a principal will accumulate at compound interest over a period of time.

annuity A series of equal payments or receipts over a period of time.

annuity schedule A table showing the growth of an annuity.

appreciation The gradual increase in value of property over a period of time.

assessed valuation The taxable value of a piece of property.

asset An owned property that has money value.

authorized stock The maximum number of shares that a corporation's charter allows the corporation to issue.

average A single number representing a group of numbers: the mean, median, or mode.

average daily balance In open-end credit, the average balance of a charge account on any day of the month.

bad debts Uncollectible credit accounts.

balance sheet A form showing a firm's assets, liabilities, and capital on a specific date.

bank discount Simple interest deducted in advance when a promissory note is cashed at a bank.

bank reconciliation statement A form used to show that a depositor's records and a bank's records are in agreement.

bankers' interest Simple interest based on a 360-day year; used by banks and business firms.

base The original value; the 100% factor; the figure to which all other numbers relate.

base year A year chosen for comparison as the 100% or "normal" year for index number calculation; the earlier year in a comparison of financial data of any two years.

beneficiary The person named to receive the value of a life insurance policy upon the death of the insured.

bond A long-term obligation of a corporation, resulting from a loan to the corporation, with interest paid to the holder.

bond quotation The price at which a bond is sold or bought, expressed as a percent of face value (e.g., 93, 103).

bond time Counting months as 30 days each for accrued interest purposes; also called 30-day-month time.

book value The current worth of a plant or tangible asset (cost minus accumulated depreciation).

cancellation method A process for calculating simple interest, in which P, R, and T are expressed in fraction form, and cancellation of fractions is carried out prior to final multiplication.

canceled check A check that has been deducted from a depositor's account by the bank and then returned to the depositor.

capital The amount invested in a firm by its owner or owners; the net worth of a business; assets minus liabilities; the claim of the owner or owners to a firm's assets.

car liability insurance Protection against loss due to injuries caused to another driver and passengers, or to property damaged in a car accident where the insured is at fault.

carrier A firm that transports goods.

carrying charge The excess of an installment price over a cash price for goods. (See also *finance charge.*)

cash discount A reduction of 1% to 3% in the amount due the seller on a purchase, granted for early payment.

cash surrender value The amount of cash that can be received for a canceled life insurance policy.

charter A formal document issued by a state, authorizing the formation of a corporation.

check An order by the drawer on the drawee to pay the payee.

clearinghouse A check-sorting center in which checks are received from a payee's bank and returned to a drawer's bank.

closed-end fund An investment fund open only to its original investors, with no additional investors allowed.

c.o.d. Cash on delivery—a term of sale.

coinsurance A practice in which the insured who carries less than an agreed amount of fire insurance shares losses with the insurance company.

collision insurance Protection against loss arising from damage to one's own car in an auto accident.

commission A wage payment that is a percentage of sales; a fee to a broker for investment services.

common stock The only, or less privileged, stock of a corporation; the voting stock of a corporation.

complement A number which, when added to another number, forms 1 or 100% (e.g., $\frac{1}{3}$ is the complement of $\frac{2}{3}$; 60% is the complement of 40%).

compound discount The term for interest in a present value problem; the difference between a future amount and a present value.

compound interest Interest paid on both principal and accumulated interest, as compared to simple interest.

comprehensive insurance Car insurance that provides protection to the insured for damage to one's car other than from an auto accident (e.g., theft, vandalism, or fire).

constant ratio formula A formula used to approximate the true annual rate of interest for a series of equal payments.

contingent liability A possible liability, arising when a note is discounted. The payee is liable to the bank if the maker defaults.

copyright The exclusive statutory right given to an author or publisher to publish and reproduce material for a specified period of time.

corporation An artificial being, created by law, that can perform business transactions.

cost The total price paid for an asset, such as merchandise or property.

cost of goods sold The amount paid for the goods sold by a trading firm; beginning inventory plus net purchases minus ending inventory.

cumulative stock Preferred stock that is entitled to unpaid past dividends (dividends in arrears).

current assets Those assets having value during the next business period only.

current liabilities Debts due in the next business period.

current ratio The comparison of current assets to current liabilities. The standard is 2:1.

decimal system A number system based on the number ten.

declining-balance method A depreciation method in which a constant rate is applied to each year's decreasing book value.

deductible In a car insurance policy, refers to the amount of any loss that the insured agrees to pay.

depreciation An expense resulting from the gradual decline in value of a tangible asset because of age or use.

depreciation schedule A table showing an asset's annual depreciation, accumulated depreciation, and book value for each year.

differential piecerate A plan of compensation providing for two or more piecerates, depending on the volume of a worker's production.

discounting a note The process of "cashing" a promissory note at a bank by paying interest deducted from the maturity value of the note at the time of discount.

dividend A share of the profits of a corporation; also, a distribution from a life insurance company or a bank.

down payment In an installment purchase, the amount of cash paid on the date of purchase.

drawee The bank on which a check is drawn.

drawer The party who writes a check.

due date The date on which a debt is to be paid; the maturity date.

effective rate The interest rate that is actually earned on an investment or paid on a loan; opposite of nominal rate.

endorsement The signature on the back of a negotiable instrument that is necessary for its transfer from one party to another.

endowment insurance Life insurance for a specified number of years (usually 20), at the end of which time the insured, if alive, receives the face of the policy.

e.o.m. End of month—a term of sale.

equivalent single rate of discount A rate that is mathematically equal to a series of trade discount rates.

estimated useful life (EUL) The projected length of time an asset will be held for use by a business.

exact interest Simple interest based on a 365-day year; used by the federal government.

execution price The actual price paid or received for a share of stock.

exemption An allowance for a person, including oneself, for whom a wage earner provides support; used to determine amounts of withholding tax.

extended term insurance The number of years and days of face value coverage to which a canceled insurance policy can be converted.

extension Quantity times price, used in invoicing problems.

face The stated amount of money on a business document, such as the face of a note, an insurance policy, or a stock.

feasibility study A study comparing alternative cost decisions.

FICA tax (Federal Insurance Contributions Act tax) A required payroll deduction, the amount of which is matched by the employer, that provides for old-age pensions, survivors' benefits, disability payments, and Medicare.

FIFO An inventory system in which the goods first purchased (first-in) are assumed to be the goods first sold (first-out).

final retail price The selling price of goods after all markups and markdowns have been completed.

finance charge The costs of an installment purchase or loan to the borrower; the carrying charge.

financial statement A formal document showing the results of business activity, such as a balance sheet or income statement.

fire insurance Protection against loss from damage to property resulting from fire or other natural causes.

fixed asset An asset whose life will extend beyond the next accounting period.

F.O.B. Free on Board—used in commercial sales to determine the point at which title passes.

F.O.B. destination Terms indicating that title passes from the seller to the buyer when the goods reach their destination.

F.O.B. shipping point Terms indicating that title passes from the seller to the buyer at the point of shipment of the goods.

frequency distribution A table used to find averages from a large number of different amounts.

fringe benefits Payments by an employer for employees beyond basic salaries (e.g., sick days, insurance, and vacations).

graduated commission A plan in which commission rates increase as sales increase.

gross proceeds The amount due to the seller of stocks or bonds before deduction of commissions and taxes.

gross profit Profit before deduction of operating expenses; sales price minus cost of goods sold.

gross profit method A method of estimating the value of an inventory.

gross wages Earnings before payroll deductions.

horizontal analysis The process of comparing accounting data from two time periods to determine rates of change or to analyze vertical percents.

incentive plan A wage plan in which earnings are based on an employee's output.

income statement A form showing the results of business activity for the past business period.

indemnity Payment for a loss.

index number An expression of a current price as a percent of a base year price.

installment buying Purchasing an item by agreeing to pay for it, plus a finance charge, in equal payments over a period of time.

insurance Protection against loss, taking the form of a policy from an insurance company for which premiums are paid.

intangible asset A long-lived, nonphysical asset, such as a patent, copyright, or trademark.

interest Money paid for the use of another's money.

interest-bearing note A promissory note that entitles the holder to interest on the face amount.

interval An arbitrary grouping of numbers in a frequency distribution.

inventory valuation The process of assigning a dollar value to goods on hand on a specific date.

invoice A business form used to record a sale or a purchase.

issued stock Stock sold by the corporation for which the stock was authorized.

kilowatt-hour The standard measure of electricity usage.

liability A debt owed to a creditor; the claim of another to a firm's assets.
LIFO An inventory system in which the goods last purchased (last-in) are assumed to be the goods first sold (first-out).
limited-payment life insurance Life insurance which provides protection for life, but requires premium payments over a specified period of time (usually 20 years).
liquid assets Those assets easily converted into cash, such as marketable securities and receivables.
list price The original, or stated, price of an article, usually found in a price list or catalog.
load fee A percentage, charged to the investor, of the offering price of a mutual fund share.
load fund A fund charging a fee for investing in it.
long-term liability A liability due after the next business period.

machine-hours method A method of depreciation involving equal expense per hour of machine time.
maker The party signing a promissory note; the giver of the note.
markdown A reduction in price from the original retail price.
markdown cancellation A reduction of a markdown.
market The cost of replacing merchandise on the date of an inventory.
markup That amount added to cost to arrive at the original retail price; expenses plus desired profit.
maturity date See *due date.*
maturity value See *amount.*
mean The arithmetic average.
median The middle number of a series of numbers arranged in increasing or decreasing order.
merchandise inventory The value of goods a firm holds for resale to customers in the normal course of business.
modal class The interval in a frequency distribution that has the highest frequency.
mode The most common number in a series.
mortgage A long-term promissory note issued in exchange for a loan on property.
mutual fund A pooling of money from individuals for the purpose of investment by a manager, with distribution of profits to be made to the investors.

negotiable Capable of being transferred from one party to another by endorsement (e.g., a check or a promissory note).
net amount to be paid The cash due to pay for a purchase after all discounts (cash and trade) have been deducted.
net asset value The value of a share of stock, or of a mutual fund share, determined by dividing the net worth of the corporation, or fund, by the number of outstanding shares.
net cost The excess of premiums paid over the cash surrender value of a life insurance policy.

net markdown Markdowns minus markdown cancellations.

net proceeds The price received after deductions for interest (in the case of a discounted note) or for taxes and commission (in the case of selling stocks or bonds).

net profit Profit remaining after deducting operating expenses; gross profit minus operating expenses.

net purchase price The price of an item bought after deduction of trade discounts; list price minus trade discounts.

net purchase price equivalent A single rate that can be applied to a list price in order to arrive at a net purchase price.

net return The excess of cash surrender value of a life insurance policy over premiums paid.

net sales Sales less sales returns.

net wages Gross wages minus deductions.

no-fault insurance Car insurance in which the policyholder's insurance company pays for damages to the policyholder's car, regardless of who is at fault.

no-load fund A mutual fund that charges no fee to investors.

nominal rate The interest rate stated in a loan or on a bond; the opposite of effective rate.

no-par A stock with no value stated on its face.

NOW account A checking account that earns interest and for which no fee is charged as long as a minimum balance is maintained.

odd lot Less than a round lot; 1 to 99 shares of a 100-unit stock.

odd-lot differential An extra charge per share added to the purchase price of a share of stock or deducted from the sales price of a share when an odd lot is traded.

offering price The price at which a mutual fund share is listed; net asset value plus load fee, if any.

open-end credit Credit in which finance charges are computed on a monthly basis, depending on the balance and the method used, as in a retail store charge account.

open-end fund A mutual fund; an investment fund with no limit on the number of investors.

ordinary whole life insurance Insurance that covers the policyholder for life and on which premiums are paid for life.

original retail price Cost plus markup.

"other" credit Credit other than open-end; credit in which repayment is made in a series of equal amounts, each being part interest and part principal.

outstanding check A check that has been deducted by the drawer but not yet received and deducted by the drawer's bank.

outstanding stock Stock in the hands of stockholders; issued stock minus treasury stock.

over-the-counter A market for the trading of stocks not listed on an exchange, at negotiated prices.

overhead The everyday costs of running a business, such as utilities, maintenance, repairs, and rent; all costs other than for materials or labor.

overtime Hours worked in excess of normal working hours.

paid-up insurance The amount of ordinary whole life insurance to which a canceled policy can be converted.

par The face value of a stock; a stock with a value printed on it.

participating stock Preferred stock that has the right to share with common stock in dividends remaining after a normal distribution to preferred and a stated distribution to common.

partnership A business owned by two or more persons.

patent A right to a process or product granted to an inventor or the inventor's assignee for exclusive use.

payee The party to whom a check or note is made out; the party who will be paid.

payroll deduction An amount taken out of an employee's gross wages by the employer.

payroll register A record of the gross earnings, payroll deductions, and net wages of a firm's employees.

per C Per hundred.

per M Per thousand.

percentage A dollar or number amount which is part of or related to the base.

periodic inventory A system of counting goods at the end of a period of time and assigning a value to those goods.

perpetual inventory A system of keeping a running inventory balance on a routine basis.

preferred stock The privileged stock of a corporation, receiving priority in the distribution of dividends or of assets in the event of a firm's liquidation.

premium A periodic payment to an insurance company for a policy.

present value The value in current dollars of a future sum.

previous balance In open-end credit, the amount due on the account at the start of a month.

principal The amount borrowed or loaned.

proceeds See *net proceeds.*

promissory note A written promise to pay a sum of money at a specified future date.

property tax A tax on the value of property owned, paid by the owner to a local or state taxing authority.

proration The process of spreading out a cost or a benefit over a period of time; also known as *allocation.*

quota An expected output (e.g., a sales representative may earn commissions only after reaching a level of sales).

rate A percent (%).

rate of yield The return on an investment, expressed as an annual percent of earnings based on cost; effective rate.

ratio A comparison of one number to another; also used in financial statement analysis.

reciprocal The arithmetic inverse (e.g., $\frac{2}{3}$ is the *reciprocal* of $\frac{3}{2}$).

recovery period See *estimated useful life.*

Regulation Z The federal Truth-in-Lending regulation, requiring disclosure of credit terms and costs to buyers and lenders.

residual value The estimated salvage or scrap value remaining at the end of the life of a tangible asset.

retail method A method of estimating the value of an inventory.

retained earnings Those profits not distributed as dividends but kept by a corporation.

r.o.g. Receipt of goods—a term of sale.

round lot An even or whole unit of trading for stocks, typically 100 shares.

salary A payment to an employee that is not related to output or hours, but rather to an arbitrary time period, such as a week or a year.

sales tax A state or local tax assessed on the value of goods or services.

savings certificate A long-term investment certificate for a stated minimum sum of money at a higher than usual interest rate.

scrap value See *residual value.*

service business A firm dealing in nonmerchandising activities.

settlement The cash payment given to the beneficiary of a life insurance policy.

short-rate table A table used in fire insurance to determine premiums on policies of less than a year and refunds on policies canceled by the insured.

simple interest Interest on principal only, as opposed to compound interest.

sinking fund A gradual accumulation of money at compound interest by equal periodic deposits.

sinking fund schedule A table showing the growth of a sinking fund.

sole proprietorship A business owned by one person.

specific identification The determination of inventory value by adding the amount of specific purchases.

standard Normal production, used as a base for incentive wage payments.

statistics The field of collecting, summarizing, and comparing numerical data for business decisions; also, a synonym for numerical data.

straight-line method A depreciation method involving an equal annual expense.

straight piecework A system providing for a minimum guaranteed pay, with piecerate payment for production above standard.

sum-of-the-years' digits method A depreciation method involving a decreasing expense, based on the sum of the individual years of an asset's life.

tangible asset A long-lasting physical asset; a plant asset.

term insurance Life insurance for a specified number of years that has no cash value at the end of the insurance period.

term of discount The time period from the date of discount to the date of maturity; the period for which a bank charges discount.

terms of sale The agreed payment conditions of a sale.

time The number of days between two dates.

trade discount A reduction of 5% or more from the list price of an item.

trade discount series More than one trade discount applied in sequence to a list price.

trademark A legal right to a name or symbol, granting exclusive use to its creator.

trading business A firm whose primary activity is the buying and selling of merchandise.

treasury stock Stock that has been reacquired by the issuing corporation.

true annual rate of interest The actual or effective rate of interest charged on an installment purchase or a loan.

turnover The rate at which an asset is replaced, as in inventory turnover or accounts receivable turnover.

unemployment tax A federal and state tax imposed on employers to provide payments to unemployed workers.

units-of-production method A depreciation method allowing equal expense per unit produced.

vertical analysis The process of determining the percent of a total represented by each individual item on a financial statement.

wage A payment based on hours worked or units produced.

weighted average The mean of data determined by considering the frequency of repetitive numbers; an inventory method in which all goods are valued at a weighted average price.

withholding tax Income tax deducted from gross wages.

yield See *rate of yield.*

ANSWERS TO *Test Your Ability* PROBLEMS

Chapter 1 page 4

1. Seventy-five
2. Two hundred seventy-five and five tenths
3. Six and seven tenths
4. Ninety-nine and ninety-nine hundredths
5. Three hundred fifty-six and two hundred five thousandths
6. One thousand nine hundred eighty-six and two hundred twenty-five thousandths
7. Ten thousand nine hundred seventy-five and one hundredth
8. Seventy-eight thousand two hundred twenty-two and sixty-five thousandths
9. One hundred twenty-two thousand nine hundred eighteen and two tenths
10. Four thousand five hundred eighty-nine and two thousandths

Chapter 1 page 5

1. 2,739
2. 1,650
3. 1,881
4. 2,091
5. 2,731
6. 23,399
7. 22,153
8. 38,777
9. 827,144

Chapter 1 page 7

1. 2,210
2. 1,780
3. 1,956
4. 1,514
5. 1,780
6. 2,274
7. 1,513
8. 2,220
9. 1,710
10. 2,220

Chapter 1 page 8

1. Vertical totals: 310, 232, 295, 249
   Horizontal totals: 220, 159, 193, 254, 260
   Grand total: 1,086
2. Vertical totals: 363, 177, 244, 264
   Horizontal totals: 162, 187, 239, 232, 228
   Grand total: 1,048
3. Dept.

Clothing	616	Counters	177
Jewelry	572	Stock	756
Pet	460	Order	1,215
Lamp	350	Grand total	2,148
Record	150		

Chapter 1 page 9

1. 76
2. 79
3. 89
4. 1,188
5. 1,806
6. 328
7. 169
8. 204
9. 1,991
10. 18,959

Chapter 1 page 10

1. Vertical totals: 339, 264
   Horizontal differences: 12, 11, 13, 19, 20
   Net total: 75
2. Vertical totals: 34,351; 23,890
   Horizontal differences: 2,222; 2,222; 2,211; 901; 2,905
   Net total: 10,461
3.

		Item	**Total Balance + Receipts**	**Balance 7/31**
Balance	397	1	128	89
Receipts	180	2	119	48
Total	577	3	65	41
Issues	342	4	55	24
Balance	235	5	117	12
		6	93	21
			577	235

Chapter 1 page 11

1. 1,170
2. 703

3. 912
4. 5,264
5. 2,592
6. 9,682
7. 26,124
8. 15,051
9. 262,444
10. 417,276

Chapter 1 page 13

1. 2,150
2. 12,520
3. 41,100
4. 130,000
5. 17,250
6. 675,000
7. 200,000
8. 54,000
9. 1,440,000
10. 2,820,000
11. 481,500
12. 450,000
13. 150,000
14. 26,400,000
15. 2,440,000

Chapter 1 page 15

1. 1,440
2. 2,040
3. 1,568
4. 2,020
5. 2,200
6. 1,201
7. 850
8. 2,000
9. 2,000

Chapter 1 page 16

1. 45
2. 6
3. 32
4. 7.5
5. 5.69
6. 355
7. 189
8. 259

Chapter 1 page 18

1. For the form used, see the Solution to Sample Problem 20, page 18. Adjusted balance = available balance = $1,043.15.
2. For the form used, see the Solution to Sample Problem 20, page 18. Bank balance − outstanding checks + late deposit. Checkbook balance − service charge − insufficient funds + collection of note. Adjusted balance = available balance = $10,290.37.
3. For the form used, see the Solution to Sample Problem 20, page 18. Bank balance − outstanding checks + late deposit. Checkbook balance − service charge − credit card advance + $18.00 error. Adjusted balance = available balance = $2,104.31.

Chapter 2 page 23

1. $\frac{8}{9}$
2. $\frac{17}{36}$
3. $\frac{19}{35}$
4. $\frac{4}{5}$
5. $\frac{1}{2}$
6. $\frac{7}{9}$
7. $\frac{5}{12}$
8. $\frac{8}{9}$
9. $\frac{7}{10}$
10. $\frac{51}{83}$
11. $\frac{27}{40}$
12. $\frac{7}{12}$
13. $\frac{1}{3}$
14. $\frac{1}{20}$
15. 1

Chapter 2 page 24

1. none
2. 5
3. 90
4. 54
5. 2
6. 2
7. 10
8. 2
9. 60
10. 25

Chapter 2 page 25

1. 6
2. 12

3. 24
4. 30
5. 6
6. 10
7. 90
8. 12
9. 80

Chapter 2 page 25

1. $2\frac{2}{3}$
2. 3
3. $2\frac{4}{7}$
4. 15
5. $2\frac{1}{5}$
6. $1\frac{22}{75}$
7. 7
8. 5
9. $3\frac{1}{5}$
10. 55
11. 40
12. $41\frac{1}{4}$
13. $41\frac{7}{15}$
14. $379\frac{1}{6}$
15. 60

Chapter 2 page 26

1. $\frac{4}{3}$
2. $\frac{29}{3}$
3. $\frac{43}{7}$
4. $\frac{35}{4}$
5. $\frac{15}{2}$
6. $\frac{69}{4}$
7. $\frac{164}{9}$
8. $\frac{71}{2}$
9. $\frac{44}{3}$
10. $\frac{367}{9}$
11. $\frac{961}{4}$
12. $\frac{251}{2}$
13. $\frac{1304}{5}$
14. $\frac{2567}{8}$
15. $\frac{5405}{6}$

Chapter 2 page 27

1. $2\frac{17}{72}$
2. $1\frac{301}{360}$
3. $2\frac{95}{168}$
4. $2\frac{53}{132}$
5. $\frac{19}{28}$
6. $1\frac{4}{21}$
7. $28\frac{7}{15}$
8. $1\frac{7}{18}$
9. $1\frac{1}{40}$
10. $44\frac{7}{72}$
11. $1\frac{1}{10}$
12. $384\frac{1}{12}$

Chapter 2 page 29

1. $\frac{1}{4}$
2. $\frac{1}{72}$
3. $\frac{25}{56}$
4. $\frac{1}{3}$
5. $\frac{7}{18}$
6. $7\frac{1}{6}$
7. $8\frac{1}{8}$
8. $66\frac{11}{42}$
9. $74\frac{1}{6}$
10. $36\frac{23}{72}$
11. $31\frac{5}{8}$
12. $1\frac{11}{12}$
13. $42\frac{35}{36}$
14. $232\frac{5}{7}$
15. $321\frac{9}{40}$

Chapter 2 page 29

1. (a) $\frac{6}{15}$
   (b) $\frac{5}{18}$
   (c) $\frac{4}{15}$
   (d) $\frac{8}{13}$
   (e) 5

2. (a) $\frac{8}{9}$
   (b) $\frac{2}{5}$
   (c) $\frac{20}{27}$
   (d) 12
   (e) $13\frac{1}{2}$

Chapter 2 page 32

1. (a) $\frac{1}{2}$
   (b) $17\frac{1}{2}$
   (c) $77\frac{11}{12}$
   (d) $7\frac{1}{2}$
   (e) $\frac{3}{10}$
   (f) $1\frac{4}{7}$
   (g) $\frac{10}{49}$
2. (a) $\frac{4}{5}$
   (b) 1:4 or .25:1
3. strawberries 120 lbs.
   ice cream 90 lbs.
   shortcake 480 lbs.
   whipped cream 30 lbs.

Chapter 3 page 37

1. 75.7
2. 18.006
3. 97.55
4. 1.00
5. 1.247
6. 56

Chapter 3 page 38

1. 1.01
2. .47
3. 3.05
4. 40.05
5. 260.30
6. 2,011.39
7. $1,291.23
8. $256.45
9. 5,680.20
10. 1.01

Chapter 3 page 38

1. 2.89
2. .79
3. 2.44
4. .82
5. .50
6. 3.65
7. 224.62
8. $220.50
9. $12,266.02
10. 1.99

Chapter 3 page 39

1. $692.10
2. 1,031.99
3. 1.16
4. .00
5. 2,245.32
6. 3,694.36
7. .00
8. .31
9. 1,213.38
10. 3.61
11. 6.66
12. 3,775
13. 1,622,820
14. 578.15
15. .00

Chapter 3 page 41

1. 5.069
2. 4.401
3. 7.572
4. 7.25
5. 20.704
6. 30,000
7. 5.644
8. 3.652
9. 800
10. 6.4
11. 827.190
12. 54.944
13. 2.005
14. 32.5
15. 4.125

Chapter 3 page 44

1. (a) .625; 62.5%
   (b) .6; 60%
   (c) .7; 70%
   (d) .556; 55.6%
   (e) .143; 14.3%
   (f) .688; 68.8%
   (g) .25; 25%
   (h) .5; 50%
2. (a) .80; $\frac{4}{5}$
   (b) .42; $\frac{21}{50}$
   (c) .875; $\frac{7}{8}$
   (d) $.06\frac{1}{4}$; $\frac{1}{16}$
   (e) $.05\frac{1}{2}$; $\frac{11}{200}$
   (f) $.85\frac{5}{7}$; $\frac{6}{7}$
   (g) .002; $\frac{1}{500}$
   (h) 2.5; $2\frac{1}{2}$

Chapter 4 page 50

1. (a) $85.00
   (b) $3.60
   (c) $227.50
   (d) $0.08
   (e) $1,310.00
   (f) $2,330.00
2. $5,600
3. 600
4.

Staff salaries	$1,890,000
Administrative salaries	240,000
Supplies	270,000
Maintenance	330,000
Transportation	120,000
Food	150,000

5. (a) 682,000
   (b) 372,000
   (c) 248,000

Chapter 4 page 52

1. (a) $120.00
   (b) $150.00
   (c) $35.00
   (d) $7,500.00
   (e) $9,000.00
   (f) $41.86

2. $27,310
3. 3,200
4. $230
5. (a) $15,500
   (b) $1,860
6. (a) $81,000
   (b) $99,000
7. (a) $28.00
   (b) $3.78

Chapter 4 page 54

1. (a) 5%
   (b) 125%
   (c) 7%
   (d) 6.2%
   (e) 200%
   (f) 116.7%
2. 83.3%
3. (a) $22
   (b) 20%
4. 40%
5. (a) 160%
   (b) 62.5%

Chapter 4 page 55

1. 19%
2. 6.5%
3. 60%
4. 7.9%
5. (a) 10%
   (b) $33\frac{1}{3}$% or 33.3%

Chapter 5 page 69

1. $369.07
2. (a) $\frac{3}{8}$
   (b) $\frac{1}{6}$
   (c) $\frac{2}{7}$
   (d) $\frac{3}{4}$
   (e) $\frac{4}{9}$
   (f) $\frac{7}{16}$
   (g) $\frac{1}{12}$
   (h) $\frac{1}{16}$
   (i) $\frac{7}{12}$
   (j) $\frac{4}{15}$

3. (a) $30
   (b) $80
   (c) $60
   (d) $40
   (e) $45
   (f) $10
4. For the form used, see Sample Problem 3, page 68. Extensions: boxes, $6.25; pads, $13.00; reams, $61.60; total, $80.85.

Chapter 5 page 73

1. $117
2. $79.80
3. $547.20
4. (a) 81%
   (b) 51%
   (c) 76%
   (d) 57%
   (e) 60%
   (f) 61.2%
5. $547.20
6. (a) 81%
   (b) 51%
   (c) 76%
   (d) 57%
   (e) 60%
   (f) 61.2%
7. (a) $327.60
   (b) $142.50
   (c) $1,332.00
   (d) $303.75
   (e) $547.20

Chapter 5 page 75

1. (a) May 7
   (b) April 30
   (c) May 10
   (d) April 26
   (e) April 16
2. (a) September 23
   (b) September 30
   (c) September 10
   (d) September 10
   (e) August 31
3. (a) $722.70
   (b) $715.40
   (c) $730.00
   (d) $708.10
   (e) $722.70
4. $282.24

Chapter 6 page 82

1. $350
2. (a) $54.00
   (b) $127.80
3. $53
4. $26
5. (a) $75.25
   (b) $60.20
6. (a) $58.75
   (b) $51.25
   (c) $15.00
   (d) $12.00
   (e) $4.50
   (f) $7.50

Chapter 6 page 86

1. (a) $50
   (b) $40
   (c) $39
   (d) 70%
   (e) 140%
   (f) 30%
2. M = $42; C = $70
3. S = $400; M = $112
4. C = $11.00; M = $1.65
5. (a) $\frac{1}{8}$
   (b) $\frac{2}{11}$
   (c) $\frac{4}{15}$
   (d) 20%
   (e) 40%
   (f) 50%
6. (a) $\frac{1}{9}$
   (b) $\frac{3}{8}$
   (c) $\frac{11}{14}$
   (d) 25%
   (e) 20%
   (f) 100%

Chapter 6 page 88

1. $18.75
2. ($1.50)
3. (a) $6.00
   (b) ($2.75)

Chapter 6 page 90

1. (a) $325
   (b) $350
   (c) $325
2. $358
3. (a) $0.32
   (b) $0.80
   (c) $1.18
   (d) $14.76
   (e) $31.71
4. $348

Chapter 7 page 100

1. $4,232.75
2. (a) $7.00
   (b) $7.22
   (c) $9.48
3. $420 ($42 each)
4. $426
5. $10.67
6. $74.55
7. $909.70
8. (a) $6,587
   (b) $6,545 ($4.675 per unit)
   (c) $6,640
   (d) $6,468

Chapter 7 page 102

1. $3,235
2. (a) $49,880
   (b) $54,540
   (c) $50,820
   (d)(c) $50,820
3. $746

Chapter 7 page 105

1. $17,415
2. $16,000

Chapter 8 page 119

1. (a) \$600
   (b) \$1,300
   (c) \$2,600
2. \$193.70
3. (a) 40
   (b) $3\frac{1}{2}$
   (c) $1\frac{3}{4}$
   (d) $45\frac{1}{4}$
   (e) \$325.80
4. \$325.80

Chapter 8 page 122

1. Carter \$240.00 minimum
   Walsh \$338.80
2. (a) \$300 minimum
   (b) \$322
   (c) \$480
3. \$1,050
4. \$191
5. \$517.50
6. \$690

Chapter 8 page 128

1. (a) \$14.12
   (b) \$28.60
   (c) \$16.22
   (d) \$16.67
2. \$11.60
3. (a) \$59
   (b) \$35
   (c) –0–
   (d) \$38
4. \$198.48
5. \$257.22

Chapter 8 page 132

1.

Crosby Repair Company Payroll
For the Week Ended August 25, 19–

Emp. No.	Exemp.	Hours Worked					Total Hours	Reg. Hrs.	OT Hrs.	OT Bonus
		M	T	W	T	F				
1	M-1	8	$7\frac{1}{2}$	$7\frac{1}{2}$	8	8	39	39	0	0
2	M-0	$8\frac{1}{4}$	8	8	$8\frac{3}{4}$	8	41	40	1	$\frac{1}{2}$
3	S-1	9	9	8	$8\frac{1}{2}$	7	$41\frac{1}{2}$	40	$1\frac{1}{2}$	$\frac{3}{4}$
4	S-1	10	$7\frac{1}{4}$	9	8	$9\frac{1}{4}$	$43\frac{1}{2}$	40	$3\frac{1}{2}$	$1\frac{3}{4}$
5	M-3	10	8	8	$8\frac{1}{2}$	$7\frac{1}{2}$	42	40	2	1
6	S-0	8	8	$9\frac{1}{2}$	$8\frac{3}{4}$	$8\frac{3}{4}$	43	40	3	$1\frac{1}{2}$
7	M-5	8	9	$9\frac{1}{2}$	$9\frac{1}{2}$	11	47	40	7	$3\frac{1}{2}$
8	M-7	10	10	8	$7\frac{1}{4}$	$8\frac{3}{4}$	44	40	4	2
TOTALS		$71\frac{1}{4}$	$66\frac{3}{4}$	$67\frac{1}{2}$	$67\frac{1}{4}$	$68\frac{1}{4}$	341	319	22	11

Emp. No.	Paid Hrs.	Rate	Gross Wages	FICA Tax	Fed. W.T.	Other Ded.	Total Ded.	Net Wages
1	39	$6.00	$ 234.00	$ 16.73	$ 21.00	$ 5.00	$ 42.73	$ 191.27
2	$41\frac{1}{2}$	5.10	211.65	15.13	21.00	8.00	44.13	167.52
3	$42\frac{1}{4}$	5.20	219.70	15.71	24.00	–0–	39.71	179.99
4	$45\frac{1}{4}$	5.60	253.40	18.12	31.00	2.50	51.62	201.78
5	43	4.40	189.20	13.53	8.00	7.00	28.53	160.67
6	$44\frac{1}{2}$	4.80	213.60	15.27	27.00	–0–	42.27	171.33
7	$50\frac{1}{2}$	4.90	247.45	17.69	11.00	10.00	38.69	208.76
8	46	4.60	211.60	15.13	2.00	4.50	21.63	189.97
TOTALS	352		$1,780.60	$127.31	$145.00	$37.00	$309.31	$1,471.29

2. (a) $4.40
   (b) $0.64
3. State $241.90
   Federal $47.20
4. (a) $250,000/mo.; $3,000,000/yr.
   (b) $22,550/mo.; $270,600/yr.
   (c) $272,550/mo.; $3,270,600/yr.

Chapter 9 page 140

1. (a) $32
   (b) $4
   (c) $3
2. (a) $475
   (b) $272
   (c) $230
3. $456
4. (a) $211
   (b) $683
   (c) $814
5. (a) $13
   (b) $487
   (c) $1
6. $92
7. (a) $18.98
   (b) $127.02
8. $140

Chapter 9 page 143

1. A $9,000
   B $18,000
   C $12,000
2. (a) $20,000
   (b) $36,000
   (c) $50,000
   (d) $10,800
   (e) $40,000
   (f) $15,000
   (g) $21,000
   (h) $63,000

Chapter 9 page 150

1. $37,500

	Class	Premium
2. (a)	22	$448
(b)	12	$133
(c)	15	$161
(d)	50	$175
(e)	26	$168
(f)	40	$290

3. (a) $122.00
   (b) $250.00
   (c) $103.00
   (d) $515.00
   (e) $380.00
   (f) $128.75

4. (a) $168.75
   (b) $620.00
   (c) $102.00
5. (a) $342.10
   (b) $160.20
   (c) $1,490.00
   (d) $1,236.00
6. $665.75

Chapter 9 page 155

1. (a) $152.10
   (b) $420.10
   (c) $180.00
   (d) $4,400.00
   (e) $238.35
   (f) $342.54
   (g) $54.27
2. (a) $77.57
   (b) $214.25
3. (c) $46.80
   (d) $1,144.00
4. (e) $21.45
   (f) $30.83
5. (g) $2.17

	Extended Term	Paid-Up
6. (a)	6 yrs., 312 days	$ 1,620
(b)	18 yrs., 192 days	$ 2,115
(c)	21 yrs., 37 days	$ 9,400
(d)	25 yrs., 224 days	$11,900

Chapter 10 page 166

1. (a) $310
   (b) $1,100
   (c) $5,825
2. (a) $500
   (b) $3,150
   (c) $2,650

3. Year	Annual Depreciation	Accumulated Depreciation	Book Value
			$3,650
1	$500	$ 500	3,150
2	500	1,000	2,650
3	500	1,500	2,150
4	500	2,000	1,650
5	500	2,500	1,150
6	500	3,000	650

4. (a) $450
   (b) $450
   (c) $1,250
5. (a) $375
   (b) $8,625
   (c) $900
   (d) $7,725
6. 1986 $ 9,600
   1987 $12,000
   1988 $ 4,400
   1989 $ 6,000

7.

Year	Hours Used	Annual Depreciation	Accumulated Depreciation	Book Value
				$40,000
1986	2,700	$13,500	$13,500	26,500
1987	2,620	13,100	26,600	13,400
1988	2,410	13,400	40,000	–0–

Chapter 10 page 172

1. (a) 45
   (b) 9
   (c) $1,800
   (d) $7,700

2.

Year	Annual Depreciation	Accumulated Depreciation	Book Value
			$9,500
1	$1,800	$1,800	7,700
2	1,600	3,400	6,100
3	1,400	4,800	4,700
4	1,200	6,000	3,500
5	1,000	7,000	2,500
6	800	7,800	1,700
7	600	8,400	1,100
8	400	8,800	700
9	200	9,000	500

3. (a) $16\frac{2}{3}$%
   (b) $6\frac{2}{3}$%
   (c) 8%
   (d) $28\frac{4}{7}$%

4.

Year	Annual Depreciation	Accumulated Depreciation	Book Value
			$25,000
1	$12,500	$12,500	12,500
2	6,250	18,750	6,250
3	3,125	21,875	3,125
4	1,125	23,000	2,000

5. (a) 1986 $930
   (b) 1987 $1,364

6.

Year	Rate	Annual Depreciation	Accumulated Depreciation	Book Value
				$30,000
1	25%	$ 7,500	$ 7,500	22,500
2	38%	11,400	18,900	11,100
3	37%	11,100	30,000	–0–

7. (a) $100 gain
   (b) $600 gain
   (c) $2,420 loss
8. $100 loss

Chapter 10 page 177

1. $1,200
2. A $60,000
   B $90,000
   C $75,000
   D $45,000
3. $0.30
4. $482.92
5. (a) $13,500
   (b) $1,710
6. 1.5%
7. (a) $54,000
   (b) $1,674
8. (a) $0.33
   (b) $11.55
   (c) $600.60
9. (a) $4,500
   (b) $3,500
   (c) $1,000 by buying

Chapter 10 page 180

1. $4,250 net profit
2. $141,250
3. $1,840 net profit

Chapter 11 page 196

1. (a) $288.00
   (b) $250.00
   (c) $262.50
   (d) $178.75
   (e) $717.75
2. (a) $163.13
   (b) $913.13

3. (a) 83 days
   (b) 120 days
   (c) 117 days
   (d) 25 days
   (e) 290 days
4. Same answers for problems 3(a)-(e).
5. (a) $8.00
   (b) $32.40
   (c) $20.40
   (d) $47.60
   (e) $16.44
6. (a) $8.11
   (b) $32.85
   (c) $20.68
   (d) $48.26
   (e) $16.67
7. (a) $29.33
   (b) $829.33

Chapter 11 page 200

1. (a) $8.25
   (b) $3.46
   (c) $20.18
   (d) $20.47
   (e) $0.09
   (f) $0.35
2. (a) $4.32
   (b) $18.00
   (c) $1.80
   (d) $32.40
   (e) $64.80
   (f) $19.80
3. (a) $6.00
   (b) $3.60
   (c) $9.60
   (d) $5.40
   (e) $15.60
   (f) $12.00
4. (a) $6.45; $651.45
   (b) $12.30; $422.30
   (c) $3.90; $653.90
   (d) $18.00; $1,218.00
   (e) $14.00; $854.00
   (f) $10.00; $760.00
   (g) $14.00; $914.00
   (h) $13.75; $613.75
   (i) $7.50; $807.50
   (j) $7.00; $1,507.00
5. $836.40

6. (a) $0.80
   (b) $1.28
   (c) $1.76
   (d) $3.36

Chapter 11 page 204

1. (a) $1.56; $501.56
   (b) $0.49; $800.49
   (c) $3.80; $1,203.80
   (d) $4.22; $754.22
   (e) $6.98; $206.98
   (f) $117.25; $3,117.25
   (g) $59.22; $8,259.22
   (h) $6.71; $2,106.71
   (i) $2.59; $452.59
   (j) $49.87; $609.87
2. (a) $1.53; $501.53
   (b) $0.48; $800.48
   (c) $3.75; $1,203.75
   (d) $4.16; $754.16
   (e) $6.88; $206.88
   (f) $115.64; $3,115.64
   (g) $58.41; $8,258.41
   (h) $6.62; $2,106.62
   (i) $2.55; $452.55
   (j) $49.19; $609.19

Chapter 11 page 207

1. (a) July 16, 1988
   (b) August 16, 1988
   (c) September 16, 1988
   (d) July 16, 1988
   (e) August 15, 1988
   (f) September 14, 1988
2. $1,425.60
3.

	Maturity Date	Discount Term
(a)	June 16	23 days
(b)	February 20	16 days
(c)	January 17	32 days
(d)	February 21	46 days
(e)	September 4	63 days

Chapter 11 page 209

1. $352.80
2. $648.45
3. $4,191.95
4. $902.58

Chapter 12 page 220

1. (a) $17.14
   (b) $1,028.29
   (c) $856,912.15
2. (a) $5,619.71; $1,619.71
   (b) $5,674.08; $1,674.08
   (c) $5,703.04; $1,703.04
3. (a) $333.01; $133.01
   (b) $1,165.82; $665.82
   (c) $1,311.09; $611.09
   (d) $5,743.49; $4,743.49
   (e) $6,333.85; $1,333.85
   (f) $11,716.59; $1,716.59
4. (a) $6,743.28
   (b) $269.73
   (c) $134,865.63
5. (a) $9,636.63
   (b) $4,964.62
   (c) $4,539.50
   (d) $1,205.84
   (e) $6,161.00
   (f) $6,578.01

6.

Year	Beg. Bal.	Annual Contrib.	Total	Int. at 7%	End. Bal.
1		$500.00	$ 500.00	$ 35.00	$ 535.00
2	$ 535.00	500.00	1,035.00	72.45	$1,107.45
3	1,107.45	500.00	1,607.45	112.52	$1,719.97
4	1,719.97	500.00	2,219.97	155.40	$2,375.37

Chapter 12 page 225

1. (a) $540.27
   (b) $27.01
   (c) $16,208.07
2. (a) $582.58; $3,417.42
   (b) $292.04; $1,707.96
   (c) $210.21; $389.79
   (d) $1,235.96; $1,764.04
   (e) $11,073.52; $8,926.48
   (f) $619.18; $230.82
3. (a) $924.96
   (b) $3,000.00
4. (a) $788.69
   (b) $15,773.75
   (c) $276.04

5. (a) $798.54
   (b) $3,400.85
   (c) $7,259.79
   (d) $1,884.17
   (e) $371.94
   (f) $6,345.20

Chapter 12 page 228

1. (a) $3.54
   (b) $177.22
   (c) $10,633.44
2. (a) $46.52
   (b) $15.44
   (c) $194.14
   (d) $42.09
   (e) $89.47
   (f) $331.56
3. (a) $177.40

(b)

Year	Amt. Beg.	Int. at 6%	Annual Contrib.	Total Incr.	Amt. End.
1			$177.40	$177.40	$ 177.40
2	$177.40	$10.64	177.40	188.04	365.44
3	365.44	21.93	177.40	199.33	564.77
4	564.77	33.89	177.40	211.29	776.06
5	776.06	46.56	177.38*	223.94	1,000.00

*rounded

4. (a) $1,900.78
   (b) $1,197.81
   (c) $440.80
   (d) $51.14
   (e) $204.17

Chapter 12 page 231

1. $553.75
2. $373.08
3. $12.03

Chapter 13 page 239

1. (a) $4.50
   (b) $–0–
   (c) $5.00
2. (a) $80.00
   (b) $228.50
   (c) $1,465.00
   (d) $1,145.00
   (e) $720.00

3. (a) $10.00
   (b) $28.50
   (c) $265.00
   (d) $145.00
   (e) $70.50
4. $448
5. (a) $50.00
   (b) $43.00
   (c) $7.12
   (d) $18.00
   (e) $9.17
6. (a) $1,080
   (b) $155

Chapter 13 page 245

1. (a) $45.03
   (b) $234.15
   (c) $11.69
   (d) $30.61
   (e) $52.68
2. (a) $30.42
   (b) $75.65
   (c) $86.12
   (d) $17.32
   (e) $180.40
3. (a) $38.71
   (b) $54.20
4.

Payment	Amount	Interest at 1%	Reduction of Principal	Principal Balance
				$650.00
1	$96.61	$6.50	$90.11	559.89
2	96.61	5.60	91.01	468.88
3	96.61	4.69	91.92	376.96
4	96.61	3.77	92.84	284.12
5	96.61	2.84	93.77	190.35
6	96.61	1.90	94.71	95.64
7	96.60	0.96	95.64	–0–

5. (a) $732
   (b) $8,784
   (c) $263,520
6. $50,000
7. $10\frac{1}{2}$%
8. 20 years

Chapter 13 page 251

1. (a) 40.00%
   (b) 12.50%
   (c) 20.00%
   (d) 12.25%
2. 12.50%
3. 12.75%
4. (a) 12.00%
   (b) 40.00%
   (c) 20.25%
   (d) 12.75%
5. (a) $25.44

   (b)

Payment	Amount	Interest at $\frac{7}{10}$%	Reduction of Principal	Principal Balance
				$100.00
1	$25.44	$0.70	$24.74	75.26
2	25.44	0.53	24.91	50.35
3	25.44	0.35	25.09	25.26
4	25.44	0.18	25.26	–0–

   (c) 8.4%
6. (a) 88.9%
   (b) 18.8%
   (c) 29.6%
   (d) 218.2%

Chapter 14 page 268

1. (a) 100,000
   (b) 88,800
   (c) 88,500
2. (a) $4
   (b) $440
3. (a) Common $1
   Preferred $7
   (b) Common $0.60
   Preferred $7
   (c) Common $–0–
   Preferred $3
4. Common $23,000
   Preferred $36,000
5. Common $26,000
   Preferred $33,000
6. (a) $27
   (b) $2

Chapter 14 page 270

1. Common $0.90
   Preferred $4.00
2. Common $1.80
   Preferred $4.80
3. Common $2.20
   Preferred $4.00
4. Common $1.50
   Preferred $17.00

Chapter 14 page 273

1. (a) $6,275.00
   (b) $6,362.50
   (c) $6,312.50
2. (a) $25.75; $7,725.00
   (b) $55.625; $4,450.00
   (c) $83.125; $4,987.50
3. (a) $25.75; $7,725.00
   (b) $55.375; $4,430.00
   (c) $82.875; $4,972.50
4. $13,612.50
5. $3,137.50

Chapter 14 page 275

1. $13,945.58
2. $4,752
3. $8,725.61

Chapter 14 page 277

1. $487.50 loss
2. (a) $1,200 gain
   (b) $720
   (c) $1,920
   (d) 5.3%
3. (a) 2.0%
   (b) 7.3%
   (c) 3.2%

Chapter 15 page 288

1. (a) 45
   (b) $45,000.00
   (c) $80.00
   (d) 8.3%
   (e) 1993
   (f) $956.25
   (g) $962.50

2. (a) $12,000
   (b) $5,610
   (c) $11,425
   (d) $15,775
3. $15,375
4. (a) $150.00
   (b) $150.00
   (c) $378.00
   (d) $191.33
   (e) $444.00
5. $170
6. (a) $3,310.42
   (b) $116,566.67

Chapter 15 page 291

1. $6,210
2. $5,300
3. $5,366.67
4. $480 gain
5. $2,640 gain
6. (a) $1,890 gain
   (b) 6.9%
   (c) 5.9%
7. (a) 6.6%
   (b) 6.9%
   (c) 5.3%
   (d) 7.8%

Chapter 15 page 294

1. (a) $10.00
   (b) $8.05
   (c) $11.06
   (d) $16.91
2. $4,052.50
3. (a) $2,800.00
   (b) $2.38
   (c) $238.00
   (d) $2,562.00
4. (a) 3,000
   (b) 5,000
   (c) 10,000
   (d) 50,000

Chapter 15 page 296

1. $0.50
2. $2,100

3. (a) $240 loss
   (b) 3.2%
   (c) 4.7%

Chapter 15 page 299

1. $33,250
2. (a) 70%
   (b) 11.7%
3. (a) $3,829.00
   (b) $4,908.00
   (c) $5,321.00
   (d) $1,674.50
   (e) $3,430.00
4. (a) $381.00 net cost
   (b) $305.10 net return
   (c) $225.00 net return
   (d) $517.60 net cost
   (e) $2,097.00 net return
5. (a) $7,390
   (b) $6,975
   (c) $415 net cost
   (d) 0.3%

Chapter 16 page 314

1. For the form used, see Sample Problem 1, page 313. Total current assets = $13,025; total fixed assets = $13,800; total assets = $26,825; total liabilities = $5,400; total liabilities + capital = $26,825.
2. For the form used, see Sample Problem 1, page 313. Total current assets = $139,720; total fixed assets = $210,000; total assets = $349,720; total current liabilities = $20,600; total long-term liabilities (bonds and mortgage) = $170,000; total liabilities = $190,600; total capital (stock + retained earnings) = $159,120; total liabilities + capital = $349,720.
3. For the form used, see Sample Problem 2, page 314. Cost of goods sold = $286,000; gross profit = $134,000; net operating profit = $14,000; net profit before taxes = $18,000; net profit after taxes = $14,040.
4. For the form used, see Sample Problem 2, page 314. Cost of goods sold = $50,100; gross profit = $11,900; net operating profit = $1,500; net profit before taxes = $1,700; income taxes = $374; net profit after taxes = $1,326.

Chapter 16 page 318

1. $22,200
2. 191
3. $252.32
4. (a) 529
   (b) 605

5. (a) 79.9
   (b) 80
   (c) 94
6. Mean 79.5
   Median 75
   Mode 75

Chapter 16 page 321

1. Mean = \$2,500 ÷ 20 = \$125.00
   Median = $\frac{20}{2}$ = 10th score = \$99.50 + $\frac{3}{6}$(\$50) = \$124.50
   Modal class \$100–\$149
2.

Interval	Midpoint	×	Frequency	=	Product
0–9	5		1		5
10–19	15		4		60
20–29	25		5		125
30–39	35		5		175
40–49	45		8		360
50–59	55		7		385
			30		1,110

   Mean = 1,110 ÷ 30 = 37
   Median = 39.5 (falls between 30–39 and 40–49 intervals)
   Modal class = 40–49

Chapter 17 page 329

1. Cash, 15.0%; accounts receivable, 5.0%; inventory, 20.0%; land, 10.0%; building, 31.0%; furniture and fixtures, 19.0%.
2. Notes payable, 4.1%; accounts payable, 16.4%; income taxes payable, 10.0%; mortgage payable, 49.1%; capital, 20.4%.
3. Net sales, 100.0%; cost of goods sold, 70.0%; gross profit, 30.0%; operating expenses, 33.0%; net operating loss, (3.0%); other income, 5.0%; net profit before taxes, 2.0%; income taxes, 0.4%; net profit after taxes, 1.6%.
4. 1989 11.0%; 1988 10.5%
   The firm's better year was 1989.
5. (a) 5.0%
   (b) (6.0%)
   (c) 50.0%
   (d) (20.0%)
   (e) (15.0%)

Chapter 17 page 332

1. (a) 2:1
   (b) 3:1
   (c) 2.5:1
   (d) 0.5:1
   (e) 0.4:1

2. (a) Firm C, 2:1; Firm D, 2:1
   (b) Firm C, 1:1; Firm D, 0.5:1
3. (a) 20.0%
   (b) 30.0%
   (c) 1.5%
   (d) (5.0%)
4. (a) 6 times
   (b) 5.9 times
   (c) 3.6 times
   (d) 9.8 times
5. (a) $88,000
   (b) 4 times
6. (a) 7.3 times
   (b) Yes, accounts are being collected every 50 days.

Chapter 17 page 335

1. (a) 150
   (b) 140
   (c) 70
2. (a) $133\frac{1}{3}$
   (b) 125
3. $60
4. No. His index is $133\frac{1}{3}$ today, while the price index is 135.

Index

A 6
B 7
C 8
D 9
E 0
F 1
G 2
H 3
I 4
J 5